BIOSEPARATION OF PROTEINS

This Is Volume I of

SEPARATION SCIENCE AND TECHNOLOGY

A reference series edited by Satinder Ahuja

BIOSEPARATION OF PROTEINS

Unfolding/Folding and Validations

Ajit Sadana
Department of Chemical Engineering
University of Mississippi
University, Mississippi

ACADEMIC PRESS
San Diego London Boston New York Sydney Tokyo Toronto

This book is printed on acid-free paper. ♾

Academic Press
a division of Harcourt Brace & Company
525 B Street, Suite 1900, San Diego, California 92101-4495, USA
http://www.apnet.com

Academic Press Limited
24-28 Oval Road, London NW1 7DX, UK
http://www.hbuk.co.uk/ap/

Library of Congress Card Catalog Number: 97-80234

International Standard Book Number: 0-12-614040-5

PRINTED IN THE UNITED STATES OF AMERICA
97 98 99 00 01 02 QW 9 8 7 6 5 4 3 2 1

CONTENTS

3 High-Resolution Fractionation Processes

4 Interfacial Protein Adsorption and Inactivation during Bioseparation

5 Protein Inactivations during Chromatographic Methods of Separation

6 Protein Inactivations during Novel Bioseparation Techniques

7 Adsorption Influence on Bioseparation and Inactivation

8 Applications and Economics of Bioseparation

9 Protein Refolding and Inactivation during Bioseparation

10 Validation of the Production of Biological Products

PREFACE

The biotechnology industry is poised for rapid growth and implementation in diverse areas. However, one major constraint is the need for a more complete understanding of protein separation processes. While progress in the technology of cloning genes to attain high and desired levels of expression has been satisfactory, large-scale production and purification of proteins has, until recently, been rather neglected.

Bioseparation stands at the very center of effective biotechnology development. Interest in protein purification has increased rather dramatically in the past few years and has become the focus of intensive research both at academic institutions and in industry. Several papers and a few books on the general area of bioseparations have recently appeared, but no book, and only a few papers, has emphasized the influence of bioseparation processes on protein inactivation. This information, which is scarce and difficult to find in the open literature, is a critical part of the bioseparation framework: One may ask of what use is a bioseparation process if information on the conformational state and the activity of the recovered protein is not presented in any detail? The aim of this book is to provide that information as part of a critical review and synthesis of the literature.

After the introductory chapter, Chapter 2 describes the three basic steps involved in the bioseparation process: cell disruption, initial fractionation, and high-resolution fractionation. The different high-resolution fractionation steps, as described in Chapter 3, are critical for meeting the stringent requirements set for product purity and effectiveness, including those of the different regulatory agencies.

The quality of the separated product is significantly affected by the inactivation of proteins at interfaces during bioseparation. Chapter 4 analyzes these interfacial protein inactivations. Until now, chromatographic processes have been utilized rather heavily during the bioseparation protocol at different stages of separation. Chapter 5 analyzes protein inactivation during chromatographic methods of separation and includes available information on the mechanistic aspects. Other techniques for effectively separating biological products of interest also need to be developed, keeping both the quantitative and the qualitative aspects in mind. Chapter 6 analyzes protein inactivations during novel bioseparation procedures.

During bioseparation, the biological product of interest must adsorb on an interface. Conformational changes accompanied by subsequent activity changes will presumably result and will significantly influence both the quantitative and the qualitative aspects of biological product recovery. Chapter 7 analyzes the influence of protein adsorption and inactivation during bioseparation. The economics of the downstream process plays a significant role in getting a biological product ready for market. This sort of information is not readily available in the open literature but is presented here in Chapter 8.

Some denaturation during bioseparation is unavoidable. Different renaturation techniques, presumably as "corrective" steps to minimize the extent of denaturation of at least some of the biological products recovered, would be extremely helpful. Some of these techniques may also be used to enhance or improve the quality of the product by facilitating the process by which the product attains the required conformational state(s). Chapter 9 analyzes protein refolding strategies and inactivation during bioseparation.

Consistency in the safety, potency, efficacy, and purity of a biological product is the manufacturer's responsibility. This consistency is the basis of governmental regulation and evaluation. Tests for impurities and contaminants are critical in the development and validation of bioseparation processes and in final product testing. Validation is the "assurance that a process is closely followed during a product's manufacture." Validation protocols or strategies provide written documentation that a process is consistently doing what its manufacturer claims it can accomplish. This validation procedure provides assurance that the process is "under control." Because recombinant techniques are used to make quite a few bioproducts, the importance of this step cannot be overemphasized. This validation process and the protocols and strategies involved therein, along with the appropriate governmental and regulatory guidelines, are presented in Chapter 10.

The text is intended for instruction at the graduate level and even at the senior undergraduate level, as well as for the industrial practitioner who, after examining the "science-based" information in the earlier chapters, will appreciate the focus on the monitoring, validation, and economics of bioseparation processes. The generalized treatment will also interest chemical, biochemical, and biomedical engineers, chemists, biochemists, and those in the medical profession who wish to better understand the fundamentals of bioseparation and its influence on protein/enzyme inactivation. Even venture capitalists will find the book of interest. Biotechnology, by its very nature, is an interdisciplinary area that requires diverse expertise. I hope that this book will foster these interactions, facilitate an appreciation of all perspectives, and help in efforts toward improved economics of bioseparation.

This text is unique in that it provides the appropriate background on bioseparation processes, while emphasizing the extent of, mechanisms of, and control of protein inactivation during these processes and their essential validation. Comparisons of protein inactivation during different bioseparation processes provide valuable information for workers in different areas who are interested in bioseparations. Readers may thus consider this a "second-level" book on bioseparations, with "first-level" books providing the fundamentals of bioseparation processes. This second-level examines and analyzes the control and validation of the product (protein) during these bioseparation processes. Second, and presumably "higher-level," books will be required to pave the way for the emergence and consolidation of protein purification as a discipline rather than as a means to an end.

LIST OF EXAMPLES

CHAPTER 1

No examples

CHAPTER 2

2.1 Describe briefly some of the advantages of protein excretion from cells (Sherwood *et al.*, 1985).
2.2 Briefly present a kinetic analysis of enzymatic lysis and disruption of yeast cell walls (Hunter and Asenjo, 1987a).
2.3 Briefly describe the principles of operation of expanded beds for particulate removal from protein solutions (Chase, 1994).
2.4 Describe briefly the recovery of proteins utilizing membranes (Martin and Manteuffel, 1988).
2.5 Provide an example where ultrafiltration has been used to clarify a fermentation broth for producing antibodies (Duffy *et al.*, 1989).
2.6 Briefly describe the purification of the IgG antibody by affinity cross-flow filtration (Weiner *et al.*, 1994).
2.7 Briefly analyze the extraction of penicillin G by an emulsion liquid membrane (ELM) process (Lee and Lee, 1992).
2.8 Briefly describe the traditional purification process for insulin production (Ladisch and Kohlmann, 1992).

2.9 Briefly analyze affinity precipitation using chitosan as a ligand carrier for protein purification (Senstad and Mattiasson, 1989).
2.10 Briefly describe an affinity precipitation method for proteins by surfactant-solubilized, ligand-modified phospholipids (Powers *et al.*, 1992).
2.11 Briefly analyze the large-scale purification of staphylococcal enterotoxin B using chromatographic procedures (Johansson *et al.*, 1990).
2.12 Briefly analyze the use of modified divinylbenzene-polystyrene resins for the separation of aspartame, phenylalanine, aspartic acid, and asparagine (Casillas *et al.*, 1992).
2.13 Briefly analyze the final fractionation steps for the recovery of SEB using chromatographic methods (Johansson *et al.*, 1990).
2.14 Briefly describe the ultrafast HPLC separation of recombinant DNA-derived proteins (Olson and Gehant, 1992).
2.15 Briefly describe the purification and characterization of lamb pregastric lipase (D'Souza and Oriel, 1992).

CHAPTER 3

3.1 Describe briefly the isolation and purification of carboxylesterase from *Bacillus stearothermophilus* (Owusu and Cowan, 1991).
3.2 Briefly analyze the processing steps for obtaining tissue plasminogen activator (tPA) from animal cell and bacterial sources with special attention to the quality of the product recovered (Datar *et al.*, 1993).
3.3 Briefly describe the purification of two endo-β-glucanases from the aerobic fungus *Penicillium capsulatum* (Connelly and Coughlan, 1991).
3.4 Briefly analyze the purification of pectin methylesterase from *Bacillus subtilis* (Pitkanen *et al.*, 1992).
3.5 Briefly analyze the purification of *Clostridium thermocellum* β-glucosidase B using ion exchange, hydrophobic interaction, and hydroxylapatite chromatography (Romaniec *et al.*, 1993).
3.6 Briefly analyze the purification of *D*-xylulokinase from the yeast *Pichia stipitis* NCYC 1541 using adsorption (hydroxylapatite column) chromatography (Flanagan and Waites, 1992).
3.7 Briefly analyze the purification of feruloyl/ρ-coumaroyl esterase from the fungus *Penicillium pinophilium* (Castanares *et al.*, 1992).
3.8 Briefly analyze the purification of chitanase from *Trichoderma harzianum* (Ulhoa and Peberdy, 1992).
3.9 Briefly describe the concerted cluster model of multivalent affinity for heterogeneous adsorption of enzymes (Dowd and Yon, 1995).

CHAPTER 4

4.11 Briefly compare the adsorption of hen lysozyme (LS2) and milk LAC on colloidal AgI (Galisteo and Norde, 1995).
4.12 Describe by appropriate modeling: (1) the principle of the replacement method, and (2) the simulation of adsorption in a well-mixed particle suspension (Cornelius *et al.*, 1992).

CHAPTER 5

5.1 Describe a procedure for the HPIEC separation of biopolymers especially suited for applications at high pH and to high-molecular weight samples (Kato *et al.*, 1984).
5.2 A method for the separation of mRNAs (van der Mast *et al.*, 1991).
5.3 Briefly describe the separation of lipase from *Pichia burtonii* (Sugihara *et al.*, 1995).
5.4 Briefly describe a process to separate basic fibroblast growth factor (bFGF) and alkaline phosphatase (PALP) from human placenta (Costa *et al.*, 1993).
5.5 Provide an example for the HPLC separation of an enzyme exhibiting microheterogeneity (Wong *et al.*, 1988).
5.6 Provide an example of protein separation using conformational differences (Regnier, 1987).
5.7 Provide an example of kinetics of denaturation of an enzyme or enzymes on a surface used in RP-HPLC (Benedek *et al.*, 1984).
5.8 Provide an analysis of the scale-up of HIC purification of the antitumor antibiotic SN-07 (Ishida *et al.*, 1989).
5.9 Provide an example for the heparin HPLC separation of proteins (Dyr and Suttnar, 1991).
5.10 Provide an example of a large-scale immunoaffinity purification of recombinant soluble human antigen (sCDS) from *E. coli* cells (Wells *et al.*, 1993).
5.11 Briefly describe a method to purify α-amylase by immunoaffinity chromatography with a cross-reactive antibody (Katoh and Terashima, 1994).
5.12 Describe the separation of enzymes and long-chain fatty acids by CPC (Cazes, 1989).

CHAPTER 6

6.1 Describe a thermodynamic analysis of the activity and stability of globular proteins in the interior of reverse micelles (Battistel *et al.*, 1988).

6.2 Provide information pertaining to (1) the amount of enzyme/protein recovered (Jolivalt *et al.*, 1990); (2) the loss of activity (Sarcar *et al.*, 1992); and (3) the structural changes, if any (Samana *et al.*, 1984) exhibited by enzymes when subjected to the reverse micelle technique.
6.3 Provide information concerning the amount of surfactant and solubilizing water required to extract a given amount of protein using reverse micellar systems (Ichikawa *et al.*, 1992).
6.4 Describe the influence of temperature on protein desolubilization from reverse micelles (Dekker *et al.*, 1990).
6.5 An analysis of the continuous extraction of an enzyme by reverse micelles (Dekker *et al.*, 1986).
6.6 Describe the effect of water content and reverse micellar extraction on protein extraction from an aqueous phase into a reverse micellar phase (Hilhorst *et al.*, 1995).
6.7 Describe an analysis for the affinity partitioning of glycoproteins in reverse micelles (Paradkar and Dordick, 1991).
6.8 Provide an example of a large-scale fermentation and separation of a recombinant protein from *E. coli* (Strandberg *et al.*, 1991).
6.9 Describe the two-phase aqueous extraction of enzymes (Kula, 1987).
6.10 Briefly describe on-line monitoring of protein activity and concentration during a two-phase aqueous extraction (Papamichael *et al.*, 1991).
6.11 Briefly describe the genetically altered charge modification utilized to enhance the electrochemical partitioning of a β-galactosidase and T4 lysozyme in aqueous two-phase systems (Luther and Glatz, 1994).
6.12 Describe a theory that helps predict the partitioning of biomolecules in two-phase systems (Diamond and Hsu, 1989).
6.13 Describe the partition behavior of the extracellular protein, lipase from *Pseudomonas cepacia* using detergent-based two-phase aqueous systems (Terstappen *et al.*, 1992).
6.14 Describe briefly a mathematical model for the metal affinity partitioning of proteins (Suh and Arnold, 1990).

CHAPTER 7

7.1 Briefly describe some of the processes that are influenced both in a favorable and in a deleterious manner by protein adsorption. Also, briefly describe some of the effects that primarily control protein adsorption (Haynes *et al.*, 1994).
7.2 Describe briefly the adsorption of blood proteins to different surfaces.
7.3 Protein adsorption on surfaces indicates quantitative as well as qualitative features (Shirahama *et al.*, 1990).

7.4 Provide applications for heterogeneous adsorption of solutes from dilute solutions (Nikitas, 1989).
7.5 There is a paradox between concentration dependent adsorption and lack of desorption in pure buffer (Kop *et al.*, 1989).
7.6 Briefly describe the competitive adsorption of HSA, IgG, and fibrinogen on silica made hydrophobic by methylation or plasma deposition of hexamethyldisoloxane (HMDSO) using *in situ* ellipsometry and TIRF (Malmsten and Lassen, 1994).
7.7 Briefly describe plasma protein adsorption onto glutathione immobilized on gold (Lestellius *et al.*, 1995).
7.8 Briefly describe the adsorption of IgG and glucose oxidase (GO_x) to highly oriented pyrolytic graphite (HOPG) as analyzed by AFM (Cullen and Lowe, 1994).
7.9 Briefly describe a macroscopic model for a single-component protein adsorption (Al-Malah *et al.*, 1995).
7.10 Develop the equations between flowing blood proteins and an artificial surface (Schaaf and Dejardin, 1987).
7.11 There are some correlations between blood protein adsorption and surface properties (Grainger *et al.*, 1989).
7.12 Describe a technique for measuring protein adsorption wherein protein molecules are not modified by the introduction of some extrinsic label that might affect the adsorption kinetics (Norde and Rouwendal, 1990).

CHAPTER 8

8.1 Provide a brief economic analysis for utilizing centrifuges for single- and multiuse facilities (Mahar, 1993).
8.2 Describe briefly the changes made by Genentech as the dosage requirements for tPA increased from 1 to 100 mg during clinical trials (Spalding, 1991).
8.3 Describe briefly the modifications made by Hoffman-LaRoche during the large-scale processing of α-interferon A.
8.4 Demonstrate the applicability of the down-scaling approach for the gel filtration of a polymeric protein mixture that has a molecular weight-size distribution between 30×10^6 and 80×10^3 Da and a mass average molecular weight of 3.98×10^6 (Naveh, 1990).
8.5 Provide economic data for the separation of tPA, monoclonal antibodies, and animal growth factors utilizing perfusion chromatography. Present three different strategies for operating chromatographic columns (Fulton *et al.*, 1992).
8.6 Provide some reasons why other bioseparation techniques have not been applied on a commercial scale. Consider a particular case, for example, two-phase aqueous systems (Huddleston *et al.*, 1992).

8.7 Provide some economic data on a technique that effectively separates relatively large amounts of monoclonal antibodies (Duffy *et al.*, 1989).
8.8 Analyze briefly some of the major cost elements in designing immunosorbent columns on a large scale (Desai, 1990).
8.9 Briefly present the costs involved in running chromatographic separations on a large scale (Peskin and Rudge, 1992).
8.10 Describe briefly the qualitative features of the Porter-Ladisch model (Porter and Ladisch, 1992) for the cost estimation of separation of α-galactosidase from soybean seeds. In other words comment on the relative costs of each purification step.
8.11 Present briefly the economics of separation of bioproducts for an *E. coli* based fermentation process (Datar, 1986).
8.12 Present briefly the process design and economics for the production of polygalaturonases from *Kluyveromyces marxianus* (Harsha *et al.*, 1993).

CHAPTER 9

9.1 Briefly mention some of the nonproteinaceous materials or additives that have been utilized to assist in protein refolding (Zardeneta and Horowitz, 1994).
9.2 Briefly present and analyze the different protein purification strategies (protocols) that have been utilized to separate proteins in the denatured state (Knuth and Burgess, 1987).
9.3 Briefly describe the effects of mutations on the aggregation of proteins (Wetzel, 1994).
9.4 Briefly describe the simulation of a folding pathway (Hinds and Levitt, 1995).
9.5 Describe the influence of the reversible and irreversible denaturation of Nase on aggregate formation (Nohara *et al.*, 1994).
9.6 Briefly describe the purification and renaturation of recombinant human interleukin-2 (IL-2) (Weir and Sparks, 1987).
9.7 Briefly describe the *in vitro* folding of glycoprotein hormone chorionic gonadotropin (Huth *et al.*, 1994).
9.8 Briefly show the influence of chaperonins and protein disulfide isomerases on the renaturation of single-chain immunotoxin (Buchner *et al.*, 1992).
9.9 Briefly describe the chaperonin-facilitated *in vitro* folding of monomeric mitochondrial rhodanese (Mendoza *et al.*, 1991).
9.10 Compare briefly the refolding of proteins by the use of assistants such as detergents, lipids, and micelles with chaperonin-assisted refolding (Zardeneta and Horowitz, 1994).
9.11 Briefly analyze cysteine to serine substitution on basic fibroblast growth factor (bFGF) IB formation during *in vitro* refolding (Rinas *et al.*, 1992).

9.12 Describe PEG-assisted refolding of three recombinant human proteins (Cleland *et al.*, 1992).
9.13 Briefly analyze the antibody-assisted protein refolding (Carlson and Yarmush, 1992).
9.14 Briefly analyze protein refolding in reverse micelles (Hagen *et al.*, 1990a).
9.15 Briefly describe the influence of environmental conditions on the refolding selectivity of insulin-like growth factor I (Hart *et al.*, 1994).

CHAPTER 10

10.1 Describe briefly some of the considerations that must be examined to set the stage for later validation work (Akers *et al.*, 1994).
10.2 Briefly describe the procedures involved in the validation of β-Urogastrone (Brewer, 1986).
10.3 Explain the concern over the removal of DNA and protein impurities in biopharmaceuticals (Briggs and Panfili, 1991).
10.4 Briefly describe the avoidance of unsafe levels of host cell protein contaminants that might lead to toxic or immunologic reactions (Eaton, 1995).
10.5 Briefly describe the validation of column-based separation processes (PDA Report, 1992).
10.6 Briefly describe the life cycle approach to analytic methods during pharmaceutical product development (Hokanson, 1994).
10.7 Briefly describe the validation procedure to purify MAbs from mouse ascites fluid (Mariani and Tarditi, 1992).
10.8 Show validation studies in the regeneration of ion-exchange celluloses (Levinson *et al.*, 1995).
10.9 Briefly describe some of the important results of cleaning validation and residue limits (Zeller, 1993).
10.10 Briefly describe chromatography cleaning validation (Adner and Sofer, 1994).

1 INTRODUCTION

I. INTRODUCTION

Advances in genetic engineering, rDNA technology, and cell fusion techniques have made it possible to produce proteins of interest; however, the technology has not kept pace with these advances in sufficient quantities. Baum (1987) emphasized the need to process biological products to a high degree of purity on a large scale. Diamond and Hsu (1989) emphasized that the separation procedures should be economical and biocompatible. The National Committee on Bioprocess Engineering (1993) identified the development of separation and purification strategies for biological products from dilute aqueous solutions as a critical need for obtaining specialty bioproducts and industrial chemicals. These dilute aqueous solutions are often obtained in processing biological materials from fermentation, plant cell culture, or whole plant material.

II. THE NEED FOR BIOSEPARATION

Furthermore, with the ever-increasing emphasis on safety with regard to regulatory agency requirements and public awareness, Lilly (1992) correctly emphasized the increasing importance of product quality, and not just the amount of product produced during a process. Lilly (1992) emphasized that to maintain product quality undesirable posttranslational changes must be either minimized or prevented. These changes may occur during both upstream and downstream processing. Also, most proteins must be folded into a specific three-dimensional conformation to express their biological activities and specificity, which complicates the process of separating and purifying them. The high cost of separation and purification coupled with the difficulty of getting highly purified products prevents some biotechnological processes with applications in medicine, agriculture, and industry from becoming viable, cost effective, and successful.

People working in the industry realized this, and subsequently many of them got involved in protein separation and purification. As a result of their research, novel and imaginative techniques sprang up. Some researchers modified existing procedures such as chromatography, electrophoresis, and precipitation. As expected, not all the techniques developed have the potential to be applied extensively. Thus, new and novel bioseparation techniques are gradually being developed and analyzed for their effectiveness. Also, Wheelwright (1989) emphasizes that even though quite a few downstream processes are in operation, there is no definite and predictable method or algorithm that one may follow to design a bioseparation protocol for a specific protein or biological product. This author emphasizes that the number of processes available and the subtle differences that exist between the different proteins make the development of a generalized algorithm for the step-by-step design of a bioseparation protocol more difficult.

Even though the generalized development of a bioseparation protocol is seemingly difficult, simplistic guidelines coupled with invaluable hands-on experience should provide the next best approach. Hopefully, the availability of more information in this area with respect to all the aspects of the bioseparation protocol should move bioseparation from an art to a science. The chapters that follow are an attempt in this direction. Also, in general, protein purification techniques should be simple, easily scalable, continuous, low cost; and, of course, should not inactivate the protein. Also, continuous processes are not always desirable. For example, high-value therapeutic proteins are produced in a batch mode for different reasons, including cost and risk factors.

III. CLASSIFICATION OF BIOSEPARATION STEPS

Cussler (1987) indicates that although a variety of bioseparation procedures exist, they can be classified into four distinct steps that include removal of insolubles, isolation of product, purification, and polishing. As is to be expected, a wide variety of bioseparation procedures are available. Because these processes contribute significantly to the cost of the product, Van Brunt (1985) emphasizes that the economic consequences of these processes must be carefully considered. Van Brunt (1985) indicates that bioseparation processes include, but are not limited to, cell disruption, centrifugation, chromatography, drying, evaporation, extraction, filtration, membrane separation, and precipitation. This author emphasizes that some of these processes are classical and their mechanisms of action are well documented in the literature. Some of the preceding processes still have to be proved, especially on the large-scale level.

The end product of interest to be obtained from these processes must meet varying, rather strict demands before it can be placed on the marketplace. For example, the product must be sterile; attain stringent quality requirements; and be free from detergents, endotoxins, proteases, etc. Curling (1985) indicates that a pure product should satisfy the demands of no immunogenic substances present, no unwanted biological activity present, no microbiological contamination, and no enzymatic activity present that is harmful to the product. For example, other proteins, modified proteins, nucleic acids, oligonucleotides, or nucleotides contribute to an immunogenic response. Enterotoxins and nonspecific activity (such as complement activation) contribute to unwanted biological activity.

In general, the end product quality requirements are largely dependent on the end use of the product. For therapeutic usage some of the requirements that are to be met include potency, identity, abnormal toxicity, nucleic acids, homogeneity, etc. (Desai, 1990). The bioseparation process or protocol that is utilized to separate the product must satisfy these requirements at the end. Huddleston *et al.* (1991) indicate that bioseparation processes are defined by the nature of the product and its application. For some cases a high degree of purity is required, whereas in others simply the absence of conflicting activity is sufficient. Huddleston *et al.* (1991) emphasize that during the initial bioseparation steps one attempts to maximize product yield even at the expense of retaining contaminants. These contaminants may be removed later using high-resolution fractionation processes. Furthermore, Huddleston *et al.* (1991) emphasize the compromise that is required in the bioseparation protocol during the harvesting, product release, clarification, enrichment, and fractionation stages. Besides, one has to be careful in the bioseparation protocol to maintain an adequate containment of any potentially hazardous by-products.

One will require a wide variety of steps in the bioseparation protocol to meet different demands on the quality of the end product. Harakas (1989), however, emphasized that one has to limit the number of steps; and one should get the most out of each step. Ideally, one should, if it is at all possible, try to restrict the bioseparation protocol to just two or three steps. Also, Harakas (1989) emphasized that one should attempt to obtain at least 90% of the product from each step. Thus, if we have two steps then the overall efficiency is 81%. If three steps are utilized, then the efficiency drops to about 73%. Note that three steps of efficiency of 80% each will eventually yield an overall efficiency of 51.2%. Thus, the need is to use as few steps as possible, and also to get as much as you can from each step.

This rapid decrease in overall efficiency has led different workers to integrate or to combine the different steps in the bioseparation protocol. This is also known as process intensification (Third International Conference on Separations for Biotechnology, 1994). Lyddiatt (1994) analyzed the use of fluidized diethylaminoethyl(DEAE)-Spherodex to combine the recovery of acidic protease with the fermentation of *Yarrowia lipolytica* cells. Also, Chang (1994) used expanded-bed adsorption for the direct extraction of glucose-6-phosphate dehydrogenase from modified yeast homogenate. This integration of steps may be either in the upstream process or in the downstream process.

Datar *et al.* (1993) have also recommended integration of unit operations. Hanson and Rouan (1994), too, have utilized the expanded-bed adsorption technique to directly recover secreted recombinant fusion protein from a crude

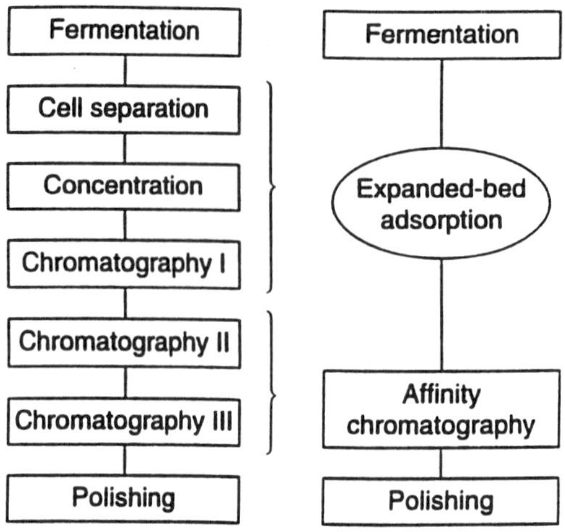

FIGURE 1.1 Integration of unit operations using genetic design of product. Left hand panel shows the classical steps involved. The right hand panel indicates the integration of cell separation, concentration, and chromatographic recovery into a single step (Nygren *et al.*, 1995).

fermenter broth. This was done without prior cell removal. The fusion protein was designed to exhibit a relatively low pI. This permitted the anionic exchange adsorption at pH 5.5. At this pH the other host proteins are not adsorbed. These authors obtained a 90% overall recovery using this procedure. Figure 1.1 shows the integration of the bioseparation steps using genetic design of this product. Nygren *et al.* (1995), too, emphasized that integrated processes may be utilized to yield biological products with high recoveries at low cost.

IV. UPSTREAM AND DOWNSTREAM PROCESSING

Traditionally, all the steps occurring in the fermentor are considered upstream processes. The processes occurring after the fermentor are considered downstream processes. During downstream processing, Dunnill (1983) indicated that one should not lose more of the product than is absolutely necessary. In other words, manipulate the downstream conditions so as to attempt to minimize loss of the valuable biological product.

The choice or selection of the process, or the sequence of steps involved in the bioseparation protocol depends on the production host, location, and physical form of the protein in the cell (Naveh, 1990). The upstream processes employed during the fermentation step will also significantly influence the composition of the bioseparation protocol. Middleberg *et al.* (1992) emphasized the importance of process interactions during the development of an optimal design and operating strategy for a bioseparation process. For example, Middleberg *et al.* (1992) stated that the separation of cellular debris and inclusion bodies is a critical step in a bioseparation process. If a coarse-grade centrifuge is used during the initial fractionation step, then this results in similar distributions for the inclusion bodies and the cellular debris. This "closeness" of the debris distribution with the inclusion body (protein) distribution will result in contamination of the protein, and will subsequently cause problems during the chromatographic steps or other suitable high-resolution fractionation steps. Middleberg *et al.* (1992) suggested that increasing the number of passes through the homogenizer will increase the fraction of cells disrupted. Also, the size of the resultant debris will be reduced. This facilitates the separation of the cell debris from the inclusion bodies. Thus, harsher homogenization conditions during the initial fractionation steps will facilitate the separation of the required protein during the high-resolution fractionation steps used in the bioseparation protocol.

Dunnill (1983) added a word of caution in that efforts to enhance the yield by greater cell rupture may lead to intractably fine debris that carries through several subsequent stages. Thus, there is a need for the design of "smart" systems. In other words, carefully analyze and examine each modification of a process suggested. Dunnill (1983) also emphasized the importance of understanding upstream (fermentation) conditions in relation to downstream processing conditions. For example, the time for harvesting has a significant effect on the cell wall strength and on the level of endogeneous proteases. An understanding of these influences will significantly influence the recovery of protein or other biological products of interest.

It is also possible to separate a biological product by different protocols or routes. Many times it may be advisable to investigate the different bioseparation routes and analyze which one is best suited to one's needs. Ajongwen *et al.* (1993) investigated the large-scale purification of *Leuconostoc mesenteriodes* NRRL 13512F dextransucrase. Figure 1.2 shows the different routes investigated by these authors for dextransucrase purification. Ajongwen *et al.* (1993) noted that routes 1-3-7-8 and 2-5 gave the highest overall purification and recovery. The 2-5 ultrafiltration procedure did give significant membrane fouling. These authors also noted that dextransucrase purification by the bioseparation routes 1-3-6-(9 or 10), 1-3-7-(9 or 10), and 1-4-(8 or 9 or 10) constrained throughput values; besides, the degree of purification and cell removal was also decreased. After a careful analysis of their process Ajongwen *et al.* (1993) noted that the best bioseparation route involved a three-stage process of continuous centrifugation, continuous ultrafiltration, and subsequent second centrifugation process. This bioseparation route yielded a 95% pure dextransucrase and the overall recovery was 60%.

Lilly (1992) highlighted some of the developments at the upstream and at the downstream stages for biological processes. Both of these have significantly contributed to enhancing the nature of biochemical engineering research. This author has succinctly emphasized the feedback element, or the interactive nature of the upstream and downstream stages, in the development of a biological process (Fig. 1.3). One optimizes the upstream and the downstream processes taken together, keeping the biological process development objective in mind (Lilly, 1992). This biological process development objective includes getting the product to the market in the quickest time possible, meeting all possible safety requirements, and making the process cost-effective and reliable. Safety factors become particularly important during scale-up (Van Brunt, 1985). One has to make sure that all liquids coming out of fermentors are free of undesirable recombinant organisms. Van Brunt (1985) emphasized that it is no small feat

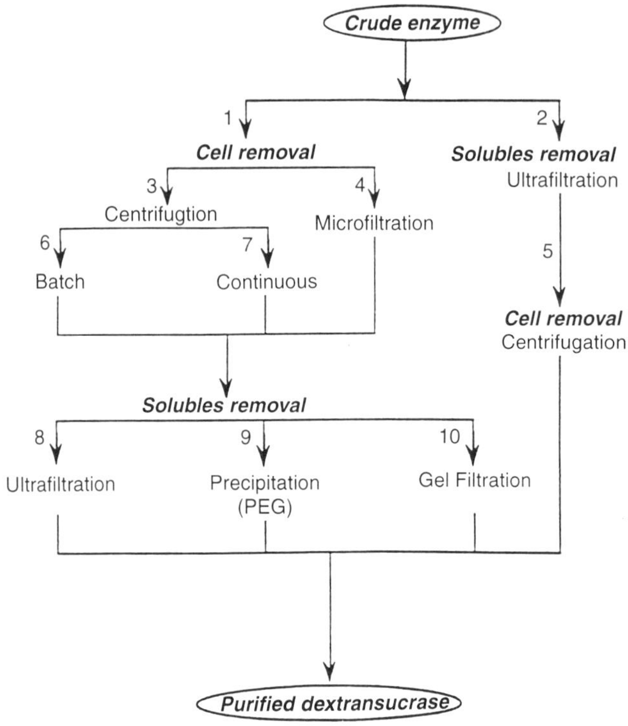

FIGURE 1.2 Different bioseparation routes for dextransucrase purification from *Leuconostoc mesenteriodes* NRRL B512F (Ajongwen *et al.*, 1993).

to heat-kill 1000-liter cultures on a regular basis. Lilly (1992) emphasized some of the significant advances that have been made to enhance the quantity and quality of a biological product. Upstream or fermentation advances include continuous media sterilization, improved agitation systems (enhance mass transfer), gas analysis by mass spectrometry (improves accuracy and facilitates the monitoring of reactions), and computerized data logging provides a good database and facilitates feedback to improve the reaction conditions (Lilly, 1992). Leser (1994) analyzed the use of an expert system to assist in the selection of protein-bioseparation processes. The database contains the physicochemical properties of the major contaminating proteins in *Escherichia coli* (Chaudhari, 1995). The database can be utilized to help select these bioseparation processes that help maximize recovery of the required protein or biological product. Wiblin (1994) also utilized a computer-based model to assist in the scale-up and optimization of expanded-bed and affinity chromatography.

Not only should one seek to improve or to modify the existing processes, but also one should actively search for novel techniques. Novel techniques are required to cope with the maturity of the different industries. These could involve bioprocessing under zero gravity (microgravity) conditions. Molecular imprinting of polymers for selective adsorption has been shown to effectively separate small molecules (Whitcombe, 1994). Chaudhari (1995) emphasized that if this method is to be used for the separation of enzymes, then methods to preserve the native structure during imprinting are required. Also, in general, gas-liquid interfaces in bioseparations should be avoided. Nevertheless, by using foams Varlie (1994) was able to separate the enzymes, trypsin, lysozyme, and catalase without significant loss in activity. These techniques along with others may be considered as emerging technologies (Third International Conference on Separations for Biotechnology, 1994). As different industries (such as biopharmaceuticals) mature, more and more sophistication will need to be employed to meet with the increasing demands of the future such as ton-scale processes for bioproducts from recombinant sources (Fulton *et al.*, 1992).

Boudreault and Armstrong (1988) indicated the advantages of near zero gravity conditions on bioprocessing. These authors emphasize that near zero gravity conditions permit containerless handling. This eliminates the vessel walls that could be a source of mechanical stress. This shear stress is disadvantageous as far as protein denaturation and quality of the protein or other biological product separated are concerned.

The Center for Space Policy (1985) in Boston has estimated that space biotechnology processing will increase to $\$15 \times 10^9$ by the year 2000. This is about 4.2% of the total estimate for biotechnology products of $\$350 \times 10^9$ by the year 2000. The emphasis will be on electrophoretic separations (Todd, 1985) and on thermodynamic phase separations (Brooks *et al.*, 1986). For example, higher throughputs (an increase by a factor of 556) for electrophoretic separations in space compared with those on the ground have been obtained. For larger macromolecules this number increases to 730 (Boudreault and Armstrong, 1988). Two-phase partitioning is also being utilized to separate biological macromolecules of interest. This technique is based on the thermodynamic principle that systems must minimize their free energy. Boudreault and Armstrong (1988) emphasized that on-the-ground sedimentation and convection lead to mixing that is larger than the natural thermodynamic demixing of a two-phase system. This effect is not present (or is minimized) under microgravity conditions. Thus, bioseparations performed under microgravity conditions exhibit the potential for the precise separation of biological macromolecules at a large-scale level.

Crystallization of proteins can be enhanced under microgravity conditions (Boudreault and Armstrong, 1988). Crystals are formed when weak interacting

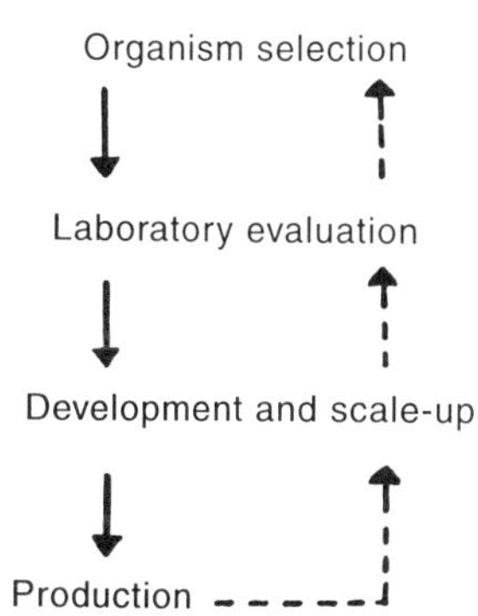

FIGURE 1.3 Steps involved during the development of an industrial fermentation process. Feedback of information denoted by dotted lines (Lilly, 1992).

forces stabilize the crystals in solution. On-the-ground sedimentation and gravity forces, which are stronger than these weak interactive forces, destabilize the crystals by producing a larger number of crystallization sites and smaller crystals. Apparently, biopharmaceutical companies were ready to pay $100,000 to $200,000 to crystallize a protein in space (Boudreault and Armstrong, 1988). The advantages and expense of near zero gravity operating conditions need to be carefully analyzed with respect to on-the-ground bioprocessing for different applications.

V. SOME FACTORS INFLUENCING BIOSEPARATION

A. Process Monitoring

Process monitoring plays an important role in the continuous search for optimizing processes for enhanced biological product recovery as far as quality and quantity are concerned. Geisow (1992) emphasized that it is helpful to monitor the structure and biological activity of different compounds during a continuous process. These continuous processes are, in general, more flexible than batch processes. Mattiasson and Hakanson (1993) stressed that to assist in the good monitoring of biological systems one should ideally use *in situ* sensors that are in direct contact with the reaction medium. This will facilitate the processing of information on a real-time basis, besides permitting a minimal response delay that assists in the control of these systems. These authors do admit, however, that biosensors have not been developed enough to meet the *in situ* requirements, and most analyses are done either off-line or on-line but outside the reactor. Thus, sampling and sample handling strategies also play an important role. If a nonfouling optical density sensor could be developed, then that would find substantial utilization in the biotechnology industry.

Geisow (1992) indicated that flow injection analysis (FIA) and synchronized peak-switching high-pressure liquid chromatography (HPLC) may be employed as semi-on-line techniques for bioprocess monitoring. In the FIA system samples are injected at intervals under computer control into a semi off-line FIA system. Samples from HPLC separations can be analyzed further by capillary electrophoresis or by mass spectrometry. On-line analysis is constained by the slower time frames for capillary electrophoresis and atmospheric pressure ionization mass spectrometry. Geisow (1992) emphasized that both capillary electrophoresis and HPLC do offer opportunities for semi-on-line monitoring of biological products.

Paliwal *et al.* (1993) addressed the importance of rapid process monitoring in fermentation and in downstream recovery processes. This is particularly important as far as quality control is concerned. These authors indicated that proteins manufactured by recombinant means must meet the strict reqirements set by the regulatory agencies. Process failures occur largely because of slightly different protein structural forms (other than the required form) that are produced during either fermentation (errors in gene expression during upstream processing) or downstream processing (posttranslational modifications). Process failures also occur when proteins from the host are coprocessed with the required protein. In both cases this is undesirable; thus, there is a need for appropriate quality control by rapid process monitioring and subsequent feedback control.

Paliwal *et al.* (1993) emphasized that the rapid detection of the different protein variants or conformational states is a difficult task. Quick corrective actions are required to help control and validate the process. These authors pointed out some of the reasons that may lead to a varying product outside the variance limits set by quality control. These reasons include aging equipment, different feedstocks, and changing parameter values during either fermentation or downstream processing. Thus, it is important to have modeling and a good database. Also, if one recognizes that one does have a poor quality product then one should: (1) make corrective actions to get the product up to requirements, for example, recycling for further purification (Paliwal *et al.*, 1993); or (2) identify and discard the product if it cannot be "fixed." Paliwal *et al.* (1993) emphasized that most of the equipment or technology either is already available or is under development.

B. Bioseparation Economics

Wesseling (1994) indicated that downstream processing makes up at least 50%, if not more, of the total cost of bioseparations. Still, one spends only about 5% of effort on this downstream processing. Furthermore, this author states that in the future marketing pressures will tend to minimize product development times. Environmental considerations, however, will tend to increase development times. These considerations are necessary to reduce the degree of waste and emissions. Wesseling (1994) demonstrated that typically to produce 0.2 m^3 of product, an average 650 m^3 of waste is generated. This represents a ratio of 1:3250 of product generated to waste produced. This ratio is more than three orders of magnitude.

Because downstream processing is such a large contributor to the total cost of producing a biological product, modifications produced either upstream or downstream that assist in the economical recovery of the product are welcome. One should, of course, carefully analyze and examine the influence of each modification. For example, by changing molecular biology or fermentation parameters, or by making appropriate modifications one can reduce or even eliminate problems encountered downstream. Uhlen and Nilsson (1985) indicated that the construction of hybrid genes may be utilized to advantage during downstream processing. Uhlen and Nilsson (1985) emphasized that one may fuse the coding of the protein of interest with the coding sequence of a polypeptide chain with a high affinity to a ligand. This facilitates the recovery of the desired protein by a single step utilizing this affinity tail technique (Uhlen, 1983, 1984; Ullman, 1984; Smith, 1984).

Heng and Glatz (1993) emphasized that the preceding charged-fusion technique is useful for solving difficult and challenging bioseparations. Heng and Glatz (1993) were able to selectively recover β-galactosidase from cell extract and noted insignificant conformational changes in the protein recovered when compared with the affinity-purified protein. This was judged by similar specific activities obtained for the recovery of the protein by the charged-fusion and the affinity-purified techniques. Although different fusion techniques have been attempted (Sassenfeld and Brewer, 1984; Uhlen *et al.*, 1983; Veide *et al.*, 1987), problems still remain, especially with regard to high-cost and large-scale operations.

C. Protein Refolding and Inclusion Bodies

Mitraki and King (1989) indicated that the presence of inclusion bodies is one of the problems associated with recombinant DNA technology. Often the recovery of the desirable protein is constrained, sometimes severely, by the presence of inclusion bodies (Pelham, 1986; Goloubinoff *et al.*, 1989). These inclusion bodies are basically aggregates of incorrectly folded and aggregrate forms of the required protein. Figure 1.4 shows one possible hypothetical folding pathway that yields these inclusion bodies (Mitraki and King, 1989). These authors indicated that these inclusion bodies are obtained from specific partially folded intermediates and not from native or fully unfolded protein chains (as Fig. 1.4 also indicates). Mitraki and King (1989) emphasized that these processes are highly specific and rather sensitive to genetic engineering modifications and environmental changes. For example, these authors suggested that if specific sites are involved in aggregate formation, then appropriate amino acid

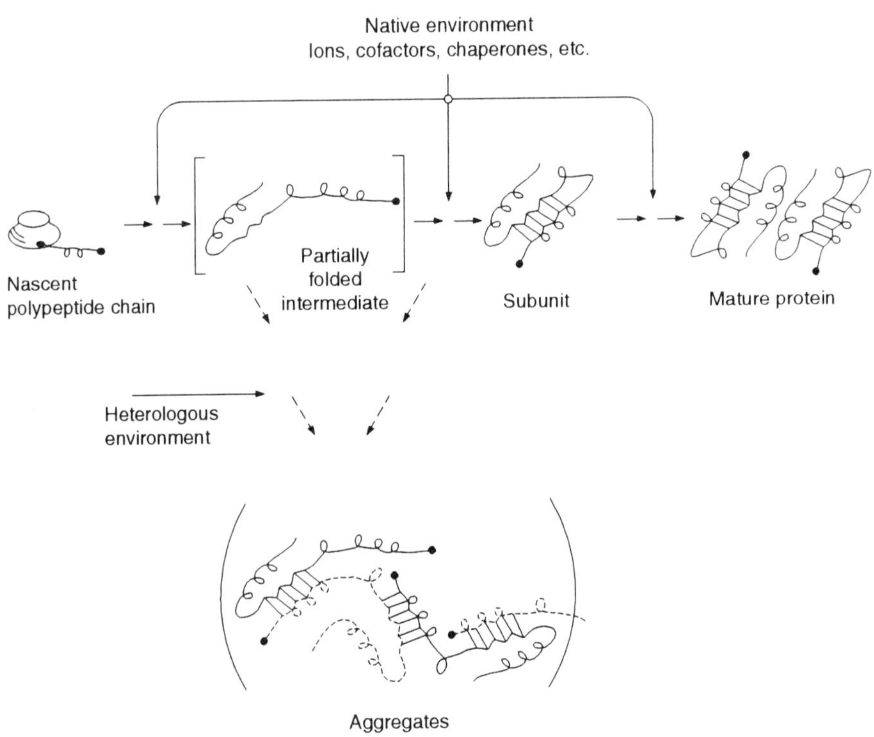

FIGURE 1.4 Hypothetical folding pathways for a dimeric protein (Mitraki and King, 1989).

changes could either increase or decrease the yield of the aggregates formed. Klibanov (1983) suggested that aggregative processes may be minimized by eliminating diffusion. Mozhaev *et al.* (1990) indicated that additional similar charged groups introduced on the protein surface by covalent modification enhance repulsion and minimize protein-protein contact. Miller (1994) analyzed the use of chaperones hsp 10 and hsp 60 to assist in the folding of the enzyme mitochondrial malate dehydrogenase *in vitro*.

Kane and Hartley (1988) indicated that there are obvious advantages to processing these aggregates or refractile bodies, especially for low-cost products. The initial purification to a relatively pure product (about 50%) is relatively straightforward. Furthermore, Kane and Hartley (1988) demonstrated that an ion- exchange step can relatively easily increase the purity of the product to 90 percent. Cheng *et al.* (1981) emphasized an additional advantage is that intracellular proteases do not attack these refractile or inclusion bodies. Thereby, proteolytic clippage that leads to a poor quality product is reduced. There are several advantages to the processing of these inclusion bodies. The major problem, of course, is that these inclusion bodies have to be solubilized, and then refolded to the correct native and active form for the protein.

The further treatment of inclusion bodies to enhance the recovery of proteins and other biological products is an area of tremendous interest with considerable effort being spent in this direction. The importance of this area is underscored by analyzing this topic in Chap. 9. The successful development of this technique will apparently depend to a large extent on understanding the mechanisms of refolding or renaturation of the protein to the correct active and stable form.

In contrast to minimizing denaturation during the bioseparation step, it may even be advantageous, in some cases, to intentionally denature a protein during bioseparation. Knuth and Burgess (1987) reviewed the principles and processes for purifying proteins in the denatured state. These authors emphasize that researchers must free themselves of the "mind block" that proteins should not be purposely denatured. These authors also indicated that many proteins can be purified in the denatured state, and subsequently renatured. These authors mentioned that many powerful techniques for the separation of proteins in the denatured state are available.

Knuth and Burgess (1987) emphasized the usage of denaturants to enhance protein bioseparations. For example, during sodium dodecyl sulfate-polyacrylamide gel electrophoresis (SDS-PAGE) the oligomers are dissociated to give each protein an equal charge. These proteins are denatured to a rodlike conformation, and electrophoretic mobility is proportional to the molecular weight. Knuth and Burgess (1987) stressed that this technique is popular as an analytic tool. The SDS-PAGE treated proteins are then renatured to the native and active form. Knuth and Burgess (1987) emphasized that a wide variety of proteins have been recovered effectively using SDS-PAGE after elution from the gels and subsequent removal of the detergent. Some of the proteins recovered include DNA topoisomerase (Hager and Burgess, 1980), estrogen receptor (Sakai and Gorski, 1984), and fructosyltransferase (Russell, 1979).

The bioseparation of different biological products of interest may be facilitated in many cases by an appropriate chemical modification. Knuth and Burgess (1987) indicated that extraction of proteins into organic solvents may be facilitated by the covalent attachment of polyethylene glycol (Inada *et al.*, 1986) or other hydrophobic agents (Criado *et al.*, 1980) to the proteins. Light (1985) indicated that the solubility of proteins into aqueous solution may be facilitated by the addition of charged groups to protein surfaces. Sadana and Henley (1986) and Inada *et al.* (1986) suggested that chemical modifications may or may not alter the specific activity and the stability of the proteins. This can, in turn, significantly influence the quantity and the quality of the protein product separated.

Burgess *et al.* (1991) indicated that the interfacial adsorption of proteins may lead to denaturation, aggregation, precipitation, foaming, and enhanced rates of chemical degradation. These authors emphasized the importance of analyzing interfacial adsorption, and how its deleterious effects on denaturation may be minimized. Leckband and Israelachvili (1993) analyzed the influence and importance of direct force measurements on protein adsorption and function. These authors emphasized that an understanding of the nonspecific protein interactions involved during the bioseparation step will significantly improve the efficiency of the bioseparation step. These authors stressed the delicate balance of the van der Waals, electrostatic, hydrophobic, and hydration forces between the protein and the surface or interface involved.

Of particular interest are the factors that lead to a deterioration of the protein quality. If these factors are known, then extreme care may be taken to help minimize these deleterious effects. For example, Kim *et al.* (1993) analyzed the factors that promoted the aggregation of albumin during ultrafiltration. At the membrane surface these authors noted rapid supersaturation of the protein. Besides, high shear promoted protein aggregation on the membrane surface. The high shear unfolds these aggregates on the membrane surface, which then facilitates protein flocculation by collisions. Protein adsorption on different surfaces and the subsequent conformational changes that result are of significant interest.

Roper and Lightfoot (1995) analyzed the separation of biomolecules using adsorptive membranes. These authors emphasized that the recovery of these fragile biological macromolecules necessitates careful attention to their unique properties. The recovery processes should, in general: (1) use mild conditions, (2) minimize process time, (3) avoid extreme pH and temperature conditions, (4) avoid exposure to air–water interfaces and shear, (5) minimize exposure to nonpolar solvents and hydrophobic adsorbents, and (6) be efficient. The mild operating conditions would help maintain the native conformation and integrity of the porduct. This helps preserve the biological activity of these macromolecules. Unnecessary long processing times lead to degradation of gene products. These authors indicated that during downstream processing (minor) variants of proteins and nucleic acids may be produced due to deamidation, oxidation, proteolysis, nicking, and aggregation. Zhang *et al.* (1991) emphasized that the fraction of degradation products increases with residence time.

Roper and Lightfoot (1995) indicated that by avoiding extreme pH and temperature conditions; and by minimizing exposure to air–water interfaces, nonpolar solvents and hydrophobic adsorbents, one may help alleviate the subsequent denaturation of many enzymes and the destabilization of biological products. These authors emphasize that the recovery process should be efficient so that they are competetive and economical. Carson (1994) emphasized that inefficient processes consume unnecessary large volumes of expensive solvents that must be either regenerated or disposed. Asenjo *et al.* (1991) indicated that solvent tankage and consumption can represent a not insignificant fraction of the total bioseparation process costs.

In a book on downstream processing by Kennedy and Cabral (1993) the central theme is product loss during the different stages of processing and how it affects the process economics. No matter how much precaution one takes, during downstream processing there will presumably be some protein or other biological product denaturation or deterioration. Thus, there is a need for corrective action, if possible. One possible corrective action is renaturation. Thus, renaturation steps are also important; and one can use them to help improve the quality of the protein separated, if need be. Knuth and Burgess (1987) reviewed the techniques involved in removing detergents from protein solutions. Furth (1980) indicated that ionic detergents may be removed by treatment on an ion-exchange column. Also, detergents with a long hydrophobic side chain may be removed by adsorption on a hydrophobic resin. Henderson *et al.* (1979) indicated that detergents may also be removed by solvent extraction. When all the factors are taken into consideration, the bottom line, as expected, makes all the difference. By considering the large sums of money that need to be spent in getting a biotechnological product ready for the market, it behooves one to pay particular attention to process economics. Gilbert (1993) indicated that the future for biotechnology is still bright. This author mentioned that 68 public, drug-related companies have raised \$6.6 billion between 1982 and 1992. There is considerable profit for the biotechnology companies with a marketable product that is in demand and meets with Federal and Drug Administration (FDA) regulations. Table 1.1 further highlights the future sales estimates of some of the most promising drugs. Only some of the drugs are presented that have estimated sales of \$1000 $\times$ 10^6 in the years 1995 and 2000 (Scrip and Rorschild Asset Management aand Other Sources, 1993).

Table 1.1 also indicates the future potential and the tremendous interest in processing biotechnological products and drugs for market consumption.

Apparently only the following four companies are really profitable in the United States: Amgen, Biogen, Genentech, and Genzyme. Gilbert (1993) mentioned that the 68 public, drug-related companies produced 17 biotechnology drugs that had a market value estimated to be \$2.4 billion in 1992 U.S. sales, and \$4.0 billion in worldwide sales. The largest selling biotechnology product in 1992 U.S. sales was Amgen's erythropoietin at \$506 million. Lilly's human growth hormone had a 1992 U.S. sales of \$430 million.

Gilbert (1993), however, cautioned about the risks that are involved in developing a product. For example, not every drug in clinical trials will be efficacious. This author mentioned that there are pitfalls and unpredictability at different stages that include drug development, clinical trials, FDA advisory committee review, and FDA approval. Furthermore, due to increasing competition, Gilbert (1993) indicated that there is further uncertainity in predicting market size, price of the drug, and market penetration. The need for improvements in both upstream and downstream processing is bound to grow along

TABLE 1.1 Estimated Pharmaceutical Drug Sales of $1000 × 10^6 and over for the Years 1995 and 2000[a]

Drug	Estimated 1995 sales (in million $)	Estimated 2000 sales (in million $)
Insulin	1000	1000–2000
Human growth hormone (hGH)	1000+	1000+
Interleukin-1 (IL-1)	1100+	
Colony-stimulating factor (CSF)	1000–2000	
G-CSF	1000–2000	1000–2500
GM-CSF	1000–3000	
Erythropoietin (EPO)	1000–2000	1000–3500
Anti-IL-2		1000+
Imaging monoclonal antibodies (MAbs)		500–2000

[a] From Scrip and Rorschild Asset Management and Other Sources (1993). *Biotechnology Suppl., 11,* May. With permission.

with: (1) increasing demands for both more biotechnology-derived drugs and (2) increasing purity requirements for these types of drugs to minimize unwanted side effects and to be more specific in their medicinal or therapeutic application.

More often than not the economics of the process (besides other general considerations) will largely determine if a company can produce a particular drug. Innovative methods ought to be explored to gain an economic edge. Hamers (1993) emphasized the need to explore a multiuse flexible facility for the manufacture of biotechnology-derived drugs. This is of particular value for small companies that cannot afford the expense required for a dedicated facility to manufacture a drug, considering the many pitfalls that abound. Note that traditionally one uses a dedicated facility to produce therapeutic products. Nevertheless, Hamers (1993) stressed that one should consider sharing a facility with other companies, especially during clinical trials where small amounts of the desired material are required. This will significantly reduce designing, building, and running costs. The improvement in process economics is possible, although at a cost.

Hamers (1993) suggested certain guidelines to follow. Particular attention, of course, must be paid to segregation (when flow paths cross), stringent validation, and careful scheduling to optimize the plant utilization and to keeping a premium on safety requirements. Other aspects of the multiuse facility that Hamers (1993) addressed include a discussion on fermentation, the primary recovery step, the purification step, the filling step, and the flow paths. Because the economics of downstream separation are important and are a significant fraction of the biotechnology-derived protein or drug processing costs, they are discussed in more detail in Chap. 8.

There is bound to be an increasing emphasis on quality control of biotechnology-derived products in the future. Also, intensified pressures from regulatory agencies, enhanced public awareness, expanded and almost fierce competition, safety factors, requirements such as more specific action of the drug with minimal side effects, and other factors will increasingly emphasize the qualitative aspects. Pharmaceutical and other biotechnology companies are well aware of this, and are beginning to pay more attention to the quality control aspects of biological product manufacturing. Information on the qualitative aspects of biological product recovery is scarcely available in the literature. This type of information is apparently available primarily in industrial sources. The reluctance of industrial sources to freely part with this information is understandable. The chapters that follow attempt to bring under one cover information on the quantitative as well as the qualitative aspects of the processing of biological products. Emphasis is placed on the downstream processing aspects.

Akers and Nail (1994) presented the top ten concerns or technical issues of importance in parenteral science. Improving the stability of unstable drugs, particularly proteins, is a major concern. These authors indicated that protein stabilization in finished formulations is a major concern to people working in biotechnological laboratories. Akers and Nail (1994) emphasized that some proteins are inherently unstable. These proteins require chemical modification or formulation additives to enhance their stability (Wang and Hanson, 1988; Ahern and Manning, 1992; Hanson and Rouan, 1992; Banerjee *et al.*, 1991).

Furthermore, Akers and Nail (1994) indicated that proteins prone to aggregation or adsorption at interfaces (particularly at dilute concentrations) may be stabilized by surface-active agents or stabilizers such as sugars, amino acids, and fatty acids. Furthermore, a careful control of pH may also significantly affect protein stability. These authors further indicated other methods to further stabilize proteins. These are the use of antioxidants to stabilize proteins containing sulfur-containing amino acids (methionine, cysteine). The damage of proteins caused by freezing or by freeze-drying may be minimized or prevented by utilizing certain sugars and amino acids. Cryoprotectants prevent or minimize damage due to freezing. Lyoprotectants minimize losses due to freeze-drying. Thus, even though there are different factors that may affect the stability or activity of a biological product, there are methods available that minimize the deleterious affects of these factors. A further study and detailed analysis of the factors that cause this denaturation, along with how its affect may be either minimized or eliminated, is of significant interest. This is one of the major themes of this manuscript, especially as it applies to the processing of different biological products of interest.

REFERENCES

Ahern, T. J. and Manning, M. C., Eds., (1992). *Stability of Protein Pharmaceuticals, Part A, Chemical and Physical Pathways of Protein Degradation,* Plenum: New York.

Ajongwen, J. N., Akintoye, A., Barker, P. E., Ganetsos, G., and Shieh, M. T. (1993). *Chem. Eng. J., 51,* B43.

Akers, M. J. and Nail, S. L. (1994). *Pharm. Technol.,* August, 26.

Asenjo, J. A., Parrado, J., and Andrews, B. A. (1991). *Ann. N.Y. Acad. Sci., 646,* 334.

Banerjee, P. S., Hosny, E. A., and Robinson, J. R. (1991). Parenteral Delivery of Peptide and Protein Drugs, In *Peptide and Protein Drug Delivery,* Lee, V. H. L., Ed., Marcel Dekker: New York, pp. 487–544.

Baum, R. M. (1987). *Chem. Engg. News,* July 20, 11.

Boudreault, R. and Armstrong, D. W. (1988). *TIBTECH, 6,* 91.

Brooks, D. E., Boyce, J., Bamberger, S. B., Harris, J. M., and Van Alstine, J. M. (1986). In *Proceedings of the Workshop on Space: Biomedicine and Biotechnology,* Ottawa, Canada, p 60.

Burgess, D. J., Longo, L., and Yoon, J. K. (1991). *J. of Parenteral Sci., 45*(5), 239.
Carson, K. L. (1994). *GEN, 14*(6), 12.
Center for Space Policy, Boston. Commercial Space Industry in the Year 2000: A Commercial Market (1985).
Chang, P. (1994). Third International Conference on Separations for Biotechnology, Society of Chemical Industry, University of Reading, Reading, UK, September 12–15.
Chaudhari, J. B. (1995). *TIBTECH, 13,* 12.
Cheng, Y. S., Kwoh, D. Y., Kwoh, D. J., Sltvedt, B. C., and Zipser, D. (1981). *Gene, 14,* 121.
Criado, M., Aguilar, J. S., and De Robertis, E. (1980). *Anal. Biochem., 103,* 289.
Curling, J. (1985). In *Proceedings Biotech '85 Europe, Geneva,* Online Publications: Pinner, Middlesex, U.K. p 221.
Cussler, E. L. (1987). Supercritical gels for protein concentration. In Protein Purification : Micro to Macro, Alan R. Liss, New York, N.Y.
Datar, R. V., Cartwright, T., and Rosen, C.-G., (1993). *Biotechnology, 11,* 349.
Desai, M. A. (1990). *J. Chem. Technol. Biotechnol., 48,* 105.
Diamond, A. D. and Hsu, J. T. (1989). *Biotechnol. Bioeng., 34,* 1000.
Draeger, N. M. and Chase, H. A. (1991). *Trans Inst. Chem. Eng. 69,* Part C, 45.
Dunnill, P. (1983). *Process Biochem.,* October, 9.
Fulton, S. P., Shahidi, A. J., Gordon, N. R., and Afeyan, N. B. (1992). *BioTechnology, 10,* 635.
Furth, A. J. (1980). *Anal. Biochem., 109,* 207.
Geisow, M. J. (1992). *TIBTECH, 10,* 230.
Gilbert, D. (1993). *Biotechnology, 11,* 654.
Goloubinoff, P., Gatenby, A. A., and Lorimer, G. H. (1989). *Nature,* (London) *337,* 44.
Hager, D. A. and Burgess, R. R. (1980). *Anal. Biochem., 109,* 76.
Hamers, M. N. (1993). *Biotechnology, 11,* 561.
Hanson, M. A. and Rouan, S. K. E. (1992). Introduction to Formulation of Protein Pharmaceuticals, In *Stability of Protein Pharmaceuticals , Part B, In Vivo Pathways of Degradation and Strategies for Protein Stabilization,* T. J. Ahern and M. C. Manning, Eds., Plenum Press: New York, pp 209–233.
Hanson, M., Stahl, S., Hjorth, R., Uhlen, M., and Moks, T. (1995) *Biotechnology, 5,* 161.
Harakas, N. K. (1989). *Biotechnology, 7,* 777.
Heng, M. H. and Glatz, C. E. (1993). *Biotechnol. Bioeng., 42,* 333.
Henley, J. P. and Sadana, A. (1984). *Biotechnol. Bioeng. 26,* 959.
Huddleston, J., Veide, A., Kohler, K., Flanagan, J., Enfors, S. O., and Lyddiatt, A. (1991). *TIBTECH, 9,* 381.
Inada, Y., Takahashi, K., Yoshimoto, T., Ajima, A., Matsushima, T., and Saito, Y. (1986). *TIBTECH, 4,* 190.
Inada, Y., Yoshimoto, T., Matsushima, A., and Saito, Y. (1986).*TIBTECH, 4,* 68.
Kane, J. F. and Hartley, D. L. (1988). *TIBTECH, 6,* 95.
Kennedy, J. F. and Cabral, J. M. S. (1993). *Downstream Processing; Chemical Engineering and Biochemistry. Recovery Processes for Biological Materials,* John Wiley & Sons: London.
Kim, K. J., Chen, V., and Fane, A. G. (1993) *Biotechnol. Bioeng., 42,* 260.
Klibanov, A.M. (1983) *Adv. Microbiol. , 29,* 1.
Knuth, M. W. and Burgess, R. R. (1987). *Protein Purification: Micro to Macro,* Burgess, R., ed., Alan R. Liss: New York, p 279.
Leckband, D. and Israelachvili, J. (1993). *Enzyme and Microb. Technol., 15,* 450.
Leser, E. (1994). Third International Conference on Separations for Biotechnology, Society of Chemical Industry, University of Reading, Reading, UK, September 12–15.
Light, A. (1985). *Biotechniques, 3,* 298.
Lilly, M. D. (1992). *Trans. Inst. Chem. Emg., 70,* Part C, 3.
Lyddiatt, A. (1994) Third International Conference on Separations for Biotechnology, Society of Chemical Industry, University of Reading, Reading, UK, September 12–15.
Mattiasson, B. and Hakanson, H. (1993). *TIBTECH, 11,* 136.
Middelberg, A. P. J., O'Neill, B. K., and Bogle, I. D. L. (1992). *Trans. Inst. Chem. Eng., 70,* Part C, 8.
Miller, A. (1994) Third International Conference on Separations for Biotechnology, Society of Chemical Industry, University of Reading, Reading, UK, September 12–15.

Mitraki, A. and King, J. (1989). *Biotechnology, 7,* 690.
Mozhaev, V. V., Melik-Nubarov, N. S., Sergeeva, M. V., Siksnis, V. A., and Martinek, K. (1990). *Biocatalysis, 3,* 179.
National Committee on Bioprocess Engineering. (1993). *Enzyme Microb. Technol., 15,* 541.
Naveh, D. (1990). *BioPharm,* May, 28.
Nygren, P. A., Stahl, S., and Uhlen, M. (1995). *TIBTECH, 12,* 184.
Paliwal, S. K., Nadler, T. K., and Regnier, F. E. (1993). *TIBTECH, 11,* 95.
Pelham, H. R. B. (1986). *Cell, 46,* 959.
Roper, D. K. and Lightfoot, E. N. (1995). *J. Chromatogr. A, 702,* 3.
Russell, R. R. B. (1979). *Anal. Biochem., 97,* 173.
Sadana, A. and Henley, J. P. (1986). *Biotechnol. Bioeng., 28,* 256.
Sakai, D. and Gorski, J. (1984). *Endocrinology, 115,* 2379.
Sassenfeld, H. M. and Brewer, S. J. (1984). *Methods Enzymol., 34,* 350.
Scrip and Rorschild Asset Management and Other Sources. (1993). *Biotechnology Suppl., 11,* May.
Smith, J.C. (1984). *Gene, 23,* 321.
Third International Conference on Separations for Biotechnology (1994). Society of Chemical Industry, University of Reading, Reading, UK, September 12–15.
Todd, P. (1985). *Biotechnology, 3,* 786.
Uhlen, M. (1983). *Gene, 23,* 369.
Uhlen, M., Nilsson, B., Guss, B., Linberg, S., Gatenbeck, S., and Philipson, L. (1983). *Gene, 23,* 369.
Uhlen, M. (1984). *Immunol.Today, 5,* 244.
Uhlen, M. and Nilsson, B. (1985). In *Proceedings Biotech '85 Europe, Geneva,* Online, Pinner, Middlesex, United Kingdom, p 171.
Ullman, A. (1984). *Gene, 29,* 27.
Van Brunt, J. (1985). *Biotechnology, 3,* 419.
Varlie, J. (1994). Third International Conference on Separations for Biotechnology, Society of Chemical Industry, University of Reading, Reading, UK, September 12–15.
Veide, A., Strandberg, L., and Enfors, S. (1987). *Enzyme Microb. Technol., 9,* 730.
Wang, Y.J. and Hanson, M. A. (1988) *J. Parenteral Sci., 42* (Suppl.), 26.
Wesseling, J. (1994). Third International Conference on Separations for Biotechnology, Society of Chemical Industry, University of Reading, Reading, UK, September 12–15.
Wheelwright, S. M. (1989). *J. Biotechnol., 11,* 89.
Whitcombe, M. (1994). Third International Conference on Separations for Biotechnology, Society of Chemical Industry, University of Reading, Reading, UK, September 12–15.
Wiblin, D. (1994). Third International Conference on Separations for Biotechnology, Society of Chemical Industry, University of Reading, Reading, Reading, UK, September 12–15.
Zhang, X., Whitely, R. D., and Wang, N. H. L. (1991). Paper presented at the American Institute of Chemical Engineers Annual Meeting, Los Angeles, CA, November 17–21.

2 STEPS IN BIOSEPARATION PROCESSES

I. INTRODUCTION

The separation of proteins and other biological products of interest performed during the bioseparation process takes place in basically three stages: disruption of cells, initial fractionation, and high-resolution fractionation. Howell (1985) emphasized that a minimum number of steps should be involved in each of these stages. If it appears that more steps are involved than are required (which is typically the case), then a reorganization or restructuring of the steps is worthwhile. This has the benefit of minimizing cost, and the risk of contamination, and makes the cleaning of equipment easier. Bear in mind that three basic goals are involved, biomass (protein or other biological product of interest) removal, initial concentration, and final purification. The combination or integration of these steps should keep these goals in mind.

Proteins and other biological products of interest may be produced by recombinant-DNA (rDNA) techniques and by mammalian cell culture techniques. Sherwood *et al.* (1985) emphasized that genetic engineering has been rather successful in producing, in general, required amounts of different types of biomass (proteins and other biological products of interest). Thus, attention has been shifted (and correctly so) to the effective separation of this biomass in a stable, pure, homogeneous, and active form in high yields. These are some of the goals of the bioseparation process. Not all these goals are met; and even if they are, they are met to different degrees. Note that the downstream separation process may dictate the type of upstream or fermentation process to be utilized to produce a biological product.

Scawen and Hammond (1989) emphasize that the biological product purification process is largely determined by the nature of the final product by its intended use. There are a large number of bioseparation processes available to separate a wide variety of biological products. Scawen and Hammond (1989) indicated that these processes must be carefully screened. One needs to select the process that is best suited for the separation of the biological product with the required and predetermined characteristics. These authors emphasized that proteins intended for therapeutic usage must by necessity be extremely pure. These therapeutic proteins must meet exacting standards set by controlling agencies to minimize unwanted side effects or immunogenic responses. There are enough examples available in the literature where deleterious side effects of drugs on humans have led to considerable suffering and long, very expensive, protacted lawsuits. On the other hand, biological products for intended diagnostic usage do not have to be very pure and need not meet these exacting standards. Note that for the preceding usage of these biological products rather small quantities of the biological product are required. Some characteristic feature of the biological product may be utilized to help separate it. For example, the major properties of proteins involved in their effective separation include size, charge, biological affinity, and hydrophobicity-solubility (Scawen and Hammond, 1989). The different bioseparation techniques utilized are based on one or more of these different properties.

Mizrahi (1986) indicated that though the advances in recombinant techniques for the production of biological products are well known, mammalian cell techniques also produce a wide variety of biological products. Some of these include animal cells as a product, enzymes, hormones, nonantibody immunoregulators, monoclonal antibodies, tumor specific antigens, etc. Edgington (1992) emphasized that because the second generation of biotechnological products is larger, more complex, more glycosylated, and contain more sulfhydryl bonds, bacterial techniques may not be suitable. A particularly vexing problem is the amount of *in vitro* folding that is required in these types of systems.

Jaenicke (1991) indicated that protein folding is constrained both by kinetics and by thermodynamics. The driving force of the three-dimensional protein structure formation is the minimization of the free energy of stabilization. There is a hierarchical nature of the three-dimensional structure formation (Janeicke, 1987). Short-ranged interactions lead to secondary structure elements. These secondary elements through the process of gradual combination and reshuffling lead to the formation of subdomains, domains, and subunit assemblies. Edgington (1992) emphasized that due to the difficult nature of the refolding techniques researchers are examining mammalian cell culture techniques more closely. Mammalian cell culture techniques have an excellent quality control system, the endoplasmic reticulum (ER). This ER eliminates the folding defects of the required proteins. Also, the proteins are correctly glycosylated and the disulfide bonds are positioned correctly. Edgington (1992) emphasized that due to the apparent zero defects of the ER system, researchers in industry and academia are beginning to place more emphasis on mammalian cell culture systems as compared with bacterial systems. The importance of *in vitro* folding techniques makes it the subject matter of Chap. 9.

The cells (bacterial or mammalian) that produce the biological products of interest have to be disrupted to be able to release the bioproducts. Howell (1985) indicated that the position of the products in the cell determine the initial bioseparation step or steps involved. The biological products may be in the cell or attached to the cell wall. This author emphasized that biological products attached to the cell wall are difficult to extract with full retention of activity. The biological products in the cell may be extracted with some problem. Thus, it is of importance to have full or as much as possible knowledge of the product to be extracted. Also, typically after the cells have been disrupted the biological product is present along with other similar and delicate substances in dilute form in solution. Therefore, initially Howell (1985) recommended that a crude and inexpensive "dewatering" step be used to help concentrate the biological product of interest. These solids may be concentrated up to about 30% solids. Thereafter, a final purification step may be utilized to concentrate the biological product to the required specifications.

One cannot overemphasize the careful selection of the initial bioseparation step in the overall scheme. The other steps or at least the quality and quantity of the bioproduct separated will be critically dependent on the initial steps selected. Some of the steps that may be utilized to initially clarify or concentrate the biological product from the cell include membrane techniques, filtration, centrifugation, or even chromatographic separations. There seems to be a continuous relation in the sequence of the steps involved in an appropriate bioseparation scheme that helps optimize the activity, stability, and quantitative yield of a bioproduct. Thus, there is a need for an integrated process for the production of a protein or a biological product of interest.

Naveh (1990) appropriately defined the process flows that occur generally in the isolation and purification of products from rDNA proteins (Fig. 2.1). The flowchart is quite complex and there is an interdependence of the different steps that may be utilized to advantage to optimize the process. Spark (1985) clearly indicated the need for integrated process control for product development and production. This author defined the protocol that is required to help optimize a process, even when it is in production. Figure 2.2 shows the four levels at which information concerning the process may be interchanged or fedback to help optimize the process. The utilization of the process control scheme presented by Spark (1985) assists in the upward as well as in the downward flow of information primarily because the data and protocols are common. The advantages of this type of system are that it is standardized, a useful database is created that is also easily accessible; and most importantly, it is live.

Wheelwright (1989)analyzed the downstream processes for the large-scale purification of proteins. This author rationalized the sequence of steps that may be utilized from among a wide variety of choices. He examined the use of heuristics or rules of thumb. Nadgir (1983) had initially classified heuristics as: (1) method (specifies the selection of a unit operation), (2) design (determines the sequence of steps utilized), (3) species (the component property largely determines the step), and (4) composition (the separation cost determines the product or feed composition). Wheelwright (1989) emphasized that while developing the design of a bioseparation process the following three factors must be kept in mind: the purity, the cost, and the speed to the market.

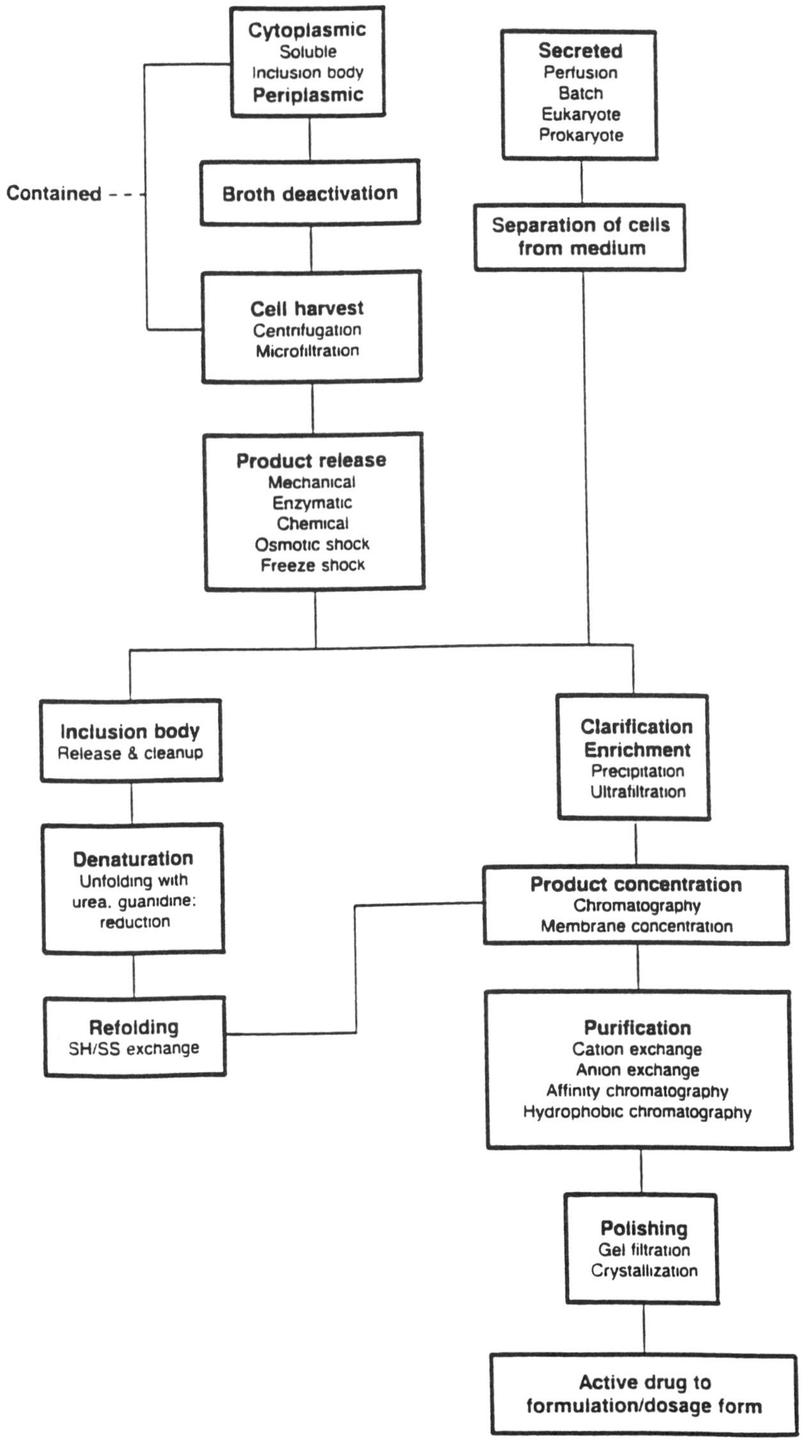

FIGURE 2.1 General overall scheme for the purification of rDNA proteins (Naveh, 1990).

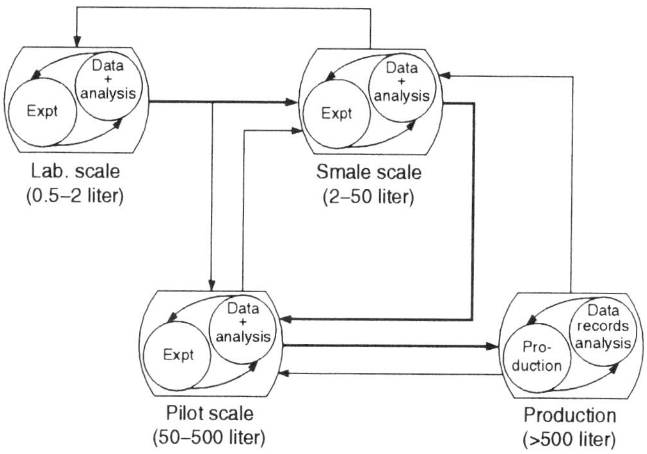

FIGURE 2.2 Interaction at four different levels during process development for bioseparation (Spark, 1985).

There is, of course, much common sense involved. There are few appropriate choices (Wheelwright, 1989). For example, either the product material is secreted or it is not. If the material is secreted, then it is to be concentrated by either ultrafiltration or centrifugation. If the product is not secreted, then the cells have to be disrupted. Then a solid–liquid separation is necessary. If the material is in the liquid phase, then use ultrafiltration or adsorption. If the product is in the solid phase, then it has to be extracted into aqueous solution. Once again, these choices emphasize the need for a complete or as much as possible knowledge about the different aspects involved in the bioseparation process.

Wheelwright (1989) recommended one should exploit that physical characteristic of the product to be separated that exhibits the greatest difference between itself and the impurities present. Also, one should examine and use different bases for separation in successive steps. Finally, Wheelwright (1989) cautioned that (1) the heuristic method of design for protein purification is rather new, and (2) exceptions will exist. Therefore, there is still a need for good common sense, and experience with these types of situations can be invaluable thanks to the high cost involved in the design of large-scale bioseparation systems. There is an urgent need for the rational and heuristic design of bioseparation systems. Nevertheless, it is anticipated that: (1) considerable expense, (2) more detailed studies, and (3) better understanding of the bioseparation principles involved during each of the bioseparation steps need to be obtained. This should be done before one can begin to predictively design a wide variety of bioseparation processes to meet the cost requirements, and the quality and quantity of bioproduct or bioproducts that match the changing market requirements.

In this chapter we present the different steps that may be utilized to help separate the product of interest. The steps to be utilized may be broadly classified into: (1) cell disruption or product excretion from the cell, (2) initial fractionation, and (3) high-resolution fractionation.

II. PRODUCT EXCRETION FROM THE CELL OR CELL DISRUPTION

Dunnill (1983) initially indicated that quite a few industrial enzymes such as proteases and amylases are naturally excreted from cells. This author predicted that as the understanding of product excretion mechanisms from cells increases it may be possible to genetically control or modulate the excretion of other useful proteins and biological products such as interferons. Dunnill (1983) cautioned, however, that not all the proteins or biological products of interest may be excreted; and even if they are, the rate of excretion may be unacceptably slow. This author also predicted the distinct advantage of excreted proteins because their contamination level is rather low, and that this would ease the further downstream processing steps. Many of Dunnill's predictions made in 1983 have been realized.

Example 2.1

Describe briefly some of the advantages of protein excretion from cells (Sherwood *et al.*, 1985).

Solution

Sherwood *et al.* (1985) indicated product excretion from cells is attractive because it does not require the breakage of cells that is both cost and energy intensive. Also, the problems of handling viscous fluids and proteinases released from the cells are avoided. Furthermore, this can act as a safety valve when high concentrations of product levels build up in the cell. High-level buildup of products in the cell may make subsequent extraction of products difficult due to, for example, the formation of inclusion bodies.

Often a signal peptide determines the secretion (Sherwood *et al.*, 1985). This signal peptide is encoded by a leader sequence to the structural gene. These authors indicated that signal peptides do have certain features in common. These authors analyzed carboxypeptidase G_2 (CPG_2), alkaline phosphatase, glyceraldehyde-3-phosphate dehydrogenase, and reduced nicotinamide adenine dinucleotide (NADH).O_2 reductase to demonstrate certain common features. For example, CPG_2 is a typical signal peptide with 22 amino acids along with a hydrophobic and hydrophilic region. The enzyme is periplasmic in location. The enzymes may be either periplasmic located or associated with the cell wall. Sherwood *et al.* (1985) indicated that about 30% of the enzymes can be secreted if the conditions are carefully controlled. These authors caution against secreting high levels of product into the fermentation medium where it may be exposed to an undesirable environment. Furthermore, these authors mentioned the attractiveness of periplasmic location of the proteins from which they can be extracted using gentle chemical or thermal treatment. Needless to say, not everybody agrees with this.

They indicated that a buildup of cloned product in the cell may lead to its leakage and may even impair certain membrane transport processes. These and other aspects of the product excretion process need to be carefully examined if one is to obtain the maximum benefit from the protein extraction process. It is to be expected that each protein will probably have associated with it its own particular or peculiar extraction and subsequent processing requirements.

However, a knowledge base builtup in this area with regard to different proteins and other biological products of interest should provide some common principles. A framework of information thus generated should significantly alleviate the problems associated with this processing step, and hopefully even attain the level of a heuristic that aids in design.

Foster (1992) indicated that quite a few biological products require cell disruption. This author indicated that cell disruption is an energy-intensive and a violent process. Therefore, not only is it important to extract the biological products of interest without destroying them, but also one should be able to contain the process within the equipment. This author emphasized that the preceding product secretion systems have often been driven by the need that cell disruption systems were difficult to contain. Furthermore, cell disruption systems should be able to: (1) be controlled, (2) be contained, and (3) be validated. Foster (1992) stressed that developments in cell disruption technologies permit this procedure to be contained when disintegrating cells from recombinant bacteria, yeast, mammalian, plant, and insect sources.

This author further indicated some of the biological products that require cell disruption in their processing include vaccines (tetanus, meningitis), enzymes (glucokinase, glycerokinase, invertase, and sacrosine dehydrogenase), toxins (enterotoxin from *Clostridium perfringens*, subcellular components (mitochondrai, chloroplasts), and intracellular constituents (DNA, RNA preparations, and virus-like particles). Also, recombinant insulin, recombinant growth hormone, and protein A and G require cell disruption. Furthermore, strict safety and containment requirements of genetically modified organisms and pathogens have significantly restricted the choice of downstream and of upstream systems. This author emphasized that all areas of biotechnology could greatly benefit by a better understanding of cell disruption systems, especially with regard to their control, containment, and validation. These factors become more significant as the scale of the process is increased. Furthermore, though the mechanisms of cell disruption by biological and chemical means are well defined, the mechanisms for cell disruption using mechanical means are far from clear.

Rehacek and Schaefer (1977) indicated that in ball mills the rotating disks induce cavitation. This cavitation causes resonance and subsequent vibration. This vibration eventually disrupts the cells. For high-pressure systems Engler and Robinson (1981) indicated that the transfer of cells from a high-pressure region to a low-pressure region leads to a disruption of cells. This process is reproducible, even though the mechanism of this disruption is not clear. It is of interest to analyze the effectiveness of cell disrupters. Also, it behooves one to test their effectiveness with respect to an acceptable criterion. An appropriate choice would be the product quality as judged by the product assay.

Foster (1992) provided a list of the required characteristics of an ideal large scale cell disrupter. He mentions that no cell disrupter actually possesses all these characteristics. The cell disrupter should be able to disrupt even tough organisms. The mechanism of cell disruption should be well understood (at least as far as possible). This assists in not disrupting the labile biological product, besides making this process reproducible. The cell disrupter should be sterilizable, and the process carried out within should be containable and validatable. The process should be amenable to being made automatic. Also, the process should be continuous, and compact, and the heat generated within should be controllable. So that the cell-disruption process is economical, the capital and the operating costs should be reasonable and inexpensive.

The preceding list of requirements that should be met by a large-scale disrupter, as suggested by Foster (1992), are all-inclusive. The list is instructive because it lets one know what is required in an ideal cell disrupter. Even if one cannot satisfy all the requirements, one should at least attempt to satisfy some of them.

The goal of providing sufficient cell disruption with minimal damage to the biological product should, of course, always be kept in mind. If this were an optimization problem, this would be a reasonable objective function. The best objective function is a monetary value of the product generated from the entire process. Modifications should continuously be made to increase the profit value from a particular process. The information provided by Foster (1992) is a step in that direction; besides it provides reasonable physical insights into the cell disruption process.

Howell (1985) indicated that cell disruption on a small scale is relatively simple. This disruption becomes rather difficult as the scale of disruption increases. This is because a great deal of energy has to be dissipated in the cells. The small cell size makes this difficult. This author indicated cell disruption devices that use ultrasonics or freeze thawing, or those that force cells through very small orifices at high pressure are successful only on a small scale. For large-scale cell disruption the Manton-Gaulin homogenizer has proved successful. This equipment pumps cells over 500 bar through a homogenizing valve. This leads to disruption of the cells by high shear and sudden pressure release on the downstream side. The detailed mechanism of cell disruption is not clear. Other factors may be involved.

On a small scale, glass or porcelain ball mills may be used to disrupt the cells (Howell, 1985). These ball mills are vibrating or rotating cylinders. About a third of the volume is taken up by the ball mills, and the remaining volume is generally only half filled with "paste."

Sauer *et al.* (1989) indicated that initially mechanical devices were used to disrupt yeasts. Some of these include the rotating disk, ball mill-type disintegrators (Kirsop, 1981; Marrfy and Kula, 1974; Rehacek and Schaefer, 1977; Schuette *et al.*, 1983, 1985), or high-pressure homogenizers (Engler and Robinson, 1979, 1981; Dunnill and Lilly, 1975; Wang *et al.*, 1979; Whitworth, 1974; Doulah *et al.*, 1975).

Sauer *et al.* (1989) indicated that these and other methods have been utilized to disrupt only laboratory-scale or small volumes of cells. They analyzed the effectiveness of a relatively new homogenizer, the Micro fluidizer. Furthermore, these authors indicated that the high-pressure Manton-Gaulin homogenizer is widely used in the industry. This homogenizer consists of a positive displacement pump, with one or more pistons that are connected to a special nozzle. Cell disruption occurs due to a combination of different mechanisms. These include shear, cavitation, and impingement.

The effectiveness of the preceding homogenizer for cell suspension concentrations ranging from 4 to 175 g dry wt/liter was analyzed (Sauer *et al.*, 1989).

The pressure was varied from 30 to 95 MPa. Up to five passes were tried. These authors noted that for recombinant and nonrecombinant (wild-type) *Escherichia coli* cells, cell disruption increased with growth rate, concentration of cells, disruption pressure, and number of passes. Furthermore, cell disruption was effectively modeled by the equation

$$\log (1 - \mathrm{CD}) = \alpha \mathrm{N}^{\beta} \mathrm{P}^{\gamma} \tag{2.1}$$

Here *CD* is the extent of cell disruption, N is the number of passes, and P is the pressure. α, β, γ are constants. Note that β is a characteristic of the system; and depends on the growth rate of cells, concentration of the cell suspension, and type of cells.

The wild-type *E. coli* cells were more difficult to disrupt than the recombinant *E. coli*. Also, at a pressure of 95 MPa and with two or three passes, 95 to 98% disruption of the recombinant *E. coli* cells was possible. Sauer *et al.* (1989) subjected the recombinant and wild-type *E. coli* cells to heat induction (at 42°C) prior to cell disruption. The heat induction resulted in a gross change in cell morphology. The cells after heat induction were linked in short chains. The links (and even the cell walls themselves) were more readily broken for the recombinant as compared with the native cells. More such studies are urgently required that shed novel physical insights into the mechanisms involved in the cell disruption process. This is of particular interest, because there is difficulty in pinpointing and in describing in any detail the mechanism or mechanisms for cell disruption. The mechanistic details obtained will greatly facilitate in controlling, validating, and making the cell disruption process more reproducible.

Engler and Robinson (1981) indicated that it is essential to clearly delineate the mechanisms of cell disruption so that one may better design and also optimize the design of such equipment. Doulah *et al.* (1975) had initially indicated that turbulent eddies, which are smaller in size than the cells, are primarily responsible for cell disruption. These turbulent eddies apparently cause the cell liquid to oscillate with sufficient kinetic energy that eventually leads to cell wall disruption. Engler and Robinson (1981) postulated that normal and shear stresses, turbulence, or stresses caused by impingement of a high velocity jet onto a stationary surface may play a significant role in cell disruption.

The role of different types of stresses on the disruption of *Candida utilis* cells in high-pressure flow devices has been analyzed (Engler and Robinson, 1981). These authors designed their experiments so that the role of the normal, shear, and impingement stresses could be analyzed independently.

Figure 2.3 shows the details of the impingement nozzle. The 80-μm inner diameter orifices created a high-velocity jet. The authors admitted that some error may have been caused in the (changing) orifice diameter by plugging and by occasional breakage. Normal stresses may be generated by rapid extrusion through an orifice. Impingement stresses were generated as the high-velocity jet struck an impingement plate near the orifice.

Based on their results Engler and Robinson (1981) concluded:

1. Normal stresses by themselves (caused by a rapid pressure release) are not sufficient to cause cell disruption.

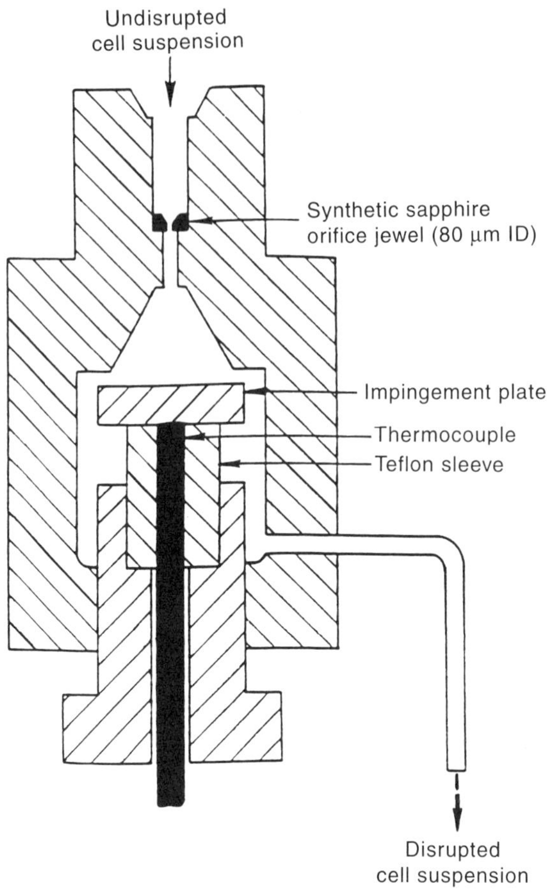

FIGURE 2.3 Impingement nozzle for cell disruption (Engler and Robinson, 1981).

2. Cell disruption by turbulent eddies requires that these eddies have a significant amount of energy. This amount of energy was not present in the eddies in the high-velocity jet produced in the experiments.
3. Their results clearly demonstrate that the high-velocity impingement of the cells on the flat plate is primarily responsible for cell disruption.

These authors noted that the fraction of cells disrupted by impingement is a first-order function of the number of passes through the disrupter. Also, there is a power law dependence on pressure over a range of pressures. The Engler and Robinson (1981) equation for cell disruption is

$$\ln (1/(1 - R)) = KNP^{\Gamma} \tag{2.2}$$

Typical K values for *Candida utilis* are 8.95×10^{-3} and 3.51×10^{-4} for the cyclic batch and the continuous culture, respectively. The values for the parameter, Γ, obtained for the cyclic batch and the continuous culture were 1.17 and 1.77, respectively. The growth rates were 0.5 and 0.1 h^{-1} for the cyclic batch and the continuous culture, respectively.

These authors suggested that because there are hydrodynamic similarities between their equipment and the Manton-Gaulin homogenizer, the cell disruption mechanisms in both cases is also similar. I am in agreement with these authors on this aspect. Nevertheless, more study on this aspect and others is required to further delineate clearly the mechanism or mechanisms involved in the cell disruption process for both recombinant and nonrecombinant (native) cells. Recognize that the cell disruption step is an early step in the entire downstream processing train. This step is bound to significantly influence the choice of subsequent downstream processing steps. Thus, it behooves the downstream processing engineer and others involved to understand and to gather as much information as possible on the cell disruption step.

The cellular products of interest may be extracted from the cells by cell wall destruction and by excretion. Another possible way is by enzymatic lysis of cell walls. Hunter and Asenjo (1987a) indicated that enzyme systems may be used to hydrolyze cell walls. These received a lot of interest due to their potential for biotechnological applications. For example, lytic systems have been used to recover a hydroxylase enzyme complex from bacterial membranes (Fish and Lilly, 1984). Hunter and Asenjo (1987a) indicated that, in general, lytic enzyme preparations contain a synergistic combination of hydrolytic activities (Phaff, 1977). The two essential activities are: (1) a lytic protease that is required to dissolve the outer part of the cell wall, and (2) an endo-β(l-3)glucanase to disintegrate the underlying glucan net. Andrews (1985) indicated that the activity of the lytic systems depends on both the source and the culture system. It would be of interest to examine the activity of lytic systems. Of particular interest would be a kinetic analysis. This analysis should be of considerable assistance in: (1) to further understanding the system, and (2) designing the small- and large-scale systems.

Example 2.2

Briefly present a kinetic analysis for enzymatic lysis and disruption of yeast cell walls (Hunter and Asenjo, 1987a).

Solution

Hunter and Asenjo (1987a) analyzed the kinetics of glucan hydrolysis, proteolysis, and lysis of brewer's yeast. They did this by using two different lytic systems with different properties. The two systems used were *Cytophaga* NCIB 9497 grown in a batch culture, and *Oerskovia xanthineolytica* LLG-109 grown in a continuous system. The *Cytophaga* system exhibits high protease activity, and the *Oerskovia* system exhibits high glucanase and low protease activities, respectively.

Hunter and Asenjo (1987a) noted the following:

1. Though the *Oerskovia* enzyme exhibited a higher initial activity compared with the *Cytophaga* enzyme, the *Cytophaga* enzyme exhibited higher conversions and higher rates of lysis at longer times. Also, the *Cytophaga* enzyme contains some inhibitors that limit its lytic ability.
2. An initial lag in yeast lysis was exhibited by both enzymes. Adsorption effects were ruled out because the *Oerskovia* enzyme adsorbs rapidly to the

yeast. Also, the *Cytophaga* enzyme does not adsorb at all. Hunter and Asenjo (1987a) attributed the lag to sequential reaction kinetics for the removal of the protein and then the glucan from the cell walls. This type of an analysis that provides physical insights into the cell wall lysis system is useful.

3. The *Cytophaga* enzyme has 10 to 20 times the lytic activity compared with the *Oerskovia* enzyme. The lysis of yeast by the *Cytophaga* enzyme does not produce long-chain proteins except at high yeast concentrations. Yeast lysis by the *Oerskovia* enzyme yields proteins at all the conditions studied. Once again, this type of analysis is useful because it sheds insights into the cell wall lysis system. This type of information should prove invaluable in the design of small- and large-scale cell wall lytic systems.

More analysis like the Hunter-Asenjo (1987a) analysis are required that further delineate the kinetic mechanisms for cell wall lysis for different types of recombinant and native (nonrecombinant) systems. The framework of data thus assembled should prove invaluable in evaluating the full potential of cell wall lysis systems as an effective bioseparation tool. There are, however, some researchers who think that this lysis system will not be practical. Also, it would be of interest to note the nature of, and the possible applications of, the products released from cells using cell lytic enzymes.

Hunter and Asenjo (1987b) also presented a simple two-step model that describes cell wall lysis. In the first step of the model the yeast cell mass is solubilized. In the next step the released protein is hydrolyzed to peptides. The Hunter-Asenjo model (1987b) was able to predict reasonably well the concentrations of soluble proteins, peptides, and carbohydrates. Initially, as the lytic enzyme attacks the cell wall, the cell components are released. The constituents are soluble proteins, peptides, and carbohydrates. The released proteins are further attacked by the proteolytic enzyme to yield peptides. The peptides and the carbohydrates are the end products, and are therefore not acted on further by the enzymes. The Michaelis-Menten form of the kinetic equation was successfully used to predict cell wall lysis and the subsequent protein breakdown. Hunter and Asenjo (1987b) indicated that their model has been verified for 0.7 to 70 g/liter yeast (dry basis), and for 4 to 40 percent crude enzyme solutions.

These authors indicated that lytic enzymes may be used to produce food grade single-cell protein from animal sources and animal feed (Jamas *et al.*, 1985), invertase (Kobayishi *et al.*, 1982), and microbial pigments (Okagbue and Lewis, 1983). Other applications are also possible.

By considering the different uses of lytic enzymes and the nonmechanical means of destruction of the cell wall possible by this technique, the initial model of Hunter and Asenjo (1987b) is useful. It lays down the basis for subsequent work with other lytic enzymes acting on different types of systems. These types of analysis should be of considerable use in the design of small- and large-scale bioseparation processes.

III. INITIAL FRACTIONATION

Typical biological product concentrations in fermentation broths range from 20 to 100 g/liter (Garcia, 1991). These include chemicals such as amino acids, alcohols, ketones, carboxylic acids, etc. The maximum concentration is much lower for recombinant proteins and some natural products, especially pharmaceuticals. Initially, a primary isolation step is required that removes the cells or cell debris. This primary operation step may include adsorption, extraction, precipitation, or even distillation (for nonlabile products). Filtration and solvent evaporation may also be used during the primary recovery-initial fractionation steps. Garcia (1991) emphasized that there should be a significant volume reduction during the early part of the separation train. This facilitates the use of subsequent low-throughput, high-performance steps such as chromatography, where high purity is required.

Howell (1985) indicated that once the cell wall has been broken one has to remove the cells from the broth. This is presuming that the biological products of interest are in the solution. Once the biological product of interest is in solution Wheelwright (1989) indicated that one needs to gradually remove the impurities and increase the relative concentration of the desired product. Howell (1985) indicated that the method to be chosen depends on the scale and on the organism used in the fermentation. The selection of the process to use from a variety of processes available is one of the major components involved in design. Because the biological products are generally expensive it is imperative to remove or to recover the entire product. This would involve significant washing stages followed by subsequent dewatering steps. For example, Howell (1985) indicated that yeasts are easy to deal with because they are generally 10 μm in diameter, and are easily flocculated or are self-flocculated. They may then be removed either by centrifugation or by filtration. Cross-flow fitration is also an important means to remove cells (Le and Atkinson, 1985). Bacteria that are smaller (around 1 μm) are more difficult to filter, and require microfilters because they pass through simple filters. Conditioning of the broth that contains the biological product of interest is important prior to carrying out any initial fractionation step.

Mosqueira *et al.* (1981) analyzed the viscosity, density, and sedimentation characteristics of mechanically disrupted baker's yeast suspensions. The analysis was performed to aid in the centrifugal separation on a laboratory and on an industrial scale. These authors indicate that methods of cell wall disruption should be such that they minimize the degradation of cell walls following rupture. This would not only minimize the proportion of cell fragments, but also restrict the fraction of viscosity increase due to colloidal cell wall glycan.

There was reasonable correlation between the laboratory results and the three pilot centrifuges (Mosqueira *et al.,* 1981). The analysis, especially the good correlation obtained between the laboratory- and the pilot-scale centrifuges, is of considerable assistance because it permits the choice of industrial centrifuges from laboratory data. An added advantage is that it permits the reasonable estimate of the separation costs at an early stage in the process development.

Howell (1985) indicated that the first crude separation or dewatering step of the final product from the clarified broth may be achieved by adsorption, precipitation, membrane separation, or even liquid–liquid extraction. Gabler *et al.* (1985) demonstrated that cell lysates may be effectively processed by cross-flow filtration (Quirk and Woodrow, 1983; Datar, 1984). Gabler *et al.*

(1985) emphasized that the concentration of both bacteria and lysates may be accomplished by the same microporous membranes. These authors indicated that ultrafiltration membranes easily concentrate protein solutions with a bare minimum of protein loss. Also, membrane processing of protein solutions is possible on a laboratory as well as on an industrial scale. An added advantage is that the biological solutions are totally contained by membrane techniques. This minimizes aerosol generation, if any.

Strandberg *et al.* (1991) indicated that aqueous two-phase partitioning has been used as a primary (or initial fractionation) step for recovering intracellular proteins from microorganisms for several years (Hustedt *et al.*, 1988). The two-phase system consists generally of polyethylene glycol (PEG) and a salt. The biological product of interest is generally collected in the PEG-rich top phase. The disintegrated cells and other nucleic acids accumulate in the salt-rich bottom phase. Kula (1990) indicated that this method is easy to scale up and it is biocompatible. Also, the method is rapid when used with centrifugal separators. Strandberg *et al.* (1991) further suggested that if the partition coefficient, K (concentration ratio of the biological product in the top phase and the bottom phase), is high, then: (1) purification and concentration of the biological product is possible, and (2) cell particles and nucleic acids are removed (Veide *et al.*, 1983, 1984). This is a good combination of the different bioseparation steps required. Of course, this presumes that high values of K are possible. However, this may not be the case. If it is not, one needs to resort to recombinant techniques to facilitate protein purification. One can therefore easily see that a wide variety of options are available, and one may need to resort to trial and error methods to improve the overall purification strategies. Experience with these types of systems would be beneficial.

Howell (1985) further indicated that the liquid-liquid extraction technique does lead to a considerable reduction in volume. An advantage of this technique is that if the cells are concentrated in the lower phase, then they are easily separated without loss of the biological product that is concentrated in the top phase. This author mentioned a Schiebel column wherein the processing of 3 kg of protein per hour per square meter (flux) is possible.

A technique that combines different functions for separation (such as clarification and initial purification) is adsorption. Chase (1994) indicated that adsorption techniques are popular methods to purify proteins (Bonerjea *et al.*, 1988; Soffer and Nystrom, 1989; Harris and Angal, 1990). A disadvantage using packed (or fixed) beds for the adsorption process is the removal of particulates, if present, from feedstocks. These particulates become entrapped in voids of the beds. This leads to excessive pressure drops and deformation of the adsorbent itself. This further increases the pressure drop.

The removal of particulates from a solution for protein purification requires at least one unit operation such as centrifugation or filtration. If not done carefully, this can lead to a significant decrease in the yield, along with further losses due to protein denaturation. However, adsorption methods are being designed to assist in the effective removal of particulates from protein recovery solutions. These procedures utilize expanded or fluidized beds.

A. Adsorption

Example 2.3

Briefly describe the principles of operation of expanded beds for particulate removal from protein solutions (Chase, 1994).

Solution

Chase (1994) reviewed the purification of proteins by adsorption in expanded beds. Figure 2.4 shows the packed bed and the stable expanded bed. As the liquid is pumped through a packed bed of adsorbent, the bed can expand, and void spaces are formed and open up. Through these open spaces particulates are allowed to pass freely (without clogging). This author indicated that at approximately twice the normal height of the packed bed, the particulates move through the bed quite freely. An upper adapter may be utilized to prevent loss of adsorbent from the top of the bed at higher liquid velocities.

The proper utilization of an expanded bed for adsorption may eliminate a step required for the removal of cells or cell debris (Chase, 1994). The adsorbed product can be eluted. This author emphasizes that fluidized beds have been utilized for the batch processing of streptomycin (Bartels *et al.*, 1958), and in the semicontinuous system for novobiocin (Belter *et al.*, 1973). Efforts have been made to use fluidized beds for the direct extraction of proteins from fermentation broths (Gailliot *et al.*, 1990; Gibson and Lyddiatt, 1990).

Chase (1994) examined how the expanded beds may be stabilized, and the equipment required for expanded-bed protocols. This author emphasized that the procedures for the scale-up of expanded beds are simple and are similar to those utilized for packed-bed procedures. In conclusion, the expanded-bed technique exhibits considerable potential for simplifying the separation of proteins from solutions containing particulates. The technique is in the early stages of development. However, the opportunity to combine clarification, concentration, and purification in a single step is well worth examining carefully.

Feedstocks containing protein cannot be directly sent to a chromatographic

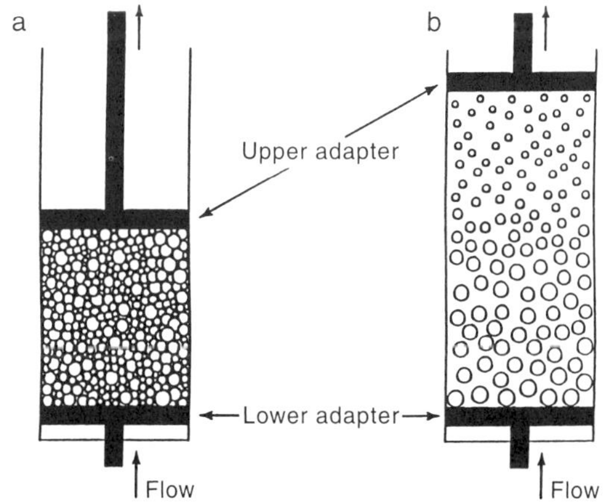

FIGURE 2.4 A packed bed and a stable, expanded bed (Chase, 1994).

bed for purification. The chromatographic bed will soon be blocked due to the buildup of cells and debris in column. There needs to be some clarification of the fermentation broth to remove the cells and debris. This is time consuming, and expensive, and some valuable product is lost. Also, as long as the protein is in contact with proteolytic enzymes, further loss of protein yield occurs. It is advantageous to remove the product quickly from the fermentation broth. Fluidized-bed adsorption, which uses high-density adsorbent particles in a upward flow fluidized bed, as indicated previously, is an option (Gailliot *et al.*, 1990; Morton and Lyddiatt, 1992). Frej *et al.* (1994) utilized the expanded-bed adsorption technique for the pilot-scale recovery of recombinant Annexin V from unclarified *E. coli* homogenate.

Frej *et al.* (1994) utilized purpose-design columns along with a novel ion-exchange adsorbent to recover Annexin V at the following three levels of scales: (1) method scouting in a small packed bed, (2) optimization in a laboratory-scale expanded bed, and (3) final pilot-scale expanded bed. These authors investigated the effect of the biomass content and viscosity of the feedstock on target protein recovery. Figure 2.5 shows the basic principles of expanded-bed adsorption.

These authors noted that a major advantage of their method was the recovery of the target protein without the use of a centrifugation or an ultrafiltration step. The cells and debris pass through the bed unhindered, whereas the target protein is adsorbed. Washing in an expanded mode followed by elution (downward flow) in a sedimented mode recovers the target protein.

The hydrodynamics of the flow and the stability of the fluidized bed were significantly affected by the viscosity and biomass content of the feedstock. Frej *et al.* (1994) investigated the effects of these parameters in some detail. These authors noted that optimum performance could be obtained for biomass content (in the feed stream) up to 5% (dry wt), and viscosities up to 10 mPa s s^{-1} at a shear rate of 1 s^{-1}. Processing of feedstocks with biomass content of 7 to 8% and viscosities of 50 mPa s s^{-1} were also feasible on proceeding with caution. They suggest reducing the viscosities of the feedstocks at the higher end of the range utilized by dilution, homogenization, and nuclease treatment. Fur-

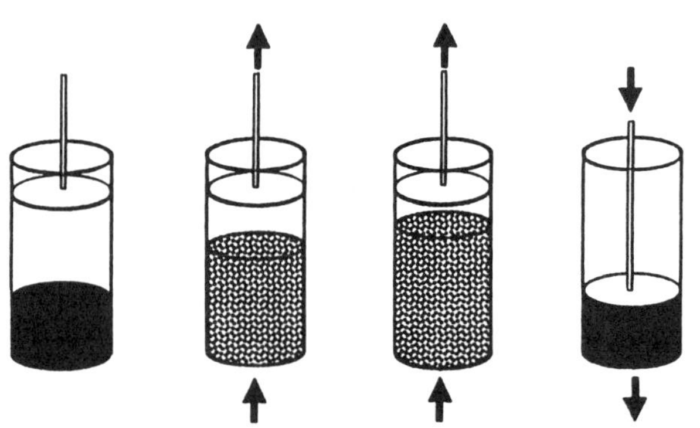

FIGURE 2.5 The different stages of operation of the expanded bed adsorption (Frej *et al.*, 1994).

thermore, the authors indicated that variations in the feedstocks significantly affect the performance of the expanded-bed adsorption column. Thus, they recommended analyzing each application carefully during method optimization. Also, the effects of viscosity and biomass content of the feedstock on the column performance should be carefully analyzed.

A particular advantage of the Frej *et al.* (1994) method is that the process was scaled up (scale-up factor of 16) from a 50-mm diameter column to a 200-mm diameter column with 95% recovery of Annexin V at both laboratory- and pilot-scale levels. The recovery was expressed as percentage of Annexin V in the eluate-Annexin V in the feedstock. These authors used an anticoagulant activity assay, and reverse-phase high pressure liquid-phase chromatography (HPLC) to determine Annexin V yields by peak area measurements.

Spalding (1991) also recommended contacting the unclarified fermentation broth with a fluidized bed for adsorption. This step functions as a chromatographic resolution step. The limitation of the blocking of bed particulates is circumvented by fluidizing. Thommes *et al.* (1995) indicated that a stable expanded bed is possible by using adsorbents with a wide distribution of particle size. Also, an inhomogeneous fluidized bed could be produced with the intention of preserving the packed-bed hydrodynamics on reducing the local mobility of the individual adsorbent particle. These authors indicated that examples of direct whole broth adsorption on a fluidized bed for bacterial as well as mammalian proteins are available (Erickson *et al.*, 1994; Hansson *et al.*, 1994).

Thommes *et al.* (1995) utilized a fluidized-bed cation exchange process to purify monoclonal antibody (mouse IgG2a) from a whole hybridoma fermentation broth. These authors utilized as matrices the commercially available Streamline material (cation exchanger), and the Bioran (commercially available controlled pore glass) material that was derivatized with sulfonic acid ligands. They applied the crude sample and the washing step in the fluidized mode, and the elution step was conducted in the fixed-bed mode. By using this process, Thommes *et al.* (1995) effectively combined the downstream processing steps of clarification, concentration, and coarse purification. These authors obtained a completely clarified eluate with purification factors between four and eight. Also, the monoclonal antibody was concentrated by a factor of more than three. Although these authors noted differences in the fluidization characteristics of the two different matrices utilized, it was noted that similar performance was exhibited by both types of columns as far as antibody binding and elution were concerned. Finally, to improve the performance of their fluidized- bed adsorption step, they are: (1) analyzing in detail the hydrodynamics of the glass material (Bioran); and (2) evaluating different types of ligands, especially affinity- and group-specific types. This should considerably assist in the purification of products from animal cell culture and from microorganisms.

Chaubal *et al.* (1995) analyzed the adsorption of three of the largest volume antibiotics, penicillin V, tetracycline, and cephalosporin C from water onto neutral polymeric sorbents. These authors indicated that for the industrial recovery of secondary metabolites (such as antibiotics) the solvent extraction technique (Belter *et al.*, 1988) is popular, followed by ion exchange for highly water-soluble antibiotics such as streptomycin (Bartels *et al.*, 1958). Neutral polymeric sorbents are also used.

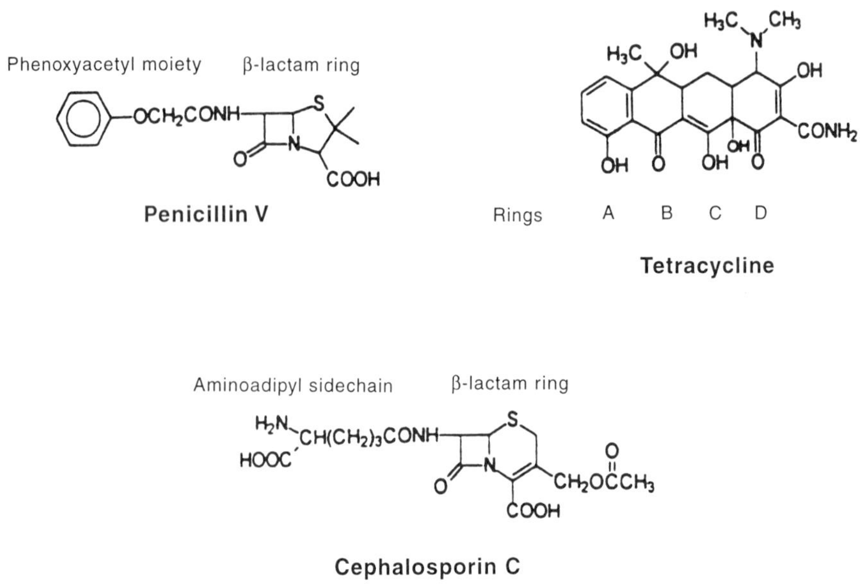

FIGURE 2.6 Structures of penicillin V, tetracycline, and cephalosporin C (Chaubal *et al.*, 1995).

Because cephalosporin C is highly water soluble and nonextractable, neutral aromatic sorbents can be effectively utilized to separate it. These authors indicated that most neutral polymeric adsorbents are copolymers of styrene (or ethylbenzene) and divinyl benzene. To obtain a better understanding of the binding of antibiotics to sorbent surfaces, ethylbenzene-divinylbenzene (Amberlite XAD-16) and an aliphatic ester (Amberlite XAD-7) were used. Figure 2.6 shows the structures of these antibiotics. They indicate that their analysis is helpful in that it provides insights into how to functionalize polymeric sorbents to improve binding affinities and selectivities.

B. Membrane Separation

Another technique that combines different functions for separation (such as clarification and initial purification) is membrane separation. Martin and Manteuffel (1988) indicated that the clarification and sterilization steps for protein solutions in the bioprocessing train often test the experience of bioprocessing engineers-chemists and other people involved in designing a process. Martin and Manteuffel (1988) indicated that microporous membrane filters are effective in clarifying and in sterilizing these solutions. These authors indicated that because many peptide and proteins are very expensive and are of value, it is undesirable to have even minimal losses of these bioproducts.

Example 2.4

Describe briefly the recovery of proteins utilizing membranes (Martin and Manteuffel, 1988).

Solution

Membrane filters have been used routinely with heat labile proteins to aid in their clarification and sterilization (Martin and Manteuffel, 1988). Because (1) proteins are known to adsorb to solid–liquid interfaces, and (2) membranes contain significant surface area up to about 500 cm^2 filtration area (Martin and Manteuffel, 1988), it is reasonable to expect adsorptive losses of proteins during membrane separation (Pitt, 1987). Membrane manufacturers have realized this, and have countered by designing the polymeric surfaces of membranes to minimize this protein adsorption (Duberstein, 1979).

The recovery of insulin, bovine serum albumin (BSA), and immunoglobulin G (IgG) following filtration through three different microporous membranes was analyzed (Martin and Manteuffel, 1988). These membranes include (1) a hydroxyl-modified hydrophilic polyamide membrane, (2) a hydrophilic polyvinylidene fluoride (PVDF) membrane, and (3) a nylon-66 membrane. Martin and Manteuffel (1988) indicated that laboratory tests that determine the effectiveness and protein loss due to adsorption in different membranes for filtration should use appropriately scaled-down throughput conditions.

These authors noted that for insulin nearly 100% recovery was attained after approximately 5ml/cm^2 of throughput volume using either a designated low-binding protein or a nondesignated low-binding protein polymeric membrane. The only differences discernible between these two types of membranes were before steady-state was attained (prior to adsorption equilibrium). These authors emphasized that although the low-binding protein membrane (as expected) binds less protein than the regular membrane, approximately 100% recovery of insulin is possible after 20 ml/cm^2 throughput volume.

Recoveries greater than 98% of both BSA and IgG utilizing two different low-binding protein membranes were realized. These recoveries were possible after a throughput volume of approximately 10 to 15 ml/cm^2 effective surface area. In addition, 100% recovery of BSA was achieved by both membranes asymptotically.

The analysis of Martin and Manteuffel (1988) demonstrates that membranes adsorb only a small amount of the proteins analyzed before steady state is achieved. They may be used for the separation of other proteins and biological products of interest. These authors emphasized that as throughputs increase, this amount of protein adsorbed initially becomes less important when compared with the final yield. Low-binding protein membranes are particularly useful when dilute protein solutions need to be filtered, and these proteins are highly expensive. These authors also cautioned that to minimize protein loss one should also examine the prefilters and the final filters. An optimization of the entire process will minimize the loss of the highly expensive biological product.

DiLeo *et al.* (1992) indicated that the production of therapeutic proteins has the risk of contamination by virus or virus-like particles, primarily if the host is mammalian in nature. These viruses may be infectious; thus there is an urgent need for their complete removal. Several treatment procedures are well known for the removal of these viruses. These treatments include thermal inactivation, chemical treatment (Horowitz *et al.,* 1985), urea or other denaturant treatment (Pocchiari *et al.*, 1988), and β-propiolactone-UV light exposure (Prince *et al.*, 1984). DiLeo *et al.* (1992) stated that these techniques have achieved validated viral clearance of 10^3- to 10^6-fold. However, efficiencies may vary depending on the type of virus (Bechtel *et al.*, 1988). DiLeo *et al.* (1992) presented a new class of membranes that have been developed specifically to remove virus from protein solutions.

The structure of the new membrane differs fundamentally from the commercially available membranes (DiLeo *et al.*, 1992). This assists in overcoming the previous shortcomings. These authors indicated this new membrane develops a refined physical separation barrier that removes virus particles based on their size. Also, an added advantage is that the minimum clearance obtained by this method is reproducible and predictable. This is true because the membrane separation is based on actual sieving properties and not on random effects.

Solution containing the virus particles could be processed continuously to remove the contaminating virus particles. Also, if additional virus clearance is required, then two membrane elements in series could be used. The particle retention characteristics of the membrane utilized increased monotonically from 3 to 8 log as the particle diameter increased from 28 to 93 nm. This membrane was able to remove 4 to 6 log overall virus particles in the size range 30 to 70 nm along with high recovery of protein. Besides, these authors presented equations to help estimate the recovery of protein as well as virus from the protein solutions. An equation for the log reduction value for the viral clearance was also presented.

The analysis presented by DiLeo *et al.* (1992) is of significant interest because it provides for a predictable, validatable, complete, and reproducible removal of virus particles from protein solutions. The protein recovery is also very high. More studies like this analysis are required that shed further physical insights into the removal of virus particles from protein solution.

Ogasawara *et al.* (1992) analyzed protein adsorption during the separation of plasma by microporous membranes. The membrane separation of plasma may be utilized to remove toxic substances with high-molecular weights for patients suffering with immunodiseases (Nose *et al.*, 1983). Blood contains different proteins, platelets, leukocytes, and red cells. These adhere to the microporous membranes as blood comes in contact with the membrane. Thereafter, there is a steady time-dependent decline in hydraulic permeability (Bauer *et al.*, 1982; Friedman *et al.*, 1983).

Ogasawara *et al.* (1992) noted an increase in the filtration resistance during the contact of BSA solutions with the microporous membranes. This also causes a lowering of pure water permeability (PWP). Furthermore, these authors indicate that the hydrophobicity and the internal structure of the microporous membrane significantly influence the amount of BSA adsorbed and the lowering of PWP. However, no information was provided by these authors on the extent of denaturation BSA undergoes during the membrane separation process.

C. Ultrafiltration

Duffy *et al.* (1989) indicated that to satisfy the present and ever-increasing demands of biological products of interest often multikilogram quantities of purified material must be produced in a cost-effective process. Considerable volumes of the fermentation broth need to be processed. For example, these authors indicated that to produce multigram quantities of antibodies requires the processing of more than 1000 liters of conditioned media. Also, the final product must continuously meet the ever-tightening FDA requirements for contaminants such as extraneous protein, DNA, and endotoxin. These authors emphasized that meeting these requirements mandates not only the careful selection of but also the continuous optimization of the different downstream processing steps.

Example 2.5

Provide an example where ultrafiltration has been used to clarify a fermentation broth for producing antibodies (Duffy *et al.*, 1989).

Solution

Monoclonal antibodies may be used as therapeutic substances and for diagnostics. Duffy *et al.* (1989) indicated that these antibodies may also be used as affinity ligands to purify other biological products of interest such as cytokines and blood-clotting factors. These authors indicated that to produce multigram quantities of antibodies requires the processing of over 1000 liters of conditioned media. They utilized cellulose triacetate membranes (surface area 50 ft^2, four in number) to effect a 50- to 100-fold concentration of several hundred liters of conditioned media in 3 to 4 hours. Their process minimized shear forces that can lead to protein denaturation. Also, because these authors used a tangential plate and frame ultrafiltration system, gel polarization was reduced. This led to acceptable flux rates.

These authors emphasized the importance of minimizing protein denaturation during the ultrafiltration step. Truskey (1987) indicated that proteins can undergo conformational changes (and subsequent denaturation) on hydrophobic membrane filter surfaces. Also, these conformational changes can lead to product instability (at interfaces). This instability can increase the product immunogenicity, which undermines the therapeutic properties. Duffy *et al.* (1989) emphasize that by utilizing hydrophilic ultrafiltration membranes the protein interactions at the interfaces can be minimized. This then maximizes the conformational integrity of the biological product of interest. This analysis is of interest because it provides some numbers for the initial purification of antibodies from solution using ultrafiltration. The emphasis by these authors on minimizing conformational changes and subsequent activity losses are of particular interest. More such studies are required that clearly examine the causes of conformational changes experienced by biological molecules during not only ultrafiltration, but also other initial purification steps employed in the bioprocessing separation train.

The influence of ionic interactions during the ultrafiltration of tryptic β-casein peptides on inorganic membranes has been analyzed (Nau *et al.*, 1995). Maubois and Leonil (1989) indicated that some peptides in β-casein may: (1) interfere in mineral nutrition (peptide β-CN(1-25), and exhibit (2) opiod (peptide β-CN(60-66)), (3) antihypertensive (peptide G-CN(177-183)), and (4) immunomodulatory (peptide β-CN(63-68)) activities. Visser *et al.* (1989) indicated that there is a gradual accumulation of peptides in an ultrafiltration reactor during the continuous production of peptide mixtures using the plasmin degradation of β-casein.

Nau *et al.* (1995) analyzed their results based on electrostatic interaction and repulsion between the peptides and the membrane, and thus on the presence of local electric fields. These authors noted that for peptides with a net charge opposite to that of the membrane, there is an electrostatic attraction between the two. This concentrates the peptide on the membrane wall, and leads to transmissions greater than one. For peptides, with the same charge as the membrane, there is repulsion prevailing that leads to lower concentrations of the peptides on the membrane than in the bulk. This leads to lower than theoretical transmissions. These authors further stated that ionic strength increases induce a small ion (with high-diffusion coefficients) screen in front of the membrane. This is detrimental to the transmission of the peptides. Now, the selectivity is based on size rather than on charge.

They emphasized that for a better understanding of the membrane transmission process, physicochemical parameters besides charge and size need to be analyzed. These authors suggested that hydrophobicity needs to be examined because it can be involved in interactions between the peptides themselves, as well as between the peptide and the membrane. In spite of not analyzing this important parameter, Nau *et al.* (1995) indicated that their model adequately, though not satisfactorily, explains the transmission of β-casein tryptic peptides during ultrafiltration on an M5 Carbosep membrane. This is true because it allows a selective separation of peptides with an enrichment in the permeate.

Two-phase aqueous extraction as a purification technique for proteins and other biological products of interest (Hustedt *et al.*, 1985a,b; Kroner *et al.*, 1982) is also of interest. Attempts have also been made to move this process toward large-scale separation (Hustedt *et al.*, 1987, 1988). Naylor (1992) indicated that solvent extraction was initially employed for the production of penicillin G. Later on it was utilized in the manufacture of other antibiotics and some nonantibiotics of animal or vegetable origin. In the original antibiotic process the fermentation broth was filtered, acidified, and then extracted with polar solvents. The isolation and purification of enzymes have been done by two-phase aqueous systems.

Richardson (1992) indicated that two-phase systems are being increasingly used for product recovery from bioreactors. The biological product should partition in the top phase, and the cell debris and unwanted products should be in the bottom phase. This is the ideal case, however. Additional work is required to make this practically feasible. The low interfacial tension present is both benefical and deleterious to process development. The low interfacial tension minimizes the denaturation of proteins and other delicate biological products of interest. However, the low interfacial tension and the small difference in density between the two phases make the separation of the two phases difficult. Furthermore, Richardson (1992) added that because the biological products of interest are present in dilute solution, a large number of stages may be required to achieve the desired separation. This may even amount to about 90% of the processing costs.

In emulsion liquid membrane systems, separation is obtained due to differences in solubility and diffusivity in the liquid film. These films generally have rather low selectivity. This selectivity can, however, be improved by adding a carrier that has an affinity for one of the components.

D. Cross-flow Filtration

Example 2.6

Briefly describe the purification of the IgG antibody by affinity cross-flow filtration (Weiner *et al.*, 1994).

Solution

Ronsohoff *et al.* (1990) indicated that the purification of recombinant proteins involves several time-consuming and expensive steps that can account for 80% of the overall costs in large-scale production processes. One way of reducing costs, as indicated earlier, is to combine steps. For example, Mattiasson and Ling (1986) combined bioadsorption and membrane separation in a single step with affinity cross-flow filtration techniques limited by the ability of suitable affinity "escort" supports (Herak and Merrill, 1989). Weiner *et al.* (1994) indicated that an ideal support should: (1) be chemically and mechanically stable, (2) possess a large specific area, (3) exhibit none or negligible nonspecific adsorption of undesired molecules, and (4) be easily modifiable to incorporate ligands.

Weiner *et al.* (1994) utilized 1- to 2-μm sized affinity microparticles for the isolation of antibody IgG from artificial IgG-human serum albumin (hSA) mixtures and from clarified hybridoma cell culture supernatants by cross-flow filtration. These authors indicated that their affinity microparticles were able to withstand the high mechanical and shear forces present in the cross-flow filtration process. Protein A was immobilized on the microparticles, and this formed a monomolecular layer on the innermost surface of the S-layer. The S-layer covered the microparticles.

These authors noted that under experimental cross-flow conditions and at a pH range of 6 to 12, there were no protein A leakage and no S-layer protein release. They were able to fit the binding adsorption equilibrium curve by a Langmuir isotherm. A maximum binding capacity, q_{max} = 13.2 μg/mg wet pellet, was obtained. The authors did note that during the washing step about 20% of the adsorbed IgG could be observed. Furthermore, the S-protein layer was susceptible to degradation by proteases, which are often present in tissue culture supernatants and serum. This could be minimized by cross-linking the cell wall fragments by glutaraldehyde, and by reducing the Schiff bases with sodium borohydride. Use of this procedure made the S-layer resistant to proteolytic degradation by a wide spectrum of proteases (Sara and Sleytr, 1987).

E. Emulsion Liquid Membrane

Example 2.7

Briefly analyze the extraction of penicillin G by an emulsion liquid membrane (ELM) process (Lee and Lee, 1992).

Solution

Lee and Lee (1992) indicated that the liquid membrane (LM) process has been utilized to concentrate weak acids and bases (Wang and Bunge, 1990; Teramoto *et al.*, 1983; Baird *et al.*, 1987), metal ions (Hirato *et al.*, 1991; Teramoto *et al.*, 1983; Goto *et al.*, 1989), and biomaterials of interest (Marchese *et al.*, 1989; Hano *et al.*, 1990). Marchese *et al.* (1989) initially applied the LM process to the extraction of penicillin G. Hano *et al.* (1990) initially applied the ELM process to the extraction of penicillin G from simulated media. Lee and Lee (1992), by using the ELM process, analyzed the influence of different parameters on the extraction of penicillin G through membranes containing an amine as a carrier. Figure 2.7 shows the schematic diagram for the facilitated transport and the extraction of penicillin G using an ELM process. The P^- is the penicillin acid anion. This reaction occurs at both interfaces (Lee and Lee, 1992)

$$\text{A (organic phase)} + \text{hydrogen ion (aqueous phase)} + P^- \rightarrow \text{AHP (organic phase)} \quad (2.3)$$

These authors utilized water–oil–water ELMs and obtained 80 to 95% degree of extraction in a batch sysyem. In the internal phase these authors noted a concentration of greater than nine times the initial concentration of penicillin G in the external phase. For the internal phase these authors theoretically calculated the pH based on the amount of penicillin transported to the internal phase. Their calculated results agree well with their experimental findings. These calculations were useful in that they permitted these authors to select a basic salt (Na_2CO_3) that gave a pH of 5 to 8. Under these conditions the penicillin was stable after the extraction process. The stability of the extracted penicillin is important. Thus, the analysis presented by Lee and Lee (1992) is of particular interest. These authors were also careful to indicate that the decomposition of penicillin G is first order in nature (Benedict *et al.*, 1945; Lundgren and Landersjo, 1970). A plot of the first-order rate constant for deactivation versus pH indicated that in the pH range of 5 to 8 penicillin G was stable. More analysis like the one presented by Lee and Lee (1992) is required that pays particular attention to the stability of not only penicillin but also other biological products of interest.

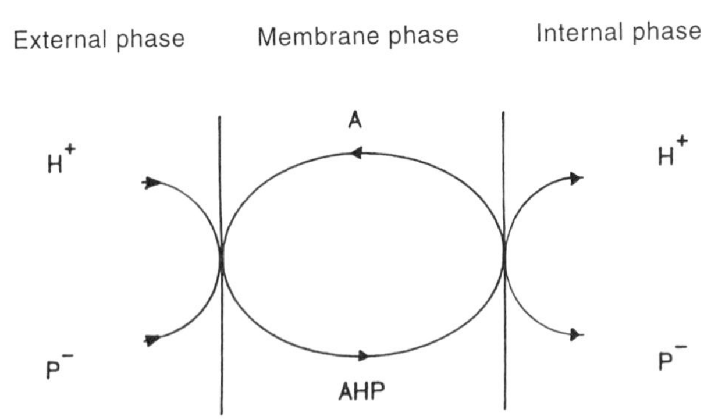

FIGURE 2.7 Schematic diagram for the penicillin G transport system (Lee and Lee, 1992).

F. Extraction

Extraction has also been traditionally utilized to extract another biological product of tremendous interest, insulin. Barfoed (1987) indicated that insulin is a polypeptide hormone. Insulin is produced in the β-cells of the pancreas. This is in response to hyperglycemia. A deficiency in insulin production in the body leads to diabetes mellitus. This author indicated that this is the largest cause of human death in industrialized nations. The next example briefly analyzes the traditional method for the production and purification of insulin.

Example 2.8

Briefly describe the traditional purification process for insulin production (Ladisch and Kohlmann, 1992).

Solution

Ladisch and Kohlmann (1992) indicated that traditionally insulin is purified from animal sources by extraction procedures followed by suitable chromatographic operations (Dolan-Heitlinger, 1982; Barfoed, 1987). The frozen pancreases of bovine or porcine origin are cut up and extracted with ethanol. Then they are acidified to pH 2. This helps remove (or inactivate) the trypsin that can degrade insulin. The extracted and acidic insulin solution is neutralized with calcium carbonate. Vacuum extraction is utilized to concentrate the insulin solution. Salt addition then precipitates the insulin. Reprecipitation of the insulin is done by redissolving the insulin followed by adjusting the pH to the isoelectric point of insulin. Ladisch and Kohlmann (1992) indicated that further purification of the insulin is undertaken by gel filtration chromatographic or ion-exchange chromatographic systems (Dolan-Heitlinger, 1982).

Large-scale purification of insulin has briefly been analyzed (Ladisch and Kohlmann, 1992). These authors mentioned that the standard methods utilized for the purification of recombinant insulin include ion-exchange, gel permeation, and reverse-phase chromatography. These methods are successful; nevertheless, Ladisch and Kohlmann (1992) suggested that further refinement in these procedures with appropriate research should be continuously undertaken to refine the procedures. This refinement and optimization of the purification is in accord with the expected demand for increasingly pure forms of insulin and other such life saving pharmaceuticals (drugs) of universal utility. These authors cited the dearth of data available in the open literature concerning complete large-scale purification sequences for the production of recombinant insulin from *E. coli*. This is not surprising considering the proprietary nature of the material and the high amount of profit that may be obtained from this information.

The information presented by Ladisch and Kohlmann (1992) is of particular value considering the scarce availability of this type of data in the open literature. What is really required is the effect of different process variables, and the sequencing of the different subprocesses on the quality and the quantity of the product separated. Eventually, the scale level of operation would also greatly impact on the stability and quality of the product separated. This type of information is, of course, not available in the open literature.

G. Precipitation

Another initial fractionation, or isolation, procedure that is used effectively for processing of proteins and other biological products of interest is precipitation. Senstad and Mattiasson (1989) indicated that traditionally affinity interactions have been used for protein purification during the later stages. These authors suggest the utilization of affinity ligands during the isolation stage in protein purification. According to these authors this concept of applying a high resolution technique at an earlier stage is attractive because it would lead to a reduction in the volume of the sample to be handled. This reduces the chemical consumption necessary during the later stages. They proposed the application of affinity precipitation using heterobifunctional ligands.

Example 2.9

Briefly analyze affinity precipitation using chitosan as a ligand carrier for protein purification (Senstad and Mattiasson, 1989).

Solution

Larsson *et al.* (1984) indicated that affinity precipitation is based on the principle of binding the soluble ligand to a target molecule that eventually leads to a precipitate. This is done by spontaneously forming aggregates with oligomeric target proteins. Heterobifunctional ligands may have more than one group on the ligand. One is for binding the target molecule. The other is to facilitate precipitation of the ligand–target protein complex. Senstad and Mattiasson (1989) utilized a soybean trypsin inhibitor (STI)-chitosan ligand for the purification of trypsin by the interaction with STI. The chitosan part was responsible for precipitation.

Figure 2.8 shows the procedure for isolating trypsin with STI-chitosan. For the purification of crude trypsin from solution Senstad and Mattiasson (1989)

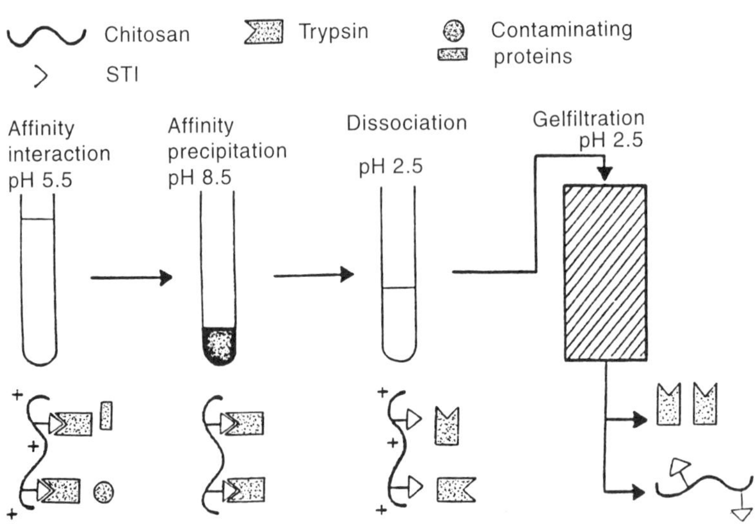

FIGURE 2.8 Isolation of trypsin with STI-chitosan (Senstad and Mattiasson, 1989).

noted a 93% yield in trypsin activity. This was obtained after isolating the precipitate, and chromatographing the soluble component under dissociating conditions. The authors also noted that the degree of purification was only 5.5. This could partly be due to the binding of chymotrypsin to the STI-chitosan. Chymotrypsin is a constituent of crude trypsin. They emphasized that their concept and model is an efficient way of recovering target molecules from solution by precipitation. We are in agreement with these authors. This concept should be tried for other systems to fully explore the potential of this technique.

Precipitation is a common purification step in isolating and recovering proteins from crude mixtures (Bell *et al.*, 1983). These authors indicated that precipitation is affected by the addition of salts and cosolvents that lower the solubility of the required protein. The addition of these salts lead to a supersaturation of the protein. This eventually leads to seed particles, crystallization, and eventual agglomeration (Bramaud *et al.*, 1995). Bramaud *et al.* (1995) analyzed the role of protein complexation in the thermal isoelectric precipitation of α-lactalbumin (α-lA) from a whey protein concentrate. Whey is a dilute liquid containing lactose, proteins, salts, and residual fat. These authors indicated that several technologies have been developed to process and utilize whey (Maubois *et al.*, 1987; Zall, 1992). Fox (1982) indicated that α-lA is one of the two components of the lactose synthesis enzyme system that catalyzes the final step in lactose biosynthesis in the lactating mammary gland. Hiraoka *et al.* (1980) initially indicated that α-lA contains 1 mol of bound calcium per mol of protein. A removal of this tightly bound calcium leads to pronounced conformational changes in α-lA. Kronman and Bratcher (1983) classified the different conformational states of α-lA based on the magnitudes of parameters of their trytophan emission spectra. Bramaud *et al.* (1995) indicated that α-lA is known to undergo well-characterized unfolding transitions. Finally, Shanbhag *et al.* (1991) characterized the conformational changes by hydrophobic partitioning as these transitions are accompanied by alteration in surface hydrophobicity.

It is important to understand the conformational changes undergone by α-lA. Bramaud *et al.* (1995) indicated that both α-lA and β-lactoglobulin (β-LG) remain entirely soluble up to temperatures of 50°C and at all values of pH between 3 and 9. At 55°C, however, some turbidity in both solutions was observed. For temperatures >50°C, and at pH between 4.2 and 4.6 (this corresponds to the isoelectric pH of α-lA), α-lA exhibited an increasing tendency to aggregate with increasing concentration. These authors emphasized that this behavior of α-lA forms the basis for the selective separation of α-lA and β-LG. Pearce (1987) attributed the reduced solubility of α-lA at the isoelectric pH to the characteristics of an unfolded-like state (intermediate and denatured form). Note that in this case, a denatured form at a higher temperature (>50°C) is actually utilized to separate and purify the α-lA.

The selective precipitation of α-lA leads to improvement of the separation parameter by centrifugation or membrane separation (Bramaud *et al.*, 1995). These authors also proposed a mechanism for the precipitation reaction. This analysis is of particular interest because it investigates the specific role of calcium on isoelectric precipitation. The precipitation depends on protein hydrophobicity, and a release of tightly bound calcium ion changes the protein hydrophobicity. More studies like these are required that provide physical insights into the precipitation process.

However, this type of precipitation is nonspecific and often, as expected, leads to an impure or a "mix" precipitate. Affinity precipitation has been developed to help improve the quality of the precipitate obtained.

Example 2.10

Briefly describe an affinity precipitation method for proteins by surfactant-solubilized, ligand-modified phospholipids (Powers *et al.*, 1992).

Solution

Powers *et al.* (1992) utilized a ligand-modified phospholipid solubilized in aqueous solution by a nonionic surfactant for the precipitation of avidin. Avidin was precipitated by contact with solutions in which dimyristoylphosphatidylethanolamine functionalized with biotin (DMPE-B) was solubilized in octaethylene glycol mono-n-dodecyl-ether ($C_{12}E_8$) solutions. These authors noted that in the avidin-DMPE-B model system, the binding interaction is specific. Figure 2.9 shows the basic principle involved in the affinity precipitation technique. Guzman *et al.* (1990) initially proposed this form of the affinity precipitation technique. Here ligand-modified phospholipids that are solubilized in aqueous solution by nonionic surfactants are contacted with multiple binding site proteins. These phospholipids apparently reduce the solubility of the target proteins thereby leading to precipitation.

A high yield as well as purity from the solutions was obtained (Powers *et al.*, 1992). For molar ratios of DMPE-B to avidin of 2, 4, and 6, the yields (the percentage of protein precipitated and which ultimately sedimented from the solution) were 32, 81, and 76% respectively. The aggregation step occurs in the order of minutes. They noted that the kinetics could be hastened by increasing the ligand:protein ratio. These authors indicate that since precipitation occurs readily below the critical micelle concentration (CMC), the two-step mechanism as shown in Figure 2.9 is involved:

1. There is an initial rapid-binding step of the ligand-modified phospholipid with the protein.
2. Thereafter, there is a slower hydrophobic interaction between the nonpolar alkyl side chains of the phospholipids that results in aggregation.

The kinetics of precipitation conform well to the Smoluchowski kinetic rate expression.

The analysis by Powers *et al.* (1992) is particularly useful for two reasons: (1) The yield and purity of the avidin precipitated is good, and (2) the predictable nature of the model conforms well with the precipitate experimentally obtained. Their analysis may be utilized to predict the precipitation of other proteins with caution. It is recommended that the analysis by these authors be extended to the precipitation of other proteins of interest. This would not only validate their model for other proteins, but also provide, if successful, a framework of a useful database. Particular care must be taken to improve the quality

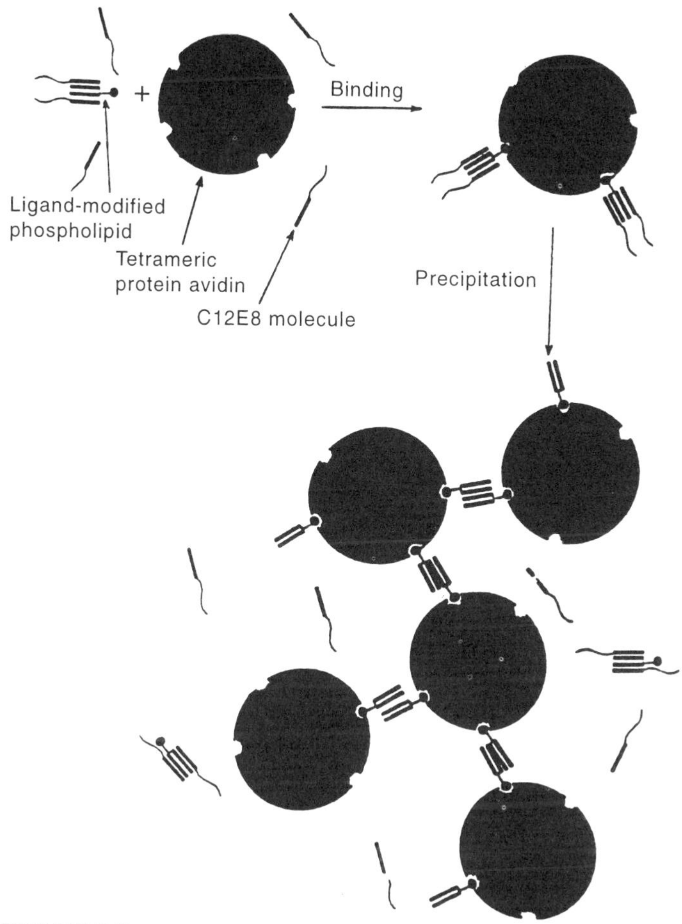

FIGURE 2.9 An affinity precipitation mechanism (Powers *et al.*, 1992).

as well as the quantity of the precipitate obtained. It would be of particular interest, for example, to be able to identify the process parameters or conditions that help improve the quality of the precipitate obtained. If this were possible, then one could control the parameters to help match the ever increasing market and governmental standards on product quality.

Affinity precipitation is an attractive technique for protein purification. However, as indicated above, it does have some drawbacks. The commercial application of this technique is constrained by the availability of stable and inexpensive bis-ligands, polyligands, and ligand carriers that exhibit high adsorption capacity as well as low degrees of nonspecific adsorption (Sun *et al.*, 1995). Sun *et al.* (1994) effectively utilized the ligand-modified polymerized liposome (PLS) for affinity ultrafiltrations. This PLS was stable, had a large hydrophobic surface area, and minimized nonspecific adsorption. It would be of interest to analyze the utilization of PLS for affinity precipitations.

Affinity precipitations occur by two mechanisms, the primary effect and the secondary effect (Morris *et al.*, 1993). The cross-linking of the protein and the subsequent network formation with ligand yield the primary effect. The secondary effect is due to the solubility decreases of the ligand carrier by changes in the pH (Senstad and Mattiasson, 1989), temperature (Galaev and Mattiasson, 1993), and salt concentration (Nguyen and Luong, 1989).

Sun *et al.* (1995) utilized STI immobilized on PLS as an affinity agent for trypsin precipitation from: (1) an artificial solution containing BSA and trypsin, and (2) a crude pancreatic extract. These authors noted that the addition of salt precipitated the PLS due to osmotic shock (Hub *et al.*, 1980). The salt could be removed by centrifugation. Thereafter, the PLS could be redispersed in water or buffer of low ionic strength.

In their analysis Sun *et al.* (1995) noted that PLS had a large specific surface area and ligand-coupling capacity. However, the adsorption efficiency did decrease on an increase in STI-coupling density due to steric hindrance. Furthermore, their experimental results indicated a lack of nonspecific adsorption, since on using PLS without immobilized STI, all the enzyme activities were detected in the supernatant.

These authors noted that from the artificial solution containing BSA and trypsin, 91% of the trypsin was coprecipitated, whereas 87% of the precipitated trypsin was recovered by elution with 0.01 M NaOH. The STI-PLS was recycled three times during the precipitation of trypsin from crude pancreatic extract. In each of these runs a purification factor of about six was obtained, although the precipitated trypsin activity decreased gradually from 93.3 to 87.7%, and then to 83.8%. They indicated that the loss of the trypsin-binding capacity may be attributed to incomplete elutions. Some denaturation may also occur, though this aspect was not considered. Overall, these authors were able to recover over 80% of the trypsin activity, and the maximum specific activity of the trypsin recovered was 1×10^4 U/mg.

They were able to precipitate and redisperse by salt addition the STI-PLS over 10 times. Thus, the PLS is an acceptable and stable ligand carrier, although its acid resistance decreased on repeated precipitation usage. These authors indicated that even though the trypsin purification has been compromised by the impurity of the STI, the PLS is a useful affinity ligand for precipitation due to the high recovery and acceptable purification factor obtained. It would be of interest to be able to determine the extent or degree of denaturation of trypsin obtained after each recycle. This might also make a contribution to the decrease in precipitation yield obtained after each recycle.

H. Chromatographic Procedures

The large-scale purification of proteins and other biological products is of considerable interest. Large-scale purification of proteins now needs to pay attention to yield and purity of the protein product separated, and also to the scale of operation. Chromatographic procedures may also be utilized, perhaps to a limited extent, as an initial fractionation technique in the purification train. The next example briefly analyzes the large-scale purification of staphylococcal enterotoxin B using three chromatographic steps. The first chromatographic step may be considered as the initial fractionation step, and the next two chromatographic steps may be considered as final fractionation steps (which are considered in the next section).

Example 2.11

Briefly analyze the large-scale purification of staphylococcal enterotoxn B using chromatographic procedures (Johansson *et al.*, 1990).

Solution

Staphylococcal enterotoxins cause food poisoning (Bergdoll, 1970). These enterotoxins are low-molecular-weight proteins (25 to 30 kD) that are similar in activity and composition. Johansson *et al.* (1990) presented a procedure for the large-scale purification of enterotoxins. The downstream procedure involves two ion-exchange steps, and one gel filtration step. For the present discussion we have placed the first chromatographic step as the initial fractionation step. The other two chromatographic procedures we place in the final fractionation category. This method helps us use the chromatographic step also as an initial purification, or fractionation, step. Johansson *et al.* (1990) emphasized that the matrices in each of the three chromatographic procedures permitted high linear flow rates at low pressure (<5 kPa).

The culture supernatant was from a fermentation of methicillin-sensitive derivative of *Staphylococcus aureus* A676. (Johansson *et al.*, 1990). The culture supernatant was received after centrifugation, sterile filtration, and adsorption of the protein. The supernatant obtained had a staphylococcal enterotoxin B (SEB) content of 0.4 mg/dm^3. The total protein content was 11 mg/cm^3, and the purity was 0.004%. The amount of culture supernatant obtained was 400 dm^3. This was diluted with 1600 dm^3 of distilled water.

This solution was then passed through a Pharmacia Column 370 KS (S Sepharose Fast Flow). During elution six 1-dm^3 samples were collected and pooled together. The total protein content decreased from 11 to 1.5 mg/cm^3 after this purification step. The SEB content increased dramatically from 0.00042 to 0.12 mg/cm^3. The purity (mg SEB/mg total protein) also increased by a factor of 2000 from 0.004 to 8. The recovery of the protein by this initial purification step using a chromatographic procedure is good. Further refinement of the protein recovered is possible using final fractionation procedures. This will be shown in a later example.

Lowe *et al.* (1992) indicated that biological molecules isolated from natural sources must meet exacting standards. For example, biologically active contaminants like DNA should be less than 10 pg per dose; and other macromolecules, viruses, and pyrogens should be less than 300 endotoxins per dose. These authors emphasized the development of gentle, but effective purification strategies for purifying different pharmaceutical products. This continuously presents new challenges to process development engineers. Lowe *et al.* (1992) emphasized the need to produce that product with the highest purity and yield. Every miniscule loss of product in the purification step leads to a loss that is multiplied by all the processing costs that have gone before (Hoare and Dunnill, 1989).

As biotechnology progresses there is an increasing demand for more selective separation methods. Traditional separation methods are being continuously modified to keep up with these demands. Reverse-phase chromatography (RPC) is one important means of isolating and purifying pharmaceutical products. The next example briefly presents an analysis of RPC removal of pharmaceutical products from dilute liquid mixtures utilizing adsorption on modified divinylbenzene-polystyrene copolymers (resins).

Example 2.12

Briefly analyze the use of modified divinylbenzene-polystyrene resins for the separation of aspartame, phenylalanine, aspartic acid, and asparagine (Casillas *et al.*, 1992).

Solution

Casillas *et al.* (1992) indicated that an ideal method for pharmaceutical purification should involve the contact of a small amount of adsorbate and resin for a short time. These authors analyzed the thermodynamics of adsorption of asparagine, aspartame, phenylalanine, and aspartic acid on modified Amberlite XAD-2 and XAD-4 resins using HPLC over a temperature range of 293 to 313 K.

They noted that above this temperature range the adsorption isotherms for the preceding components were linear, and may be characterized by temperature-dependent adsorption equilibrium constants characteristic of the resin-adsorbate system. For a modified XAD-2 resin the ΔS values were -14.8, -12.9, -13.13, -28.7 kJ/mol for asparagine, aspartic acid, phenylalanine, and aspartame, respectively. Similarly, in the same order, the ΔH values were 4.37, 4.15, 9.47, and 10.83 kJ/mol, respectively.

Furthermore, the pore size diameter and the polarity of the functional group of the resin (along with the side chain, or amino acid ring, or dipeptide) strongly influenced the adsorption equilibrium. These authors indicated that based on their results an efficient separation of the smaller amino acids from their larger homologues should be possible based on pore size selectivity of the resin. The differences in adsorption equilibria between phenylalanine (the most expensive of the four studied) and the others should permit the effective separation of this compound from a fermentation mixture using these resins. The preceding study is particularly useful because it is anticipated that the demand for artificial sweeteners is bound to grow in the future.

Different methods for the initial purification or isolation of the proteins or other biological products of interest have been briefly presented. Mattiasson and Ling (1988) indicated that in the future high selectivity may be possible by two-phase extraction methods. For the present, however, chromatography is the only choice for industrial separations.

IV. HIGH-RESOLUTION FRACTIONATION

Naveh (1990) indicated that only chromatography has the power to provide the high resolution required to obtain an ultrapure product. He emphasized that the previous purification steps only prepare the biological product for purification by chromatographic procedures. These previous procedures provide the necessary volume reduction, clarification, and adjustment of the crude extract. The chromatographic steps utilized at this stage could include cation exchange, anion exchange, affinity chromatography, and hydrophobic chromatography. This author emphasized that a train of chromatographic steps may be required to attain the high specificity required.

A. Chromatographic Procedures

Lowe *et al.* (1992) indicated that affinity chromatography is being given a second look because this technique is predictable and exhibits a rational character. This should simplify process design. However, affinity chromatography is being looked at seriously due to: (1) nonspecific adsorption, (2) expensive and labile ligands utilized, and (3) difficulties associated with sterilization of relatively unstable adsorbents. These authors emphasized that new developments in affinity chromatography are alleviating some of the previous drawbacks. Biomimetic ligands that selectively pick-up target proteins are being developed thanks to molecular modeling techniques combined with binding and crystallographic studies. They indicated that naturally occurring ligands are not only expensive but also retain much less than full activity when immobilized. Use of biomimetic ligands or pseudo-ligands, is one means to circumvent these difficulties.

These authors emphasized that the use of "designer" ligands for different biomolecules of interest will open a new era in purification. Nevertheless, these improvements will still be constrained by ligand leakage and subsequent product contamination. Thus, as in these and generally in all purification steps, a complete or overall picture should always be kept in mind. Developments for the different aspects of a purification process should keep pace with the others. Otherwise, the purification process will only be as good as the weakest link.

The large-scale purification of therapeutic proteins is of tremendous interest. Naveh (1990) emphasized that in these cases it is desirable to remove low levels of impurities, to reduce pyrogens, and to exchange into buffer systems suitable for formulation. This author indicated that gel filtration chromatography has an order of magnitude lower productivity than other chromatographic steps. Nevertheless, it is utilized for the preceding applications. In the next example we carry on our discussion of the purification of SEB. Please see example 2.11 for the initial purification steps.

Example 2.13

Briefly analyze the final fractionation steps for the recovery of *SEB* using chromatographic methods (Johansson *et al.*, 1990).

Solution

The final fractionation step involves an ion-exchange step and a gel filtration step (Johansson *et al.*, 1990). Table 2.1 neatly summarizes the recovery percentages of SEB during each stage in the purification process. The numbers for the initial fractionation steps are provided again, so that one may obtain a clear picture as SEB progresses along the purification train. The SEB content

TABLE 2.1 Large-Scale Purification of SEB[a]

Step	Starting volume, dm^3	Protein content, mg/cm^{-3}	SEB mg/cm^{-3}	Percent recovery	Percent purity
Culture supernatant	400	11	0.00042		0.004
Initial step (cation exchange)	2000	1.5	0.12	86	8
Intermediate step (cation exchange)	6	2.3	0.99	87	43
Final step (gel filtration)	0.63	0.4	0.41	99	102

[a] Johansson *et al.*, 1990.

(mg/cm^3), and the purity (mg SEB/mg total protein) is included. The starting volume prior to each step is also given. The table dramatically brings out the reduction (in general) in volume of the protein solution processed as it goes along the bioprocessing train.

The overall recovery was 74%. This number may be obtained by multiplying the recovery percentages obtained for each stage ($0.86 \times 0.87 \times 0.99 = 0.74$). These authors emphasized that the final product obtained yields a single band on sodium dodecyl sulfate (SDS) polyacrylamide gel electrophoresis (PAGE). Also, a single peak was obtained when subjected to analytic gel filtration.

The analysis presented by Johansson *et al.* (1990) is of particular interest because it demonstrates conclusively an improved way of purifying large amounts of SEB. What is more important and encouraging is the utilization of new media and equipment to help develop this large-scale purification process. By continued efforts in this direction where newer media, equipment, and procedures may be utilized to develop and optimize large-scale purification procedures, one will begin to see real improvements in developing biotechnological processes, in general. These types of efforts should be continuously encouraged.

Procedures are required that assist in the development of large-scale processes. Olson and Gehant (1992) indicated that rapid HPLC is quickly becoming a major tool that assists in the development of large-scale processes to purify recombinant proteins.

Example 2.14

Briefly describe the ultrafast HPLC separation of recombinant DNA-derived proteins (Olson and Gehant, 1992).

Solution

Olson and Gehant (1992) analyzed applications of ultrafast HPLC during the process development of recombinant-derived proteins. These authors indicated that rapid HPLC is gradually becoming an important tool during the development of large-scale processes to purify recombinant proteins. Nugent (1990a) indicated that though this is a relatively new technique, it has already been put to use in the development of robust manufacturing processes.

Furthermore, these authors indicated that two parameters are especially important for rapid HPLC development: instrumentation and resins. Nugent (1990b) indicated that instrumentation has been developed specifically for rapid HPLC. This permits the selectivity and sensitivity required for rapid HPLC. The column packing material for rapid HPLC should permit a rapid (kinetics wise) association and dissociation, and the geometry of this material also plays a significant role. The protein should not be trapped in the pores of the particle; this facilitates the rapid association and dissociation required. The wide-pore HPLC resins permit a high-resolution analysis of some complex samples for (e.g., α- and β-subunits of a recombinant chimeric protein with an analysis time of less than 5 min from sample to sample). Samples of greater than 97% purity were obtained.

The rapid (or fast) HPLC permits one to: (1) obtain quantitative estimates of product recoveries at different steps, and (2) visualize the wide variety of host cell impurities. An estimate of these host cell impurities is particularly important for biotechnology, where during the production of required proteins a host of variants are produced by posttranslational modification such as by deamidation, proteolysis, and glycosation, etc. (Garnick *et al.*, 1988; Frenz *et al.*, 1990; Anicetti *et al.*, 1989).

The development of rapid HPLC is of particular significance because it would: (1) permit additional monitoring of a protein bioseparation process, (2) permit optimization of protein purity, and (3) aid in the development of large-scale bioseparation processes. The optimization could be done considering both the quality and the quantitative aspects. The rapid and accurate monitoring of biotechnology processes is of paramount importance, considering the complexities and heterogeneities involved. Real-time information obtained by different channels, along with the rapid processing of such data, will go a long way in controlling (by feedback or by other mechanisms) and manipulating the quality, quantity, and stability of a wide range of biotechnology products.

Now that we have briefly described the steps that are involved in the processing of proteins and other biological products of interest, it would be worthwhile to provide an example that describes the entire processing train. Of particular interest would be the changes in the specific activity and yield as the biological product goes along the processing train.

Example 2.15

Briefly describe the purification and characterization of lamb pregastric lipase (D'Souza and Oriel, 1992).

Solution

D'Souza and Oriel (1992) described the purification of lamb gastric lipase. Pregastric lipases (PGL) play an important role in fat digestion in newborn animals (Ramsey and Young, 1961; Hamosh *et al.*, 1981). These PGLs from calf and lamb are used to enhance cheese food flavors (Anti, 1969; Nelson, 1972; Kilara, 1985).

D'Souza and Oriel (1992) indicated that the initial purification steps include acid precipitation, pepsin digestion, and Econo-Pac blue affinity chromatography. These steps were undertaken to remove the nonlipase protein constituents. The Econo-Pac chromatography was included to remove an albumin-like contaminant. These authors noted that following the initial purification steps, the pregastric lipases eluted as a broad activity peak from the Superose 12 (gel filtration) column. The overall purification scheme presented in Table 2.2 is of interest, especially the sequence of numbers on the specific activity and the yield as the SEB progresses along the "separation train." PGL2 was purified 61-fold with a yield of 0.2%. Initial crude enzyme activity was 21 mU/mg.

The sharp fall in the yield with increasing purity (or specific activity) is dramatically brought out in this example. Similar numbers (and pattern of numbers) are to be expected for the separation and purification of other proteins and biological products of interest. Perhaps this helps us understand why, in general, ultrapure proteins and other biological products of interest are so expensive.

Because high-resolution fractionation is the topic of the next chapter only three typical examples are provided here to give a flavor of what to expect in this procedure. More detailed examples will be provided in the next chapter, which also includes the polishing step. The polishing step may be used to either enhance the purity of the product by, for example, the removal of pyrogens from therapeutic proteins or enhance the specific activity of the protein itself.

In this chapter we briefly present, by using appropriate examples, the different procedures utililized to separate a wide variety of proteins and other biological products of interest. There is a continuous need to improve the separation of the various products during the different stages in the bioseparation train. One should be cautioned that just because these procedures exist, it does not mean that one will always attain a high level of success. The comments provided by Geisow (1992) are of interest and should help place the above discussion in proper perspective.

Geisow (1992) indicated that chromatographic media are frequently used during different stages in the bioseparation train. However, these chromatographic media that may be found suitable at the laboratory-scale level may not be suitable at the pilot–plant level. Perhaps they are suitable at the pilot–plant level, but then again they may be unsuitable at the production level. Also, just

TABLE 2.2 Purification Strategy for PGL2[a]

Purification step	Specific activity (mU/mg)	Yield (%)
Acid precipitation	64	46
Pepsin digestion	138	34
Econo-Pac blue cartridge effluent	379	25
Superose column	658	6.5
PGL2	1292	0.2

[a] From D'Souza, T. M. and Oriel, P. (1992). *Appl. Biochem. Biotechnol.*, *36*, 183. With permission.

like catalysts, these chromatographic media may not have a lot-to-lot consistency. In designing a production-scale process, this author emphasizes that one has to compromise between efficiency (cost-effectiveness) and the quality of the final product (governmental regulations). Furthermore, it takes about a year or two to help satisfy this compromise or equation. Thus, one is continuously faced with problems, and experience with process development at the production level often can be invaluable.

This author emphasizes that both the time for process development and the processing costs should be reduced so that one may satisfy the increasing demands for recombinant therapeutic proteins (such as bovine somatotropin, albumin, and hemoglobin). Finally, economic considerations almost make it mandatory to utilize the lowest possible (but suitable) cost media, the minimum number of purification steps (without compromising product quality, stability, etc.), and the fastest throughput possible. Because economic considerations play such a predominant role in bioseparation processes (they may represent 50% of the final product costs), they form the basis of Chap. 8.

V. CONCLUSIONS

The three (broad) stages that are involved during the bioseparation of proteins and other biological products of interest presented in this chapter indicate the interdependence of one on the other. There is a choice of the process that one can use in separating a protein or other bioproduct. An initial choice determines to a large extent future choices of separation procedures. Basically one has to utilize high throughput and comparatively inexpensive procedures in the beginning. After there has been a substantial reduction in the solution volume, one can use more expensive and low throughput procedures (such as chromatography) to attain the required quality (regulations set by the market or governmental agencies), quantity, and stability of the protein or bioproduct.

Apparently, bioseparation is still more of an art and less of a science. Thus, there is a need to carefully integrate the different pieces of information available in the open literature. This is one of the goals of this book. Information on different aspects of the bioseparation process is presumablly available to a large extent at different industrial locations. The reluctance of those in the industry to part with their knowledge and information (especially of the proprietary type) is clearly understandable. Nevertheless, this is the type of information that is required (especially at the higher production levels) to better understand the bioseparation process for different proteins and other biological products of interest. The problem is further exacerbated by the fact that the separation of each different type of protein or biological product requires one to almost start from square one in each instance. There is nothing like experience to help design bioseparation processes, especially at the higher industrial-scale levels. It is hoped that in the future as the knowledge base on bioseparation processes for different proteins and biological products builds up, one can see bioseparation gradually move from an art to a science. The presentation of information in the chapters that follow is an effort in this direction.

REFERENCES

Andrews, B. A. (1976). Ph.D. Thesis, University of London, London.
Anicetti, V. R. Keyt, B. A., and Hancock, W. S. (1989). *TIBTECH, 7,* 342.
Anti, A. W. (1969). *Food Prod. Dev., 3,* 17.
Baird, R. S., Bunge, A. L., and Noble, R. D. (1987). *AIChE J., 33,* 43.
Barfoed, H. C. (1987). Insulin Production Technology, *Chem. Eng. Prog., 83,* 49.
Bartels, C. R., Kleiman, G., Korzun, N., and Irish, D. B. (1958). *Chem. Eng. Prog., 54,* 49.
Bauer, H., Chimel, H., Stroh, N., and Walita, E. (1982). *J. Membr. Sci. 11,* 321.
Bechtel, M. K., Bagdasarian, A., Olson, W. P., and Estep, T. N. (1988). *Biomater., Artif. Cells Artif. Organs, 16,* 123.
Bell, D. J., Hoare, M., and Dunnill, P. (1983). *Adv. Biochem. Eng. Biotechnol., 26,* 1.
Belter, P. A., Cunningham, F. L., and Chen, J. W. (1973). *Biotechnol. Bioeng., 15,* 533.
Belter, P. A., Cussler, E. L., and Hsu, W. S. (1988). Bioseparations. Downstream Processing for Biotechnology. Wiley-Interscience, New York.
Benedict, R.G., Schmidt, W.H., Coghill, R.D., and Oelson, A.P. (1945). *J. Bacteriol., 49,* 85.
Bergdoll, M. S. (1970). Enterotoxins, In *Microbiological Toxins,* Montie, T.C., Kadis, S., and Ajly, S. J., Eds. Academic Press: New York, 265-326.
Bonerjea, J., Jackson, J., Hoare, M., and Dunnill, P. (1988). *Enzyme Microb. Technol., 10,* 357.
Bramaud, C., Aimar, P., and Daufin, G. (1995). *Biotechnol. Bioeng., 47,* 121.
Casillas, J. L., Addo-Yobo, F., Kenney, C. N., Aracil, J., and Martinez, M. (1992). *J. Chem. Technol. Biotechnol., 55,* 163.
Chase, H. A. (1994). *TIBTECH, 12,* 296.
Chaubal, M. V., Payne, G. F., Reynolds, C. H., and Albright, R. L. (1995). *Biotechnol. Bioeng., 47,* 215.
Datar, R. (1984). *World Biotechnol. Rep., 2,* A15.
DiLeo, A J., Allegrezza, Jr., A. E., and Builder, S. E. (1992). *Biotechnology, 10,* 182.
Dolan-Heitlinger, J. (1982). In *Recombinant DNA and Biosynthetic Human Insulin, A Source Book,* Eli Lilly: Indianapolis, IN.
Doulah, M. S., Hammond, T. H., and Brookman, J. S. G. (1975). *Biotechnol. Bioeng., 17,* 845.
D'Souza, T. M. and Oriel, P. (1992). *Appl. Biochem. Biotechnol., 36,* 183.
Duberstein, R. (1979). *J. Parenteral Sci. Technol., 33,* 250.
Duffy, S. A., Moellering, B. J., Prior, G. M., Doyle, K. R., and Prior, C. P. (1989). *Pharm. Technol. Int., 1*(3), 46.
Dunnill, P. (1983). *Process Biochem., 10,* 9.
Dunnill, P. and Lilly, M. D. (1975). *Biotechnol. Bioeng. Symp., 3,* 97.
Edgington, S. M. (1992). *Biotechnology, 10,* 1413.
Engler, C. R. and Robinson, C. W. (1979). *Biotechnol. Bioeng., 21,* 1861.
Engler, C. R. and Robinson, C. W. (1981). *Biotechnol. Bioeng., 23,* 765.
Erickson, J. C., Finch, J. D., and Greene, D. C. (1994). Direct Capture of Recombinant Proteins from Animal Cell Culture Media Using a Fluidized Bed Adsorber. In *Animal Cell Technology: Products for Today, Prospects for Tomorrow,* Griffiths, B., Spier, R. F., and Berthold, W., Eds., Butterworth and Heinemann: Oxford, 557–560.
Fish, N. M. and Lilly, M. D. (1984) *Biotechnology, 2*(7), 623.
Foster, D. (1992). *Biotechnology, 10,* 1539.
Fox, P. F. (1982). *Developments in Dairy Chemistry,* Applied Science: London.
Frej, A. K., Hjorth, R., and Hammarstrom, A. (1994). *Biotechnol. Bioeng., 44,* 922.
Frenz, J., Hancock, W. S., and Henzel, W. J. (1990). Reversed Phase Chromatography in Analytical Biotechnology of Proteins, In *HPLC of Biological Macromolecules-Methods and Applications,* Gooding, K. M. and Regnier, F. E., Eds., Marcel Dekker: New York.
Friedman, L. I., Hardwick, R. A., Daniels, J. R., Stromberg, R. R., and Ciarkoski, A. A. (1983). *Artif. Organs, 7,* 345.
Gabler, R., Ryan, M., Zahka, J., Case, D., and Hatch, P. (1985). *Biotech '85 Europe,* Online Publications: Pinner, Middlesex, UK, 733.
Gailliot, F. P., Gleason, C., Wilson, J. J., and Zwarick, J. (1990). *Biotechnol. Prog., 6,* 370.
Galaev, I. Y. and Mattiasson, B. (1993). *Biotechnol. Bioeng. 41,* 1101.
Garcia, A. A. (1991). *Biotechnol. Prog., 7,* 33.

Garnick, R. L., Solli, N. J., and Papa, P. A. (1988). *Anal. Chem., 60,* 2547.
Geisow, M. J. (1992). *Trends Biotechnol., 10,* 373.
Gibson, N. B. and Lyddiatt, A. (1990). In *Separations for Biotechnology,* Verrall, M. S. and Hudson, M. J., Eds., Ellis Horwood: Chichester, UK, 217–224.
Goto, M., Kondo, K., and Nakashio, F. (1989). *J. Chem. Eng. Jpn., 22,* 71.
Guzman, R. Z., Kilpatrick, P. K., and Carbonell, R. G. (1990). In *Downstream Processing and Bioseparation: Recovery and Purification of Biological Products,* Eds., Hamel, J. F., Hunter, B. J., and Sikdar, S. K., ACS Symposium Series, 419, American Chemical Society: Washington, DC, pp 212–236.
Hamosh, M., Scanlon, J. W., Ganot, D., Likel, M., Scanlon, K. B., and Hamosh, P. (1981). *J. Clin. Invest., 67,* 838.
Hano, T., Ohtake, T., Matsumoto, M., Ogawa, S., and Hori, F. (1990). *J. Chem. Eng. Jpn., 23,* 772.
Hansson, M., Stahl, S., Hjorth, R., Uhlen, M., and Moks, T. (1994). *Biotechnology, 12,* 285.
Harris, E. L. V. and Angal, S. (1990). *Protein Purification Methods. A Practical Approach,* IRL: Oxford.
Herak, D. C. and Merrill, E. W. (1989). *Biotechnol. Prog. 5,* 9.
Hirato, T., Kishigami, I., Awakura, Y., and Majima, H. (1991). *Hydrometallurgy, 26,* 19.
Hiraoka, Y., Segawa, T., Kuwajima, K., Sugai, S., Murai, N. (1980). *Biochem. Biophys. Res. Commun., 95,* 1098.
Hoare, M. and Dunnill, P. (1989). *Philos. Trans. R. Soc. London, Ser. B. 324,* 497.
Horowitz, B., Wiebe, M. E., Lippin, A., and Stryker, M. H. (1985). *Transfusion, 25,* 516.
Howell, J. (1985). *Process Biotechnol. (Suppl),* February, p iv.
Hub, H. H., Hupfer, B., Koch, H., and Ringsdorf, H. (1980). *Angew Chem. Int. Edn. Engl., 19,* 938.
Hunter, J. B. and Asenjo, J. A. (1987a). *Biotechnol. Bioeng., 30,* 471.
Hunter, J. B. and Asenjo, J. A. (1987b). *Biotechnol. Bioeng., 30,* 481.
Hustedt, H., Kroner, K. H., and Kula, M. R. (1985a). Applications of Phase Partitioning in Biotechnology. In *Partitioning in Aqueous Two-Phase Systems,* Walter, H. and Brooks, D. E., Eds., Academic: London, 529.
Hustedt, H., Kroner, K. H., Menge, U., and Kula, M. R. (1985b). *Trends Biotechnol., 3,* 139.
Hustedt, H., Borner, B., Kroner, K. H., and Papamichael, N. (1987). *Biotechnol. Tech., 1,* 49.
Hustedt, H., Kroner, K. H., and Papamichael, N. (1988). *Process Biochem., 23,* 129.
Jaenicke, R. (1987). *Prog. Biophys. Mol. Biol., 49,* 117.
Jaenicke, R. (1991). *Biochemistry, 30,* 3147.
Jamas, S., Rha, C. K., and Sinskey, A. J. (1985). Directed Biosynthesis of Yeast Glucans with Known Structure-Function Properties, Presented at the American Chemical Society 190 National Meeting, Chicago, IL, September 9–13.
Johansson, H. J., Pettersson, T. N., and Berglof, J. H. (1990). *J. Chem. Technol. Biotechnol., 49,* 233.
Kilara, A. (1985). *Process Biochem., 20,* 35.
Kirsop, B. (1981). *Chem. Ind., 4,* 418.
Kobayashi, R., Miwa, T., Yamomoto, S., and Nagasaki, S. (1982). *Eur. J. Appl. Microbiol. Biotechnol., 15,* 14.
Kroner, K. H., Hustedt, H., and Kula, M. R. (1982). *Biotechnol. Bioeng., 24,* 1015.
Kronman, M. J. and Bratcher, S. C. (1983). *J. Biol. Chem. 258,* 10875.
Kula, M. R. (1990). *Bioseparation, 1,* 181.
Ladisch, M. R. and Kohlmann, K. L. (1992). *Biotechnol. Prog., 8,* 469.
Le, M. S. and Atkinson, T. (1985). *Process Biochem., 20*(1), 26.
Larsson, P. O., Flygare, S., and Mosbach, K. (1984). *Methods Enzymol., 104,* 364.
Lee, S. C. and Lee, W. K. (1992). *J. Chem. Technol. Biotechnol., 55,* 251.
Lowe, C. R., Burton, S. J., Burton, N. P., Alderton, W. K., Pitts, J. M., and Thomas, J. A. (1992). *Trends Biotechnol., 10,* 442.
Lundgren, P. and Landersjo, L. (1970). *Acta Pharm. Suec., 7,* 509.
Marchese, J., Lopez, J. L., and Quinn, J. A. (1989). *J. Chem. Technol. Biotechnol., 46,* 149.
Marrfy, F. and Kula, M. R. (1974). *Biotechnol. Bioeng., 16,* 623.
Martin, J. M. and Manteuffel, R. L. (1988). *Biopharm,* November, 20.

Mattiasson, B. and Ling, T. C. L. (1988). Extraction in Aqueous Two-Phase Systems, In *Separation for Biotechnology,* Verrall, M. S., and Hadson, M. J., Eds., Ellis Horwood: Chichester, UK, 281.

Mattiasson, B. and Ling, T. G. I. (1986). In *Membrane Separations in Biotechnology,* Ed., McGregor, W. C., Marcel Dekker, New York.

Maubois, J. L., Pierre, A., Fauquant, J., and Poit, M. (1987). *Int. Dairy Fed., 212,* 145.

Maubois, J. L. and Leonil, J. (1989). *Lait, 69,* 245.

Mizrahi, A. (1986). *Process Biochem., 8,* 108.

Morris, J. E., Hoffman, A. S., and Fisher, R. R. (1991). *Biotechnol. Bioeng., 41,* 991.

Morton, P. H. and Lyddiatt, A. (1992). In *Ion Exchange Advances,* Slater, M. J., Ed., Elsevier Applied Science: New York.

Mosqueira, F. G., Higgins, J. J., Dunnill, P., and Lilly, M. D. (1981). *Biotechnol. Bioeng., 23,* 335.

Nadgir, V. M. and Liu, Y. A. (1983). *AIChE J., 29,* 926.

Nau, F., Kerhere, F. L., Leonil, J., and Daufin, G. (1995). *Biotechnol. Bioeng. 46,* 246.

Naveh, D. (1990). *Biopharm, 3,* 28.

Naylor, A. (1992). *J. Chem. Technol. Biotechnol., 55*(3), 294.

Nelson, J. H. (1972). *J. Am. Oil Chemists Soc., 49,* 559.

Nguyen, A. L. and Luong, J. H. T. (1989). *Biotechnol. Bioeng. 34,* 1186.

Nose, Y., Kambic, H. E., and Mastubaria, S. (1983). In *Plasmapheresis: Therapeutic Applications and New Techniques,* (eds. Nose, Y., Malchesky, P. S., Smith, J. W., and Krakauer, R. S., Eds., Raven: New York, 1.

Nugent, K. and Olson, K. (1990). *Biochromatography, 5,* 101.

Nugent, K. (1990a). Ultrafast Protein Analysis: A Powerful Technique for Recombinant Protein Process Monitoring, In *Current Research in Protein Chemistry,* Villafranca, J. J., Ed., Academic: San Diego, 233–244.

Nugent, K. A. (1990a). *Am. Biotechnol. Lab., 8*(7), 24.

Ogasawura, K. I., Tsuda, S., Ozawa, K., and Sakai, K. (1992). *Chem. Eng. J. 48,* B1.

Okagbue, R. N. and Lewis, M. J. (1983). *Biotechnol. Lett., 5,* 731.

Olson, K. C. and Gehant, R. L. (1992). *Biotechnol. Prog., 8,* 562.

Pearce, R. J. (1987). *Aust. J. Dairy Technol. 42,* 75.

Phaff, H. J. (1977). Enzymatic Yeast Cell Wall Degradation, In *Proteins: Improvement through Chemical and Enzymatic Modification,* Vol. 60, Feeney, R. J. and Whitaker, J. B., Eds., American Chemical Society: Washington, DC.

Pirotta, M. (1982). *Angew. Makromol. Chem., 109/110,* 197.

Pitt, A. M. (1987). *J. Parenteral Sci. Technol., 41,* 110.

Pocchiari, M., Macchi, G., Peano, S., and Conz, A. (1988). *Biomater., Artif. Cells, Artif. Organs, 16,* 123.

Powers, D. D., Willard, B. L., Carbonell, R. G., and Kilpatrick, P. K. (1992). *Biotechnol. Prog., 8,* 436.

Prince, A. M., Stephen, W., and Brotman, B. (1984). *Rev. Inf. Dis., 5,* 92.

Quirk, A. and Woodrow, J. R. (1983). *Biotechnol. Lett. 5*(4), 277.

Ramsey, H. A. and Young, J. W. (1961). *J. Dairy Sci., 44,* 2227.

Rehacek, J. and Schaefer, J. (1977). *Biotechnol. Bioeng., 19,* 1523.

Richardson, J. F. (1992). *J. Chem. Technol. Biotechnol., 55*(3), 304.

Ronsohoff, T. C., Murphy, M. K., and Levine, H. L. (1990). *Biopharm, 3,* 20.

Sacco, D. and Dellacherie, E. (1984). *Anal. Chem., 56,* 1521.

Sara, M. and Sleytr, U. B. (1987). *J. Membr. Sci. 33,* 27.

Sauer, T., Robinson, C. W., and Glick, B. R. (1989). *Biotechnol. Bioeng., 33,* 1330.

Scawen, M. D. and Hammond, P. M. (1989). *J. Chem. Technol. Biotechnol., 46,* 85.

Schuette, H., Kroner, K. H., Hustedt, H., and Kula, M. R. (1983). *Enzyme Microb. Technol., 5,* 143.

Schuette, H., Kraume-Fluegel, R., and Kula, M. R. (1985). *Chem. Ing. Tech., 57,* 626.

Senstad, C. and Mattiasson, B. (1989). *Biotechnol. Bioeng., 33,* 216.

Shanbhag, V. P., Johannsson, G., and Ortin, A. (1991). *Biochem. Int., 24,* 439.

Sherwood, R., Minton, N., Quirk, A., and Atkinson, T. (1985). *Biotech '85 Europe,* Online Publications: Pinner, Middlesex, UK, 161.

Soffer, G. K. and Nystrom, L. E. (1989). *Process Chromatography. A Practical Guide,* Academic: New York.

Spalding, B. J. (1991). *Bio/Technology* 9, 229.
Spark, L. B. (1985). *Biotech '85 Europe,* Online Publications: Pinner, Middlesex, UK, 203.
Strandberg, L., Kohler, K., and Enfors, S. O. (1991). *Process Biochem.*, 26, 225.
Sun, Y., Yu, K., Jin, X. H., and Zhou, X. Z. (1995). *Biotechnol. Bioeng.* 47, 20.
Sun, Y., Yu, K., and Zhou, X. Z. (1994). Polymerized Liposome as an Affinity Support for Enzyme Affinity Ultrafiltration, In *Better Living through Innovative Biochemical Engineering. Proceedings of the Third Asia-Pacific Biochemical Engineering Conference,* Teo, W. K., Yap, M. G. S., and Oh, S. K. W., Eds., Continental, Singapore, 613–615.
Teramoto, M., Sakai, T., Yanagawa, K., Ohsuga, M., and Miyake, Y. (1983). *Sep. Sci. Technol.*, *18*(8), 735.
Teramoto, M., Yamashiro, T., Inoue, A., Yamomoto, A., Matsuyama, H., and Miyake, Y. (1991). *J. Membr. Sci.*, *58*, 11.
Thommes, J., Halfar, M., Lenz, S., and Kula, M. R. (1995). *Biotechnol. Bioeng.* 45, 205.
Truskey, G., (1987). *J. Parenteral Sci. Technol.*, *41*(6), 180.
Veide, A., Smeds, A. L., and Enfors, S. O. (1983). *Biotechnol. Bioeng.*, *25*, 1789.
Veide, A., Lindback, T., and Enfors, S. O. (1984). *Enzyme Microb. Technol.*, *6*, 325.
Visser, S., Noorman, H. H., Slangen, C. J., and Rollema, H. S. (1989). *J. Dairy Res.*, *56*, 323.
Wang, C. C. and Bunge, A. L. (1990). *J. Membr. Sci.*, *53*, 71.
Wang, D. I. C., Cooney, C. L., Demain, A. L., Dunnill, P., Humphrey, A. E., and Lilly, M. D. (1979). *Fermentation and Enzyme Technology,* John Wiley & Sons: New York, p 238.
Weiner, C., Sara, M., Dasgupta, G., and Sleytr, U. B.(1994). *Biotechnol. Bioeng.* 44, 55.
Wheelwright, S. M. (1989). *J. Biotechnol.*, *11*, 89.
Whitworth, D. A. (1974). *Biotechnol. Bioeng.*, *16*, 371.
Zall, R. R. (1992). Whey and lactose processing, In *Sources and Composition of Whey and Permeate,* Zadow, J. G., Ed., Elsevier: London.

3 HIGH-RESOLUTION FRACTIONATION PROCESSES

I. INTRODUCTION

The bioseparation of a wide variety of biological products is undertaken today using a wide variety of protocols. There are only a few basic procedures available for the bioseparation of different biological products. However, one has to apply them in an appropriate sequence (or protocol) depending on the nature of the product being separated and its eventual utilization. For example, Scawen and Hammond (1989) indicated that proteins intended for therapeutic or medicinal use need by necessity to be extremely pure to minimize side effects. If these same proteins were to be used for diagnostic purposes then the purity level would not need to be as stringent. Another important parameter that significantly influences the protocol to be used is the scale of operation. Depending on the scale of operation the various bioseparation procedures to be used may be utilized in different order, and also more than once in the same process. This depends on the final product requirements as set by the industry, and state or federal governmental regulations.

In the previous chapter we analyze the cell disruption processes along with the initial fractionation processes for separating bioproducts of interest. In this chapter we analyze the role of final fractionation or high-resolution fractionation processes. These treatment steps will be the last processing steps before the product is ready for the customer. Whatever changes have to be undertaken to match whatever requirement that is necessary have to be performed in these high-resolution steps. The previous steps may have yielded a product that is not quite ready for the market. The high-resolution or final fractionation steps must get the product ready for the market. Some of the changes that may be undertaken at this step include getting the product free from impurities, salts, pathogens, endotoxins, contaminating proteins, etc. The protein or other biological product may not be in the required correct structural form. Then the final fractionation processes must be designed or selected keeping these and other factors (or required changes) in mind. Some proteins or enzymes may be separated together until this stage. Then this stage must be utilized to separate the required protein or enzyme. For example, Ulhoa and Perbedy (1992) utilized gel filtration on Sephadex G-100 to free chitanase from chitobiase.

The high-resolution fractionation processes are, in general, low volume as well as expensive procedures. These include chromatographic procedures and crystallization. These are the procedures that are most frequently employed as the final or polishing and confectioning steps. In general, these steps will entail some sort of a drying step (Cussler, 1987). This, of course, does not mean that other bioseparation procedures may not be the last step that the biological product undergoes before it reaches the market. Once again, depending on the market requirement other more suitable procedures may be employed as the last step before the product reaches the market. In the next section we will analyze the use of gel filtration and other chromatographic procedures, crystallization, and other techniques as final or high-resolution fractionation techniques. Special attention will be paid to (1) the amount of product denaturation or activity loss that occurs, and (2) the processes that lead to the preceding losses during these final fractionation procedures. Particular attention will also be paid to the quality of the bioproduct separated. The causes or the identification of the processes that leads to these activity losses or deterioration in product quality would then assist in minimizing these activity losses.

II. CHROMATOGRAPHIC PROCEDURES

Different chromatographic procedures have been utilized as the final fractionation step during the bioseparation of a wide variety of biological products of interest. We will analyze the effectiveness of these different procedures by examining the different examples of their utility available in the literature. Some of these procedures, like gel filtration chromatography, have traditionally been used more frequently than other chromatographic procedures. However, there is no hard and fast rule that gel filtration chromatography should be used as the final fractionation step in all or even most bioseparation trains. The primary determinant of choice of the final fractionation technique should frequently be the product that is separated and its required quality and quantity.

A. Gel Filtration Chromatography

Separation of bioproducts in gel filtration chromatography is based on molecular size. A mobile solvent phase surrounds an inert stationary phase that is highly rigid and porous (Scawen and Hammond, 1989). As the sample that contains the bioproduct of interest is applied to the column, the bioproduct partitions between the pores in the beads and the solvent. These authors indicated that if one is interested in large-scale operations, then the rigidity of the beads plays an important role. They mention some traditional gel filtration materials that have been used. These include Sephadex, Superose, and Sephacryl (Pharmacia LKB, Sweden), Biogel P (a polyacrylamide made by BioRad laboratories, U.S.A), Trisacryl (IBF Biotechnics, France), Cellulofine (Amicon, U.S.A), and Fractogel (Toso Haas). The importance of the compatibility of steps that follow in sequence is emphasized (Scawen and Hammond, 1989). For example, a gel filtration step (which causes a dilution) should follow an ion exchange step (which concentrates product). These authors recommended using the gel filtration step when the volume of product is low. We will demonstrate the effectiveness of gel filtration as a final fractionation technique. Particular attention will be paid to the product losses as they occur and the reasons that probably cause them during the final fractionation step and during the previous bioseparation steps. So that we may present a complete picture it is essential to present the activity losses not only during the final fractionation step but also during any previous bioseparation step undergone by the bioproduct of interest.

Example 3.1

Describe briefly the isolation and purification of carboxylesterase from *Bacillus stearothermophilus* (Owusu and Cowan, 1991).

Solution

Esterases are required for the stereospecific hydrolysis and synthesis of esters (Brookes and Lilly, 1987; Tombo et al., 1987). Owusu and Cowan (1991) purified a thermostable carboxylesterase from *Bacillus stearothermophilus* utilizing centrifugation, DEAE-Sephacel ion-exchange chromatography, and gel filtration chromatography. Polyacrylamide gel electrophoresis (PAGE) under dissociating and nondissociating conditions of the samples was also undertaken.

Owusu and Cowan (1991) noted a single low-mobility band when nondenaturing PAGE of cell-free extract samples was performed. After purification by precipitation and anion-exchange chromatography these authors obtained two bands of higher mobility. Further purification by gel chromatography revealed just a single band of high mobility. The gel filtration and the sodium dodecyl sulfate (SDS)-PAGE results suggest that the purified G18A7 esterase is a monomeric protein with a molecular weight of approximately 40,000.

The single, double, and single bands observed by Owusu and Cowan (1991) are of interest. These authors suggested the dissociation of the high-molecular-weight esterase during the initial purification step (ion-exchange chromatography) followed by separation of the two monomers during the gel filtration step. Krish (1971) and Barker and Jencks (1969) suggested that esterases may occur in multiple forms in microorganisms. These multiple forms may have different chromatographic and electrophoretic properties. Owusu and Cowan (1991) indicated that the multiple forms may arise from the association of active monomers. The equilibrium of this reaction is affected by pH, salt concentration, and enzyme dilution. The status of the amount of monomers or multimeric enzyme forms would affect the activity and stability characteristics of this enzyme.

Owusu and Cowan (1991) suggested the following explanation to account for the significant differences in thermostability between the crude and the purified esterase forms: (1) thiol-containing compounds in the crude cell extract may lead to cysteine-containing protein denaturation; (2) high levels of proteolytic activity in the crude cell extract may increase activity loss due to proteolysis; (3) coprecipitaion with less stable compounds present in the crude cell extract may enhance activity loss (these less stable compounds are not present in the purified samples); and (4) conformational changes encountered during the processing steps. These authors indicated that steps (1) and (4) are primarily responsible for activity losses. We are in agreement with them on this, and would like to place more emphasis on step (4) compared with step (2). This underscores the importance of "gentleness" during processing of carboxylesterase, proteins, and other biological products of interest.

Table 3.1 shows the effect of processing on the specific activity, yield, and purification of esterase as it proceeds along the bioseparation train (Owusu and Cowan, 1991). These authors were able to obtain a 56-fold purification of esterase along with a 48% recovery. This is a relatively high value of the yield. One contributing factor to this high yield is the not so high purification factor value obtained for this enzyme. Furthermore, they suggested that the main cause of activity losses was due to the changes in aggregation of the enzyme state during the processing. The analysis by these authors is of particular interest because they provide some appropriate reasons for activity losses undergone by the enzyme during the processing steps. More such types of analysis are required in the literature that clearly delineate the causes of protein activity losses during the bioseparation steps. The identification of the causes of protein activity losses during bioseparation would permit the control and possible rectification of the processes to help minimize these activity losses.

Brumeanu *et al.* (1995) develped a two-step chromatographic procedure to purify antigenized immunoglobulins derivatized with monomethoxypolyethylene glycol. Polyethylene glycol (PEG) is nontoxic and nonimmunogenic, and is approved by the U.S. Food and Drug Administration (FDA) for internal use in humans. PEGylated proteins preserve their biological activity. The extent of this depends on the degree of PEGylation. Clinical trials have been conducted for m(mono) PEG-derivatized enzymes, cytokines, and monoclonal antibodies (Kawashima *et al.*, 1991; Nho *et al.*, 1992; Lisi *et al.*, 1982). Brumeanu *et al.* (1995) pointed out, however, that during the PEGylation process, various de-

TABLE 3.1 Purification Protocol for G18A7[a]

Bioseparation step	Specific activity (U/mg)	Yield (%)	Purification factor
Cell extract	0.2	100	1
Ammonium sulfate	0.22	73	1.1
DEAE-Sephacel chromatography	4.63	66	23
TSK G3000 chromatography	11.2	48	56

[a] From Owusu, R. K. and Cowan, D. A. (1991). *Enzyme Microb. Technol.*, *14*, 236. With permission.

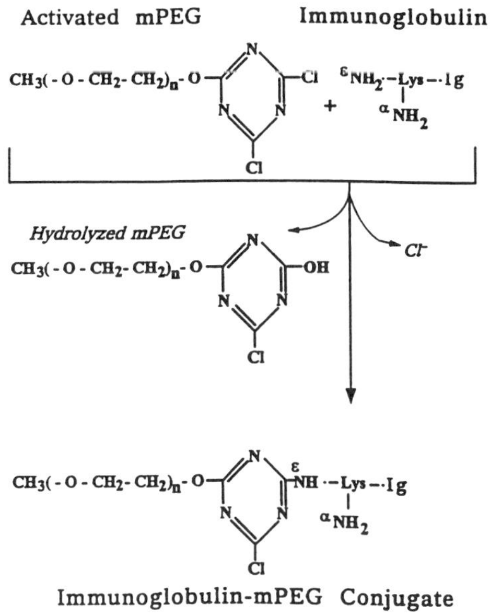

FIGURE 3.1 Chemical reactions for the attachment of mPEG to AIgs [From Brumeanu, T. D. *et al.*, (1995). *J. Chromatogr.*, *696*, 219].

grees of derivatization occur due to microheterogeneity of the protein, the distribution of both the number and the portion of attachment of PEG polymers, the inherent polydispersity of PEG polymers, etc. These authors emphasized that protein-mPEG preparations are likely to contain species of highly PEGylated proteins as well as unreacted materials. For biological usage a relatively pure population of mildly PEGylated antigenized immunoglobulins Igs (AIgs) are required.

Figure 3.1 shows the chemical reactions for the attachment of mPEG to AIgs. Brumeanu *et al.* (1995) utilized a two-step procedure to purify mPEG-derivatized AIgs (AIgs-mPEG). Size-exclusion chromatography using ammonium hydrogen carbonate as a buffer system was utilized to remove the hydrolyzed mPEG polymers. AIg-mPEG preparations were applied to a Ultrogel AcA-44 gel column. These authors indicated that during the conjugation process significant amounts of hydrolyzed mPEG were produced. Two peaks were obtained. The first peak contained AIgs-mPEG as well as unconjugated AIgs. Free hydrolyzed mPEG eluted as the second peak.

Then the fractions of the conjugates free of hydrolyzed mPEGs were chromatographed again on a Q300 anion-exchange column. This step permitted the removal of un-PEGylated AIgs from AIg-mPEGs. However, mildly PEGylated AIgs were required. On the anion-exchange high-pressure liquid chromatography (HPLC) three peaks were obtained. The third peak eluted at the same location as unconjugated control Ig-HA. The first peak represented the highly PEGylated AIgs that could not bind to the matrix. Peak 2 contained the mildly PEGylated Ig-hemaglutinin (HA). This was revealed by SDS-PAGE analysis.

Finally, Brumeanu *et al.* (1995) emphasized that they were able to isolate relatively pure populations of 6 to 8% degree of derivatization as indicated by an electrophoretic assay. These authors further indicated that mildly PEGylated Ig-HA exhibited a long half life in blood circulation and induced strong T-cell activation *in vivo*.

It would be of interest to analyze the sequence of steps required to process a biological product of interest obtained from animal and bacterial sources. Ultimately, the process economics is of interest. Nevertheless, this is also an excellent academic exercise because it permits one to compare the different bioprocessing steps required at the different stages of getting the product ready for the market. Also, the same bioseparation step may be utilized more than once at different stages of the bioseparation protocol.

Example 3.2

Briefly analyze the processing steps for obtaining tissue plasminogen activator (tPA) from animal cell and bacterial sources with special attention to the quality of the product recovered (Datar *et al.*, 1993).

Solution

Datar *et al.* (1993) analyzed the processing steps for producing tPA from animal cell and bacterial sources. These authors indicated that recombinant tPA sells for $2200 per dose or $22,000 per gram. This is about 20 times the cost of its competitor, streptokinase. Conversations with a couple of medical doctors indicate that streptokinase may be used only once on a single individual (due to an immunogenic response). However, tPA may be used more than once on the same individual. These authors emphasized that low production costs will become increasingly important for the development of second generation therapeutics such as bovine growth hormone. Thus, they have analyzed the processing implications of obtaining tPA from the preceeding two sources. tPA is selected as a model product for study because: (1) it is regarded as a flagship product for the young biotechnology industry, and (2) it was the first product made ready for the market using genetically engineered mammalian cells instead of recombinant bacteria.

An annual production rate of 11,000 g of tPA per year was selected based on an analysis of the market requirments (Datar *et al.*, 1993), and the authors based their fermentation capacity on this amount. They compared the pilot-scale production runs of tPA in *Escherichia coli* and Chinese hamster ovary (CHO) cells. tPA from both of these sources was purified using variations of similar purification protocols. The steps involved in the protocols included affinity chromatography, ion-exchange chromatography, and amino acid-Sepharose and gel chromatography. Besides these steps, ultrafiltration, centrifugation, microfiltration, solubilization, cleavage, and refolding were also used. Figure 3.2 shows the process flow sheet for purifying tPA from *E. coli* and CHO cells. Table 3.2 shows some of the important parameters involved including the sequence of recovery operations. Two points are worth noting. The product concentration was 33.5 and 460 tPA mg/L for the CHO and the *E. coli* processes, respectively. Also, the overall process yield was 47 and 2.8% for the CHO and the *E. coli* processes, respectively. By using these yield percentages of 47 and 2.8%, Datar *et al.* (1993) estimated the plant capacity

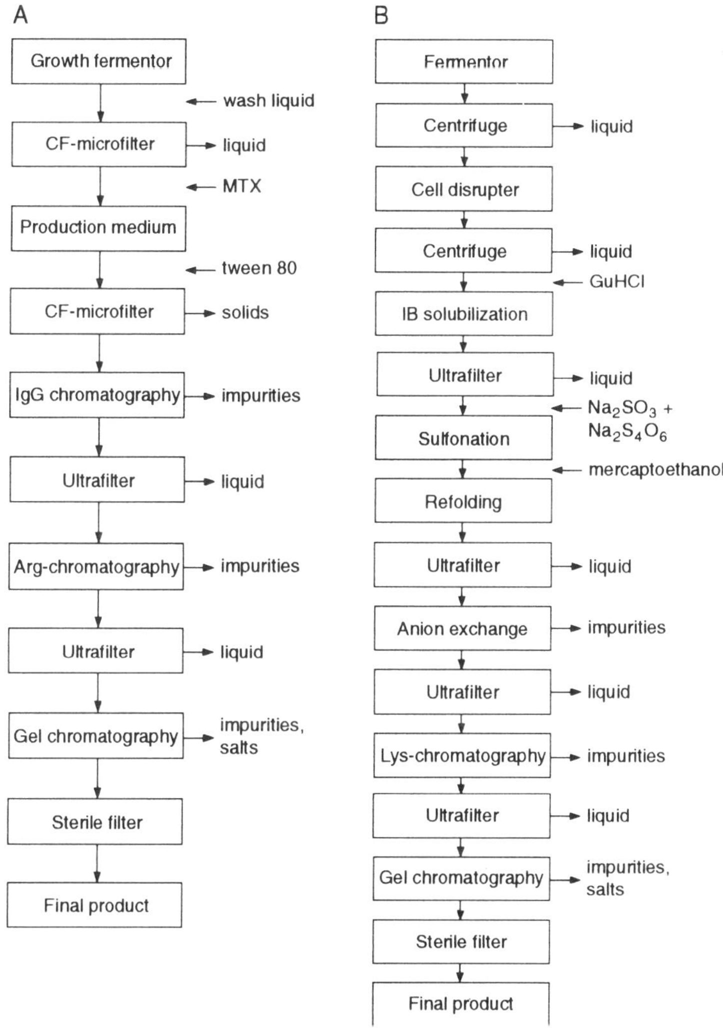

FIGURE 3.2 Process flowsheet for purifying tPA from *E. coli* and from CHO Cells [Datar, R. V. *et al.* (1993) *Biotechnology, 11,* 349].

requirements to be 700,000 and 865,000 liters/year for the CHO and the *E. coli* processes, respectively.

The CHO process with only five steps is not very recovery intensive. The *E. coli* process with 16 steps in the purification protocol is recovery intensive. It is worth noting that in both processes affinity chromatography and gel chromatography have been used toward the end of the purification process (as high-resolution fractionation processes) to help remove impurities and salts.

The purity of tPA obtained by both the sources was greater than 99.5% as judged by SDS-PAGE and by gel filtration chromatography (Datar *et al.*, 1993). The tPA produced had a single chain form. Some of the impurities represented fragments of tPA. In the case of *E. coli* special care has to be taken

TABLE 3.2 Purification Protocol for tPA from Chinese Hamster Ovary (CHO) Cells and from *E. coli*[a]

	CHO cells	*E. coli*
Bioseparation step	Microfiltration Ultrafiltration Affinity chromatography Gel chromatography	Centrifugation Ultrafiltration Solubilization Cleavage Refolding Affinity chromatography Gel chromatography
Fermentor size	28,000 liter (growth) 14,000 liter (production)	17,300 liter
Product concentration	33.5 mg tPA/liter	460 mg/liter
Yield	47%	2.8%

[a] From Datar, R. V. *et al.* (1993). *Biotechnology, 11,* 349. With permission.

to obtain a clearance greater than 99.999% for endotoxin. Another concern is the presence of incompletely renatured forms that have the potential to act as immunogens. These causes of concern for the *E. coli* process combined with the inherent difficulty of undertaking renaturations at the large-scale level with the present level of knowledge act as deterents as far as this route is concerned.

It is of interest to note the drop in the product yield as the product proceeds along the purification protocol. Datar *et al.* (1993) indicated that for the *E. coli* process the final yield of 2.8% reflects a 20% yield for the refolding operation and a 56% yield for the ultrafiltration step that follows. In this case the drop in product yield for the final fractionation steps that include affinity and gel chromatography is only a few percent. Also, the critical step, as indicated previously is increasing the yield of the refolding step. There is not much drop in the product yield during the final fractionation steps because the levels of product prior to utilizing these steps is so low (a few percent). If process improvements can be made in the future that will permit the treatment of higher levels of product yields during the final fractionation steps, then one may note higher levels of product losses. Until then one has to concentrate on improving tPA yields during the primary and secondary stages of purification, and primarily during the refolding steps as suggested by Datar *et al.* (1993).

The analysis by Datar *et al.* (1993) is particularly interesting because it shows the "bottlenecks" involved during the processing of tPA. Furthermore, if these bottlenecks can be removed, then a "sensitivity" analysis suggests a considerable improvement of the product yield. Similar studies on tPA and other biological products of interest are required in the open literature to help push bioseparation processes from an art to a science.

Often during the processing of proteins or enzymes two slightly different forms of the enzyme are produced. Then in the final stage these two slightly different forms need to be separated to obtain the required protein or enzyme in a homogeneous and pure form.

Gilbert *et al.* (1992) purified and characterized xylanases from the thermophilic ascomycete *Thielavia terrestris* 255B. Ward and Moo-Young (1989) indicated that xylans are the major hemicellulose in angiosperms. They account for 20 to 30 percent of the dry weight of woody tissue. Gilbert *et al.* (1992) indicated that xylans are heterogeneous polysaccharides. The major constituent of these polysaccharides are linear chains of D-xylosyl residues. Xylanases are required for the hydrolysis of these xylans. Gilbert *et al.* (1992) emphasized that xylanotic systems involve the hydrolysis of xylans to fermentable products, and the further removal of hemicelluloses and lignin from bioleaching pulp treatments.

The three separation steps employed by these authors include an ultrafiltration step, an anion exchange step (Q-Sepharose column), and a gel filtration (Superose 12 column) step. The purification protocol, the specific activity after each step, and the yield are given in Table 3.3. There is a considerable increase in the specific activity of the xylanase from 46 to 2154 U/mg, which represents a purification factor of 46.8. However, as expected this is obtained at the cost of a very low yield of 4.7%. In this case, we can see that there is a considerable drop in the yield (from 50 to 4.7%) during the gel filtration step. The gel filtration step was required to separate xylanase I (xyl I) from xylanase II (xyl II). The xylanase II peak eluted last from the gel filtration column and was purified to at least 99% homogeneity.

The analysis of Gilbert *et al.* (1992) is of significant interest. These authors emphasized that the xylanases interact with the gel filtration column due to the presence of a high tyrosine content. This assists in the final purification step. They did not make any comment on the possible conformational changes that may result due to the interactions of the xylanases with the gel filtration column. These conformational changes could also contribute to the low yield values. In that case the interactions with the gel filtration column are a mixed blessing. In one way they facilitate the separation. In the other case the interactions may lead to conformational changes and possible subsequent activity or yield losses of these xylanases. It would be helpful to explore and to analyze this aspect further to shed insights into what really causes the yield losses during the gel filtration or any other high-resolution fractionation step.

It would be useful to note and to compare the purification of xylanases by different workers. It is of interest to note the different bioseparation procedures used, the actual bioseparation protocol utilized, and the effect of these on the quality and quantity of the xylanase produced.

TABLE 3.3 Purification Protocol for Xylanase II from *Thielavia terrestris* 225B[a]

Bioseparation step	Activity units	Specific activity (U/mg)	Yield (%)
Cell culture	128,600	46	100
Pellicon retentate	78,400	176	61
Anion exchange chromatography	64,500	462	50
Gel filtration chromatography	6,032	2,154	4.7

[a] From Gilbert, M. *et al.* (1992). *Appl. Biochem. Biotechnol., 34/35,* 247. With permission.

TABLE 3.4 Purification Protocol for Xylanase from *Streptomyces* A451[a]

Bioseparation step	Specific activity (U/mg)	Yield (%)	Purification factor
Cell extract	4.6	100	1.0
NH_4SO_4 precipitation	14.8	81	3.2
DEAE-Sepharose chromatography	51.4	62	11.2
CM-Sepharose chromatography	201.4	54	43.8
Sephadex G75 gel filtration chromatography			
Xylanase I	672	21	146
Xylanase II	383	18	83.2

[a] From He, L. *et al.* (1993). *Enzyme Microb. Technol.*, *15*, 13. With permission.

He *et al.* (1993) purified two isoenzymes of xylanase (1,4 β D-xylan xylanhydrolase, E.C. 3.2.1.8) from *Streptomyces* sp. A451 using ammonium sulfate fractionation and DEAE- and CM-Sepharose chromatography. The high resolution fractionation step was gel permeation chromatography on Sephadex G75. These authors utilized the following steps in sequence: ammonium sulfate addition and precipitation, centrifugation, desalting using Sephadex G25, treatment by a DEAE-Sepharose FF column and by a CM-Sepharose column, concentration by ultrafiltration (Amicon), gel filtration on a Sephadex G75 column, and finally freeze-drying of samples at − 70°C.

The two isoenzymes of xylanase, xyl I and xyl II, were purified to homogeneity by the preceding bioseparation protocol and exhibited specific activities of 672 and 383 U/mg, respectively. The final fractionation step (gel filtration) was required to resolve xyl I and xyl II. SDS-PAGE indicated that the two isoenzymes were homogeneous. The yields of xyl I and xyl II were 21 and 18%, respectively. Table 3.4 shows the purification steps, the specific activity, the percentage yield, and the purification factor after each bioseparation step. Xyl I and xyl II were purified by a factor of 146 and 83, respectively.

The purification protocol utilized by He *et al.* (1993) exhibits yields of about 39%. This is a relatively high number, and is not unexpected if one considers that the purified enzymes exhibit high levels of stability. The authors emphasized that the two isoenzymes exhibit similar stability profiles. They indicated that greater than 90% of the activity was retained in the temperature range 45 to 60°C. Also, xyl II retained about 90 percent of its activity at 70°C. At 40°C, the enzymes exhibited a reduced level of activity of 80% (of control) after a 24-h period. It would be of interest to find the causes of activity loss. These would then help minimize the loss of xyl I and xyl II activity during each step of the bioseparation protocol. He *et al.*(1993) state that the purification of the two different xylanase forms supports the concept by Wong *et al.* (1988) that the actinomycetes secrete multiple forms of xylanase to assist in the solubilization of the heterogeneous substrate. These authors indicate that the enzyme-substrate interactions may also exhibit subtle differences.

The large-scale production of artificial sweeteners from corn starch represents one possible breakthrough for biotechnology compared with their manufacture by the acid hydrolysis of starch (which gave low yields). The present enzymatic process yields greater than 95% conversion of starch to glucose with a minimium of bioproduct formation (Goldstein, 1990).

Brown *et al.* (1993) purified a highly thermostable glucose isomerase from a thermophilic bacterium *Thermotoga maritima*. These authors indicated the advantages of operating temperatures of 95 to 100°C where syrups of approximately 55% fructose could be produced. Glucose isomerase stable at these high temperatures may be isolated from thermophilic bacteria such as *Bacillus stearothermophilus* (Suekene *et al.*, 1978), *Thermus aquaticus* (Lehmacher and Bisswanger, 1990), and *Thermus thermophilus* (Dekker *et al.*, 1991). The enzyme *D*-glucose (*D*-xylose) isomerase (EC 5.3.1.5) isomerizes glucose syrup and fructose [high-fructose corn syrup-HFCS). Brown *et al.* (1993) utilized the following bioseparation procedures to purify the glucose isomerase: centrifugation, Q-Sepharose column treatment, Phenyl-650M hydrophobic resin column treatment, Q-Sepharose HP anion exchange column treatment, and HiLoad Superdex 200 gel filtration column treatment.

Table 3.5 indicates the purification protocol and shows eventually a 130-fold increase in the purity of the enzyme after the high-resolution (gel filtration) step. Brown *et al.* (1993) indicated that a second anion exchange step was utilized to remove a contaminating protein that coeluted in the previous column. SDS-PAGE of the final sample exhibited a single band with a molecular weight of 45,000. These authors stated that the glucose isomerase from *Thermotoga maritima* has a homotetrameric structure. They further indicated that this is the most common oligomeric state of the glucose isomerase.

The overall yield of the glucose isomerase after the gel filtration is reasonably high (17%). This is in agreement with the stability of the enzyme during incubation at elevated temperatures (100°C). Brown *et al.* (1993) emphasized that the *T. maritima* glucose isomerase demonstrates significant potential for useful application in the industrial isomerization process. About two-thirds of the glucose isomerase activity was lost after treatment by the first two chromatographic steps. It would also be of interest to examine and to analyze the causes of activity losses during each of the bioseparation processes involved. No comments were provided by these authors for the loss in glucose isomerase activity during each of the bioseparation procedures utilized.

TABLE 3.5 Purification Protocol for Glucose Isomerase from *Thermotoga maritima*[a]

Bioseparation step	Specific activity (U/mg)	Yield (%)	Purification factor
Cell-free extract	0.15	100	1.0
Q-Sepharose FF anion exchange chromatography	0.47	60	3.1
Phenyl-650M hydrophobic interaction chromatography	4.0	35	27
Q-Sepharose HP anion exchange chromatography	5.0	27	33.3
Superdex 200 gel filtration chromatography	19.5	17	130

[a] From Brown, S. H. *et al.* (1993). *Biotechnol. Bioeng.*, *41*, 878. With permission.

Example 3.3

Briefly describe the purification of two endo-β-glucanases from the aerobic fungus *Penicillium capsulatum* (Connelly and Coughlan, 1991).

Solution

Connelly and Coughlan (1991) purified two endo-β-glucanases from the solid-state cultures of the aerobic fungus *Penicillium capsulatum.* Glucanases catalyze the depolymerization of glycans. Glycans are polysaccharides that cause problems in the beer industry by forming precipitates, hazes, and gels in stored beer (Rivere, 1977). They also cause "gumminess" in poultry diets that restricts the food intake (Jeroch *et al.*, 1988).

Connelly and Coughlan (1991) utilized the following sequence of bioseparation procedures: freeze-drying of the crude extract, gel filtration on a Sephacryl S-300 column, ion exchange on a DE-52 cellulose column, ultrafiltration, and finally electrophoresis. Table 3.6 shows the purification protocol, the increase in specific activity, and the corresponding decrease in yield as the endo-β-glucanases proceed down the bioseparation train. There is about a 19.2 fold purification increase of the β-glucanase specific activity after the ultrafiltration step along with a 55% yield. The yield of the enzymes obtained is relatively high. Nondenaturing electrophoresis was required to separate the two endo-β-glucanases. There was an increase in the specific activity after the electrophoresis step for β-glucanase A, but a decrease in the specific activity for β-glucanase B. The yields dropped drastically from 55 to 23 and 21.7% for β-glucanase A and B, respectively. At least in this case, rather similar structure enzymes may be separated but with a significant decrease in yield. A single high-resolution bioseparation step cuts the yield by more than one-half.

Some points about the purification protocol are worth mentioning. The gel filtration step is used earlier in this purification protocol. These authors mentioned that the crude extract is more stable than the purified enzyme. Apparently the purification protocol removes a stabilizing substance or substances

TABLE 3.6 Purification Protocol for Endo-β-Glucanases from *Penicillium capsulatum*[a]

Bioseparation step	Specific activity (U/ml)	Yield (%)	Purification factor
Cell extract	46	100	1
Freeze-dried crude extract	56.1	96	1.2
Gel filtration	139.8	79.2	3.0
Freeze-drying	161	79	3.5
Ion-exchange chromatography	773.1	56	16.8
Ultrafiltration	883.2	55	19.2
Electrophoresis	920	23	18.5 β-glucanase A
	789	21.7	18.5 β-glucanase B

[a] From Connelly, I. C. and Coughlan, M. P. (1991). *Enzyme Microb. Technol., 13,* 462. With permission.

TABLE 3.7 Purification Protocol for β-Glucosidases from *Clostridium thermocellum*[a]

Bioseparation step	Specific activity (U/mg)	Yield (%)	Purification factor	
Cell free extract	0.06	100	1.0	
DEAE-Sepharose chromatography				
Fraction A	0.09	43	1.5	
Fraction B	0.19	30	3.2	
Mono Q (fraction A)	1.5	15	25	
Mono Q (fraction B)	1.5	18	25	
Mono P (fraction A)	37.5	8	625	
Mono P (fraction B)	7.5	15	125	
Superose 12 gel filtration chromatography				
	0.06	8	2083	fraction A
	0.12	10	1383	fraction B

[a] From Katayeva, I. A. (1992). *Enzyme Microb. Technol.*, *14*, 407. With permission.

(as yet unidentified). They indicated that the two endo-β-glucanases were purified to homogeneity as judged by electrophoresis and by isoelectric focusing. Both the enzymes increased the fluidity of solutions and have been classified as endo enzymes, or more corectly as endo-β-1,4 glucanhydrolase (EC 3.2.1.4).

Clostridium thermocellum excretes β-glucosidases that metabolize cellobiose. The degradation of crystalline cellulose by a cellulase complex yields cellobiose as a major end product. Katayeva *et al.* (1992) purified two β-glucosidases from the anaerobic bacterium *C. thermocellum*. The cells of *C. thermocellum* were broken by sonication and the crude extract was centrifuged. These authors then subjected the preparation to DEAE-Sepharose CL-6B chromatography from which the two peaks of β-glucosidase A (first peak) and β-glucosidase B (second peak) were obtained. Further purification was undertaken on an ion-exchange chromatographic, high-performance Mono Q HR 10/10 column. Thereafter, the sample was concentrated by ultrafiltration. Chromatofocusing was then performed on a Mono P HR 5/20 column. Finally, high-resolution fractionation was carried out on a gel filtration 12 HR 10/30 column. Both the enzymes obtained after the purification protocol were homogeneous. The specific activities of β-glucosidase A and B obtained after the preceding bioseparation protocol were 125 and 83 U/mg, respectively. Table 3.7 shows the steps involved in the bioseparation protocol, and indicates the more than three orders of magnitude of the purification factor obtained for both the enzymes. The yields, as expected, for such a high purification factor are low (8 and 10% for β-glucosidase A and B, respectively). A large fraction of the enzyme is lost during the first chromatographic step (DEAE-Sepharose; 57 and 70% for the A and B forms, respectively).

It would be of interest to better select a column for the first step, if possible, to minimize the considerable loss of activity (of 57 and 70%) during this stage. The authors did not provide any reasons for the loss of enzyme activity during each of the bioseparation stages and did not suggest how it may be minimized.

Such information would provide considerable assistance in preserving the activity of both of these enzymes during the bioseparation protocol.

Sometimes, one may require more than one type of high-resolution fractionation step (e.g., gel filtration) to help attain the desired purity of the product to free it from other contaminating proteins. The next example describes a purification protocol where two different gel filtration steps are utilized.

Example 3.4

Briefly analyze the purification of pectin methylesterase from *Bacillus subtilis* (Pitkanen *et al.*, 1992).

Solution

Pitkanen *et al.* (1992) purified *Erwinia chrysanthemi* B374 pectin methylesterase (PME) from *Bacillus subtilis.* Pectin is a major structural component of plant cell walls. Pectinases are required to degrade the pectin and play a major role in the texture changes of fruit and vegetables during ripening and storage. PMEs are required for the clarification of cider (Rombouts and Pilnik, 1986), and for the deesterification of pectin to methanol and pectate.

Centrifugation, ultrafiltration, and two gel filtration steps were utillized to purify PME (EC 3.1.1.11) to homogeneity (Pitkanen *et al.*, 1992). A purification factor of 11 and a yield of 12% were obtained. Table 3.8 shows the purification protocol. The significant decrease in the yield in the second row of the Table 3.8 (41%) was mainly due to the ultrafiltration step. Pitkanen *et al.* (1992) emphasized that the loss of protein was mainly due to leakage in the ultrafiltration membranes. This was in spite of the fact that the molecular mass of PME is 36,000 Da (Heikinheimo *et al.*, 1991), which is about four times the molecular cutoff (10000 Da) of the membranes. If a better retention of activity is required, then other means of concentrating the sample must be selected. Either obtain a different membrane with a lower cutoff or utilize some other process so that PME losses during this step may be minimized. These authors indicated that the contaminating flagellin proteins of *B. subtilis* were removed in the void volume during the first gel filtration step. After the second gel filtration step, the PME was homogeneous as judged by SDS-PAGE. There is about a threefold drop in yield (41 to 12%) on subjecting the PME to the second gel filtration step. Furthermore, Pitkanen *et al.* (1992) indicated that a low activity of the enzyme in the absence of salts is due either to nonspecific binding or to

TABLE 3.8 Purification Protocol for PME from *Bacillus subtilis*[a]

Bioseparation step	Specific activity (U/mg)	Yield (%)	Purification factor
Cell free extract	108	100	1.0
Bio-gel P-100 chromatography	632	41	5.8
Superose 12 gel chromatography	1190	12	11

[a] From Pitkanen, K. *et al.* (1992). *Enzyme Microb. Technol.*, *14*, 832. With permission.

aggregation of the enzyme. It is worthwhile maintaining a 20 to 150 mM level of monovalent or divalent cations concentration to realize the four-fold increase in activity (obtained for the purified enzyme) during the bioseparation steps.

Chromatographic procedures have been utilized in the bioseparation protocol as the primary separation technique, as a secondary separation technique, and as a high-resolution fractionation technique. Other chromatographic procedures that have been utilized as a high-resolution fractionation technique include hydroxylapatite chromatography, ion-exchange chromatography, adsorption chromatography, Phenyl-Sepharose chromatography, and hydrophobic interaction chromatography. The preceding chromatographic techniques will be demonstrated as effective high-resolution fractionation techniques by considering examples where they have been used as the last bioseparation step in the purification protocol.

B. Hydroxylapatite Chromatography

Adsorption chromatography may be utilized as a high-resolution fractionation step. This high-resolution fractionation may be performed on a hydroxylapatite column.

Example 3.5

Briefly analyze the purification of *Clostridium thermocellum* β-glucosidase B using ion exchange, hydrophobic interaction, and hydroxylapatite chromatography (Romaniec *et al.*, 1993).

Solution

Romaniec *et al.* (1993) presented a purification protocol for *C. thermocellum* β-glucosidase B. Woodward and Wiseman (1982) indicated that β-glucosidase (β-*D*-glucoside glucohydrolase, EC 3.2.1.21) catalyzes the hydrolysis of β-glucoside linkages between glucose and alkyl, aryl, or saccharide groups. Romaniec *et al.* (1993) demonstrated that β-glucosidase in cellulolytic microbes hydrolyze cellobiose and cellodextrins to fermentable glucose.

Romaniec *et al.* (1993) carried out the bioseparation protocol at 4°C and under aerobic conditions. These authors utilized the following bioseparation procedures: ammonium sulfate precipitation, Q-Sepharose FF column treatment, butyl-Sepharose column treatment, and finally hydroxylapatite column treatment. The hydroxylapatite column basically uses adsorption chromatography to help purify the component or components. Table 3.9 shows the bioseparation procedures employed; and the specific activity, yield, and purification factor after each step. These authors indicated that the β-glucosidase was purified 51-fold with a yield of 9%. This is a relatively moderately high yield considering the purification factor obtained. The enzyme was purified to homogeneity as analyzed by SDS PAGE, which revealed just a single band. In this case note that the high-resolution fractionation step (the hydroxylapatite chromatography) was utilized twice to attain the desired purity. The second hydroxylapatite chromatographic step improved the purification by about 57%. No reasons were provided by Romaniec *et al.* (1993) to explain the losses of β-glucosidase activity during each of the bioseparation procedures.

TABLE 3.9 Purification Protocol for β-Glucosidase from *Clostridium thermocellum*[a]

Bioseparation step	Specific activity (U/mg)	Yield (%)	Purification factor
Crude cell extract	8.6	100	1.0
Ammonium sulfate precipitation	15	79	1.7
Q-Sepharose FF column chromatography	25	38	2.9
Butyl-Sepharose column chromatography	97	24	11.3
Hydroxylapatite chromatography	275	17	32
Hydroxylapatite chromatography	441	9	51.3

[a] From Romaniec, M. P. M. *et al.* (1993). *Enzyme Microb. Technol.*, *15*, 393. With permission.

Furthermore, these authors noted that the purified enzyme was relatively stable at 45°C (half-life of 10 h), and its stability was increased by the addition of divalent cations. At 60°C, the half-life of the enzyme was 40 min. These authors also indicated that the purified β-glucosidase B is a broad specificity β-glucosidase.

Silica gel is frequently used as the packing material for HPLC separations. Knox and Kaur (1988) emphasized that for good separations or results it is essential the column packings have regular shapes, for example, spherical with uniform size in the range 3 to 10μm in diameter. Some organic polymers along with silica gel are able to satisfy the preceding requirements. Honda *et al.* (1995) indicated that silica gel does have disadvantages in that its uncoated surface is dissolved by alkaline solution and bonds to alkyl groups that tend to be cleaved in acidic solution. It would be of interest to explore the utilization of other materials as column packing materials, especially, those materials that exhibit unique adsorption characteristics for the separation of substances that have not been successful using the standard silica-based packings. Honda *et al.* (1995) suggested that some of these alternative materials could be titania (Trudinger *et al.*, 1990), zirconia (Blackwell and Carr, 1992), and polypeptides (Hirayama *et al.*, 1990). However, for utilization as HPLC column packings these materials must be shaped (into spheres) and have a proper uniform size.

Lewis and Smith (1967) indicated that sorbitol (*D*-glucitol) is an acyclic polyol that occurs in plants and fruits. In industry, it is used as a sweetener and food additive, and in pharmaceutical applications. Buetler (1984) suggested that sorbitol dehydrogenase oxidizes sorbitol to *D*-fructose in the presence of nicotinamide adenine dinucleotide (NAD). This reaction may be utilized to determine the amount of sorbitol in foods. Schneider and Giffhorn (1991) purified sorbitol dehydrogenase from *Pseudomonas* sp. to homogeneity. These authors indicated that this dehydrogenase is not affected by xylitol, and thus its interference in the assay is minimal.

Schneider and Giffhorn (1991) utilized $(NH_4)_2SO_4$ precipitation, chromatography on Q-Sepharose, Procion-blue affinity chromatography, and finally hydroxylapatite chromatography. In between the Q-Sepharose column and the Procion-blue affinity chromatography, the eluted fractions were concentrated by ultrafiltration. This step could have contributed to the decrease in sorbitol dehydrogenase yield (78 to 54%). Table 3.10 shows the bioseparation protocol followed, along with the specific activity, the yield, and the purification factor after each step. The sorbitol dehydrogenase was purified by a factor of 172.1 and the yield was 22%. The final purified enzyme was homogeneous as judged by analytic-PAGE and by SDS-PAGE, each of which yielded only a single protein band. Both the yield (22%) and the purification factor (172.1) are relatively high.

These authors indicated that the purified enzyme is quite stable, and its stability is increased by the addition of sucrose. They analyzed the stability of the purified enzyme in standard buffer at 4, 20, 30, 40, and 50°C. Samples were tested in the absence and in the presence of sucrose (numbers in parentheses). The half-lives of the enzyme were (in the same order) 4 days (17 days), 32 h (96 h), 5 min (24 min), and 1 min (2 min) of incubation. The addition of sucrose stabilized the enzyme anywhere from 100 to 450 percent for the temperature ranges and experimental conditions utilized previously.

Example 3.6

Briefly analyze the purification of *D*-xylulokinase from the yeast *Pichia stipitis* NCYC 1541 using adsorption (hydroxylapatite column) chromatography (Flanagan and Waites, 1992).

Solution

Flanagan and Waites (1992) purified D-xylulokinase from the pentose-fermenting yeast *P. stipitis* NCYC 1541 using a two-stage adsorption chromatographic (hydroxylapatite column) process at 4°C. These authors indicated that *D*-xylose is the second largest amount of sugar that is naturally available. The enzyme *D*-xylulokinase catalyzes a critical step in the conversion of *D*-xylose to ethanol (Lachke and Jeffries, 1986). This step is the phosphorylation of *D*-xylulose to *D*-xylulose-5-phosphate. Flanagan and Waites (1992) indicated that once the *D*-xylulose-5-phosphate is formed, it then passes into the pentose phosphate cycle.

These authors sonocated the cells. Thereafter the cell debris was removed

TABLE 3.10 Purification Protocol for Sorbitol Dehydrogenase from *Pseudomonas* sp[a]

Bioseparation step	Specific activity (U/mg)	Yield (%)	Purification factor
Crude extract	0.52	100	1.0
Ammonium sulfate precipitation	0.77	98	1.5
Q-Sepharose chromatography	9.4	78	18.1
Procion-blue chromatography	45.1	54	45.1
Chromatography on hydroxylapatite	89.5	22	172.1

[a] From Schneider, K. H. and Giffhorn, F. (1991). *Enzyme Microb. Technol.*, *13*, 332. With permission.

TABLE 3.11 Purification Protocol for D-Xylulokinase from *Pichia stipitis*[a,b]

Bioseparation step	Specific activity (U/mg)	Yield (%)	Purification factor
Crude cell extract	0.93	100	1.0
First hydroxylapatite chromatography	7.35	56	8.0
Second hydroxylapatite chromatography	21.37	49	23.0

[a] From Flanagan, T. and Waites, M. J. (1992). *Enzyme Microb. Technol., 14,* 975. With permission.

[b] Flanagan and Waites (1992) indicated that the results are averages of four runs.

by centrifugation. Further purification of the *D*-xylulose was performed on a Biorad hydroxylapatite adsorption column in a two-stage process. In between the two stages the sample was concentrated by ultrafiltration (10,000 Da molecular mass membrane). SDS-PAGE performed on the purified sample gave a single band indicating the homogeneity of the purified sample.

Table 3.11 shows the purification protocol followed. The specific activity, the yield, and the purification factor are given after each of the two hydroxylapatite steps. The yield is relatively high, and the actual purification factor obtained is relatively low. The major activity loss occurs in the first adsorption column. Flanagan and Waites (1992) indicated that the partially purified enzyme is relatively unstable at 4°C, where it loses all its activity within 24 h. These authors emphasized that this activity was retained when the enzyme was stored at −20°C. Stryer (1981) suggested that a divalent metal ion is required as a cofactor for all kinase reactions. Flanagan and Waites (1992) indicate that Mg^{2+} is the best cofactor for this enzyme. A 3 mM concentration of this cofactor is the optimum concentration as far as retention of the enzyme activity is concerned.

Other chromatographic techniques have also been used as the final fractionation or high-resolution fractionation techniques.

C. Hydrophobic Interaction Chromatography

Example 3.7

Briefly analyze the purification of feruloyl/*p*-coumaroyl esterase from the fungus *Penicillium pinophilium* (Castanares *et al.,* 1992).

Solution

Castanares *et al.* (1992) purified feruloyl-*p*-coumaroyl esterase from the fungus *P. pinophilium.* These authors indicate that xylans are major constituents of woods and agricultural residues. *p*-Coumaric esterase esterifies the *L*-arabinose residues that are in the backbone of the xylan polysaccharides. Wong *et al.* (1988) indicated that the selective modification of the xylans may be industrially important.

Table 3.12 shows the bioseparation procedures utilized to purify feru-

TABLE 3.12 Purification Protocol for Feruloyl–p-Coumaroyl Esterase from *Penicillium pinophilium*[a]

Bioseparation step	Specific activity (U/mg)	Yield (%)	Purification factor
Crude extract	0.08	100	1
Amicon (ultrafiltration) concentrate	0.17	48.5	2.3
DEAE-Sepharose chromatography	3.60	39.5	45
Phenyl-Superose chromatography	28.4	11.5	355

[a] From Castanares, A. *et al.* (1992). *Enzyme Microb. Technol., 14,* 875. With permission.

loyl-p-coumaroyl esterase from *P. pinophilium.* The specific activity of the enzyme, the yield, and the purification factor after each step are also given. No numbers of the yield after each step were provided by the authors. However, these yield values can easily be calculated. The yield is simply given by total activity after each step divided by the total activity present in the crude enzyme extract. The crude enzyme solution was extracted by an Amicon membrane (ultrafiltration), followed by anion-exchange chromatographic purification on a DEAE-Sepharose CL 6B column. This was followed by purification on a hydrophobic interaction column. The enzyme was purified by a factor of 355 and the specific activity increased from 0.08 to 28.4 U/mg. The calculated yield is 11.5%. This is a relatively high yield considering the relatively high purification factor value obtained.

These authors indicated that the purified enzyme does have a broad specificity. It is of interest to note that the hydrophobic interaction chromatographic treatment increases the purification of the feruloyl-p-coumaroyl esterase by a factor of about eight. The authors were able to obtain a pure enzyme as noted by a single component by SDS-PAGE. This component had a molecular mass of 57,000 Da. Castanares *et al.* (1992) further indicated that the purified enzyme exhibits good stability characteristics in the temperature range from 37 to 55°C. For example, at 37°C, the enzyme retained about 60% of its activity after a 25-h incubation. The enzyme activity retained dropped to 30% for this same time period at 50°C. The high stability of the enzyme may contribute to its eventual relatively high yield on purification.

D. Affinity Chromatography

Example 3.8

Briefly analyze the purification of chitanase from *Trichoderma harzianum* (Ulhoa and Peberdy, 1992).

Solution

Ulhoa and Peberdy (1992) purified extraceullar chitanase from *T. harzianum.* These authors indicated that chitanases are endoglycosidases. These enzymes are poly-β1-4(2-acetamido-2-deoxy)-D-glucoside glucanohydrolases,

EC 3.2.1.14. Rosenberger (1979) suggested that in fungi, chitanases probably have a physiological role.

The bioseparation protocol was carried out by Ulhoa and Peberdy (1992) at 4°C. These authors utilized the following bioseparation procedures: ammonium sulfate precipitation, ion-exchange chromatography on a Q-Sepharose column, and gel filtration on a Sephadex G-100 column followed by hydrophobic interaction chromatography on a Phenyl-Sepharose CL 4B column. The bioseparation protocol, the specific activity, the purification factor, and the yield after each step are given in Table 3.13. The chitanase obtained is purified by a factor of 76 and the yield is 10%. During the hydrophobic interaction chromatography step the purification factor increases substantially from 17 to 76; and, as expected, the yield drops from 26 to 10%. The yield value is relatively moderately high considering the high purification factor obtained. The gel filtration step was the next to the last step, and the hydrophobic interaction chromatography step was the last step.

These authors indicated that ammonium sulfate precipitation was used because many workers used this technique as the first step to recover proteins (Usui *et al.*, 1987; Zarain-Herzerg and Arroyo-Begovich, 1983). Besides, in general, one obtains a 70 to 80% enzyme activity recovery using this technique. Ulhoa and Peberdy (1992) recovered 61% of chitanase using this technique (see Table 3.13), besides removing contaminants such as polysaccharides. Furthermore, contaminating proteins were removed by the ion-exchange chromatographic step on the Q-Sepharose column. The gel filtration on the Sephadex G-100 column was utilized to free chitanase from chitobiase. Further purification of the chitanase was undertaken on a Phenyl-Sepharose CL-4B column. The purified enzyme was homogeneous with no trace contaminants as noted by SDS-PAGE that revealed the enzyme migrating as just a single band. These authors noted that the purified enzyme did not exhibit any chitobiase activity using *p*-nitrophenyl-*β*-*N*-actylglucosamine as substrate. This chitobiase activity was exhibited in the crude enzyme extract.

The numbers presented by Ulhoa and Peberdy (1992) and the statements made are of interest, especially the statements that indicate the specific contaminants removed during each stage and the corresponding drops in enzyme recovery yield after each bioseparation step. Numbers like these for this and other protein-bioseparation protocol systems are required to help move biosepara-

TABLE 3.13 Purification Protocol for Chitanase from *Trichoderma harzianum*[a]

Bioseparation step	Specific activity (U/mg)	Yield (%)	Purification factor
Ammonium sulfate precipitation	3.47	100	1.0
Q-Sepharose chromatography	13.76	61	4.0
Sephadex-G100 chromatography	55.10	26	17
Affinity chromatography	245.0	10	76

[a] From Ulhoa, C. J. and Peberdy, J. F. (1992). *Enzyme Microb. Technol.*, *14*, 236. With permission.

tion from an art to a science. It is hoped that in the future such numbers become more widely available in the open literature.

Theoretical modeling that describes the binding of the analyte to the affinity ligand on the chromatographic column is of interest. This permits one to estimate, control, and improve the binding (and eventual bioseparation) of the analyte or product of interest.

Example 3.9

Briefly describe the concerted cluster model of multivalent affinity for heterogeneous adsorption of enzymes (Dowd and Yon, 1995).

Solution

The bioseparation of biological molecules may be carried out by the adsorption on suitable materials. The selectivity of the purification step is often aided by attaching suitable ligands to the matrix. Dowd and Yon (1995)indicated that quite a few theoretical models have been utilized to analyze the binding kinetics of proteins having several identical binding sites with ligands immobilized on insoluble stationary phases (Hethcote *et al.*, 1983; Liapis, 1989; Yon, 1988; Dowd and Yon, 1992). Dowd and Yon (1995) emphasized that for multivalent interaction to occur the ligands (two or more) should be in a cluster with suitable geometry to facilitate the binding during the time it takes the protein molecule to make the required two or more contacts. These authors further indicated that cooperativity between the ligands in the cluster is required due to proximity and entropic considerations. Dowd and Yon (1995) analyzed the binding of rabbit muscle dehydrogenase to the biomimetic dye Cibacron Blue immobilized on cellulose. The experimental analysis was performed to permit a better understanding of the clustering hypothesis.

These authors considered only single and paired ligands in their analysis. They ignored higher order clusters. This was done for statistical and accessibility reasons. Holbrook *et al.* (1975) initially indicated that lactate dehydrogenase has four coenzyme binding sites. These bind the free dye with identical microscopic binding constants. The analysis of Dowd and Yon (1995) indicates for Cibacron Blue cellulose the existence of a small proportion of high affinity sites for binding lactate dehydrogenase. In terms of the cluster model (Yon, 1988) these include ligand pairs as well as single ligands. Dowd and Yon (1995) emphasized that the values of the association constants obtained by them are comparable or close to the published values for both the immobilized and the free dye. However, these authors emphasize that the values of the association constants obtained by them for ligand pairs are only about 100-fold greater than the constants for single ligands. According to the model they should be at least seven orders of magnitude higher. Apparently the experimentally determined cooperativity is much lower than that predicted by the cluster model. These authors indicated that geometrical-steric constraints presumably lead to lower estimated values of the association binding constant. Note that a simultaneous good fit of both ligands is required.

Finally these authors emphasized their analysis supports the premise that the cellulose matrix is rigid enough to permit the permanent existence of ligand clusters. The Dowd and Yon (1995) analysis is of significant interest because it provides physical insights into the analyte-ligand binding process. A better understanding of this process would eventually lead to a better bioseparation process. This understanding would also assist in other areas where such a type of binding is critical, for example, immunoassays.

In the next two examples we analyze the effectiveness of ion-exchange chromatography as a high-resolution fractionation technique. These two examples have been selected at random, and we hope that they demonstrate, as in the previous cases, the potential of this method as a high-resolution fractionation technique.

E. Ion-Exchange Chromatography

Example 3.10

Briefly analyze the purification of κ-carrageenase from *Pseudomonas carrageenovora* (Ostgaard *et al.*, 1993).

Solution

Carrageenans constitute a major component of cell wall structure in red algae (Ostgaard *et al.*, 1993). These authors indicated that there is an increasing demand for cell wall degrading enzymes in seaweed technology so that one may effectively obtain the protoplasts. Cell wall degrading enzymes (such as carrageenases) also assist in understanding the structure of carrageenans. They indicate that carrageenases have been obtained from different sources including *Pseudomonas*, *Cytophaga*, and others (Le Gall *et al.*, 1990; Sarwar *et al.*, 1987).

Ostgaard *et al.* (1993) purified κ-carrageenase from *P. carrageenovora* on a large scale. For smaller amounts of enzyme production these authors utilized $(NH_4)_2SO_4$ precipitation followed by gel filtration on a Sephadex column. The final step was ion exchange on a CM-Sepharose CL-6B column. The gel filtration column assisted in desalting. For large-scale purifications, they utilized ultrafiltration followed by cation exchange on a S-Sepharose FF column. Table 3.14 shows the bioseparation steps involved; and the specific activity, yield, and purification factor after each step.

TABLE 3.14 Purification Protocol for κ-Carrageenase from *Pseudomonas carrageenovora* [a]

Bioseparation step	Specific activity (U/mg)	Yield (%)	Purification factor
Small scale			
Concentrated broth	1.8	100	1.0
Precipitation–desalting	6.9	75	3.9
CM-Sepharose CL-6B ion-exchange chromatography	51	30	28
Large scale			
Dialyzed broth	2.3	100	1.0
S-Sepharose FF cation-exchange chromatography	55	91	23

[a] From Ostgaard, K. *et al.* (1993). *Enzyme Microb. Technol.*, *15*, 326. With permission.

These authors further indicated that the enzyme was purified at room temperature without loss of activity due to thermal effects. They stated that the carrageenase activity in the crude supernatant was stable within $\pm 5\%$ at room temperature for about 20 h in the pH range 5.5 to 6.5. There was a small amount of activity loss when ultrafiltration was used. This is not surprising (as in this case) when the globular protein has a molecular weight about three times the nominal membrane cutoff at 10,000 Da. For small-scale applications Ostgaard *et al.* (1993) demonstrated that ammonium sulfate precipitation should give a yield of about 90% (Greer, 1984). The subsequent desalting by the dialysis reduces the yield further to 75%. The treatment by the CM-Sepharose column further reduced the yield to 30%.

SDS-gel electrophoresis was utilized to test the purity of the enzymes separated. The authors obtained a characteristic double band (32,000 and 34,000 Da *Mr*). Similar double bands have been observed for this enzyme (Greer, 1984). Ostgaard *et al.* (1993) indicated that the enzyme obtained by them may be considered pure considering the state of seaweed and phycocolloid studies. These authors admitted that higher purity enzymes may be required for structural studies of carrageenans (Knutsen, 1991). Nevertheless, the yields obtained by these authors for both small-scale and large-scale purification are relatively high. One possible explanation could be the not so high purification value obtained.

It is reasonable to assume that subsequent processes that yield higher purification factor values of the enzyme (required possibly for structural studies of carrageenans) would eventually give lower yields of the enzyme. Besides, the enzyme is relatively stable at room temperature. In any case, the analysis and numbers provided by Ostgaard *et al.* (1993) are of interest because these authors indicated the possible reasons for the quantitative loss of enzyme activity during each bioseparation step. Similar studies are required for the purification of other enzymes from seaweed and other sources.

Example 3.11

Briefly analyze the production of blood proteins using the ion-exchange technique (Cueille and Tayot, 1985).

Solution

Cueille and Tayot (1985) utilized silica particles (Spherosil column) as the ion-exchange technique to purify albumin from human placental blood and from human plasma. These authors indicated that the Spherosil particles have good mechanical properties (rigidity, incompressibility, and generally exhibit a low-pressure drop). This permits their possible application on an industrial scale. They emphasized that these Spherosil particles have a large pore diameter. This permits the easy diffusion of the proteins or other biological macromolecules of interest in the pores. Also, these particles may be coated by suitable polymers. This permits suitable interaction with the different biological products, besides minimizing irreversible adsorption.

Figure 3.3 shows the steps that are involved to purify human albumin from placental blood. Cueille and Tayot (1985) coated the Spherosil beads with suitable polysaccharides. This permitted the treatment of 20,000 liters per day of protein solution. The authors were able to obtain 140 kg of protein using these

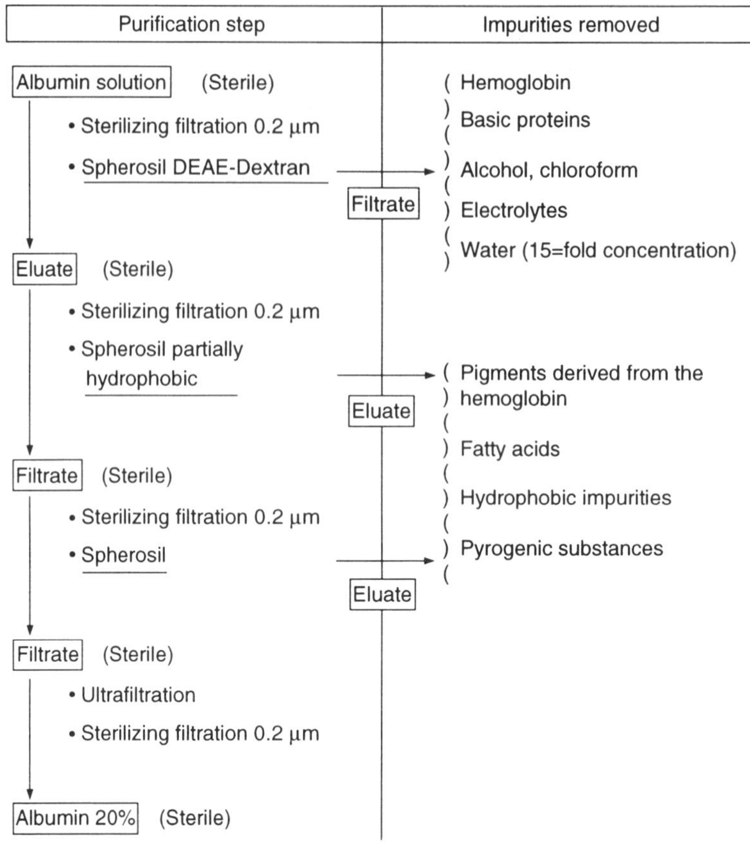

FIGURE 3.3 Process flowsheet for purifying human albumin from placental blood [From Cueille, G. and Tayot, J. L. (1985) *Biotech '95 Proceedings,* Online Publications: Pinner, Middlesex, UK, p. 141].

flow rates. In the first step, the authors used an electro-positive support Spherodex (Spherosil-DEAE Dextran). This permitted the fixing of the negatively charged albumin. Alcohols and other impurites were not fixed and were permitted to flow through the column. Chloroform, hemoglobin, and other electrolytes were also removed in this column. These authors indicated that after this step the albumin is still contaminated with pigments (derived from hemoglobin), fatty acids, and hydrophobic impurities.

The pigments are removed in the second step. Here the eluant from the first column is treated by a partially hydrophobic Spherosil column that has a large affinity for the pigments. In the third chromatographic step the fatty acids, hydrophobic impurities, and pyrogenic substances are removed. These authors emphasized that between the chromatographic steps the processed albumin solution is stored in sterilized and refrigerated tanks. No information was provided concerning the loss of albumin activity during each bioseparation step. Nevertheless, these authors indicated that the albumin purified was extremely pure (100% in cellulose acetate electrophoresis) and pyrogen free, and met all the Pharmacopeia requirements. A similar modified process may be utilized to purify albumin from human plasma. These authors further indicated that ion-exchange and affinity chromatographic methods were also being developed to purify polio virus from cell culture supernatants, factor VIII and IX from human plasma, cholera and tetanus toxin from solutions, etc. These methods were for both laboratory and industrial-scale levels. Such types of efforts need to be strongly encouraged so that one may meet the ever increasing demands for these products at a reasonable cost. Besides, such studies will provide the much required information to develop a database to help move bioseparation of various biological products of interest from an art to a science.

Vijayalakshmi and Thomas (1985) pointed out some parameters that help minimize blood protein denaturation during their bioseparation. These authors indicated that the addition of sucrose and albumin helps minimize the denaturation of blood coagulating factors (fragile macromolecules) during their bioseparation. This step is advisable prior to concentration of the protein (blood coagulating factor) solutions by a mild technique such as ultrafiltration. When affinity chromatography was used to purify the blood proteins, these authors emphasized the need for caution during the removal of competitive ligands. Extensive dialysis or diafiltration would lead to the denaturation of the proteins. Also, when metal-chelating agents were used, these authors suggested the utilization of mild elution conditions to minimize metal leakage. They further advised not to dialyze out the metal from the purified product.

III. CRYSTALLIZATION

Crystallization is often used as a high-resolution, polishing, or confectioning step during the bioseparation of biological macromolecules of interest. Proteins and antibiotics are often purified by crystallization. Lang *et al.* (1992) described the procedure for the crystallization of bacterial luciferase from marine bioluminescent bacterium *Vibrio harveyi* MAV. These authors utilized the following steps:

1. Homogeneous luciferase was concentrated by microfiltration.
2. The sitting-drop vapor diffusion method with crystallization plates was used (McPherson, 1990). Each drop consisted of 5 μl of 10 to 50 mg/ml of the protein solution and 5 μl of a precipitating solution containing the polymer or salt (precipitating agent). Lang *et al.* (1992) indicated that the drops were equilibrated against 1 ml of the corresponding solution.
3. The BIOMEK 1000 automated laboratory system (Beckman, Palo Alto, CA) was utilized by these authors to enhance the accuracy, reproducibility, and precision.

It would be of interest to analyze the effectiveness of the crystallization procedure to produce ultrapure bioproducts, and to note the extent of product losses, if any.

Example 3.12

Briefly analyze the large-scale purification and crystallization of lipase from *Geotrichum candidum* (Hedrich *et al.*, 1991).

Solution

Hedrich *et al.* (1991) developed a large-scale procedure for the purification of lipases from the fungus *G. candidum.* Lipases have a high specificity for the fatty acids having at least one cis-Δ9 double bond. These authors were able to prepare 260 mg of the pure enzyme within 2 days. They utilized a two-step purification protocol that employed a Q-Sepharose column and a Phenyl-Sepharose column. Table 3.15 shows the purification protocol, the specific activity of the lipase, the yield, and the purification factor after each step. It is of interest to note that the separation on the Q-Sepharose column produces only a marginal decrease in the yield (2%) while producing a purification factor of 6.4. The subsequent treatment on the Phenyl-Sepharose CL-4B column gave a substantial decrease in yield (42% after treatment) with a purification factor of 13.2. These authors indicate that the final product was homogeneous as noted by SDS-PAGE treatment.

For the crystallization experiments the two-step isolation procedure was slightly modified. A DEAE-Sepharose FF column was used instead of the Q-Sepharose column as the first step. For crystallization the hanging-drop method was used at 6°C using multichamber plates. For example, the *G. candidum* lipase after the two-step isolation procedure was crystallized in the presence of PEG. These authors suggested that the size of the crystals depends on the molecular weight of the crystallization agent. Using 11% PEG at a pH of 4 to 5.5, Hedrich *et al.* (1991) noted that the lipase formed thin plates of 30 × 30 μm in size.

Furthermore, they were able to obtain five different crystals depending on the amount of glycosylation. The crystals obtained by these authors are pure enough to obtain X-ray diffraction data relating structural influences on the amount of glycolysation. No information was provided by Hedrich *et al.* (1991) on the lipase activity losses during the crystallization procedure.

It would be of interest to note the purification and crystallization of lipase from another organism. One would like to compare the different bioseparation protocols utilized. The next example briefly examines the purification and crystallization of bacterial luciferase from *V. harveyi* MAV.

Example 3.13

Briefly analyze the purification and the crystallization of lipase from *Vibrio harveyi* (Lang *et al.*, 1992).

TABLE 3.15 Purification Protocol for Lipase from *Geotrichum candidum*[a]

Bioseparation step	Specific activity (U/mg)	Yield (%)	Purification factor
Crude enzyme	80	100	1.0
Q-Sepharose chromatography	508	98	6.4
Phenyl-Sepharose chromatography	1052	42	13.2

[a] From Hedrich, H. C. *et al.* (1991). *Enzyme Microb. Technol.*, *13*, 840. With permission.

TABLE 3.16 Purification Protocol for Luciferase from *Vibrio harveyi* MAV[a]

Bioseparation step	Specific activity (light units[b]/mg) × 10^{-11}	Yield (%)	Purification factor
Osmotic lysis	0.457	100	1.0
Q-Sepharose fast flow chromatography	3.35	79.2	7.3
Phenyl-Sepharose CL4B chromatography	4.76	47.2	10.3
Hydroxylapatite chromatography	10.20	30.5	22.2
Ω-Aminohexyl-agarose chromatography	15.6	29.2	34.4

[a] From Lang, D. *et al.* (1992). *Enzyme Microb. Technol., 14,* 479. With permission.
[b] One light unit corresponds to the emission of 1 quantum per second.

Solution

Lang *et al.* (1992) utilized the following bioseparation protocol to purify bacterial luciferase from *V. harveyi* MAV: ion-exchange chromatography, S-Sepharose fast flow (cation exchange), hydrophobic interaction chromatography on Phenyl-Sepharose CL-4B, adsorption chromatography on hydroxylapatite, and affinity chromatography on Ω-aminohexyl agarose. By utilizing this bioseparation protocol these authors were able to obtain a yield of 29% and a 34-fold purification factor. The yield is relatively high. This is understandable considering the relatively low purification factor value obtained (Table 3.16).

These authors noted that the luciferase could be crystallized in two different forms (needles and rhombics) depending on the precipitating agents used. For example, bundles of needles were obtained in ammonium sulfite, whereas rhombic crystals were obtained in ammonium sulfate. Lang *et al.* (1992) obtained pure crystals that were suitable for X-ray diffraction studies. No information was provided by these authors on the loss of luciferase activity during the crystallization procedure. However, in separate temperature studies they noted that about 44 and 78% of the luciferase activity was lost in about four weeks at 4 and 25°C, respectively. Because the crystallization was performed at room temperature for a period of 1 to 4 days, this time period represents about a 2.8 to 11.3% loss in luciferase activity. Activity losses during the actual bioseparation procedures are really required.

Crystallization may be utilized as a high-resolution fractionation step to purify antibiotics. Antibiotics inhibit the multiplication of various microorganisms by interfering with cell metabolism or cell wall development. Sollmann (1957) initially indicated that broad spectrum antibiotics are effective against a wide range of bacteria. Bienkowski *et al.* (1988) indicated that the three most important classes of antibiotics are penicillins, cephalosporins, and tetracyclines. These authors further indicated that, in general, antibiotic manufacturing plants contain a fermentation section, and a rather large, energy-intensive separation-purification system (Figure 3.4). The purification protocol may con-

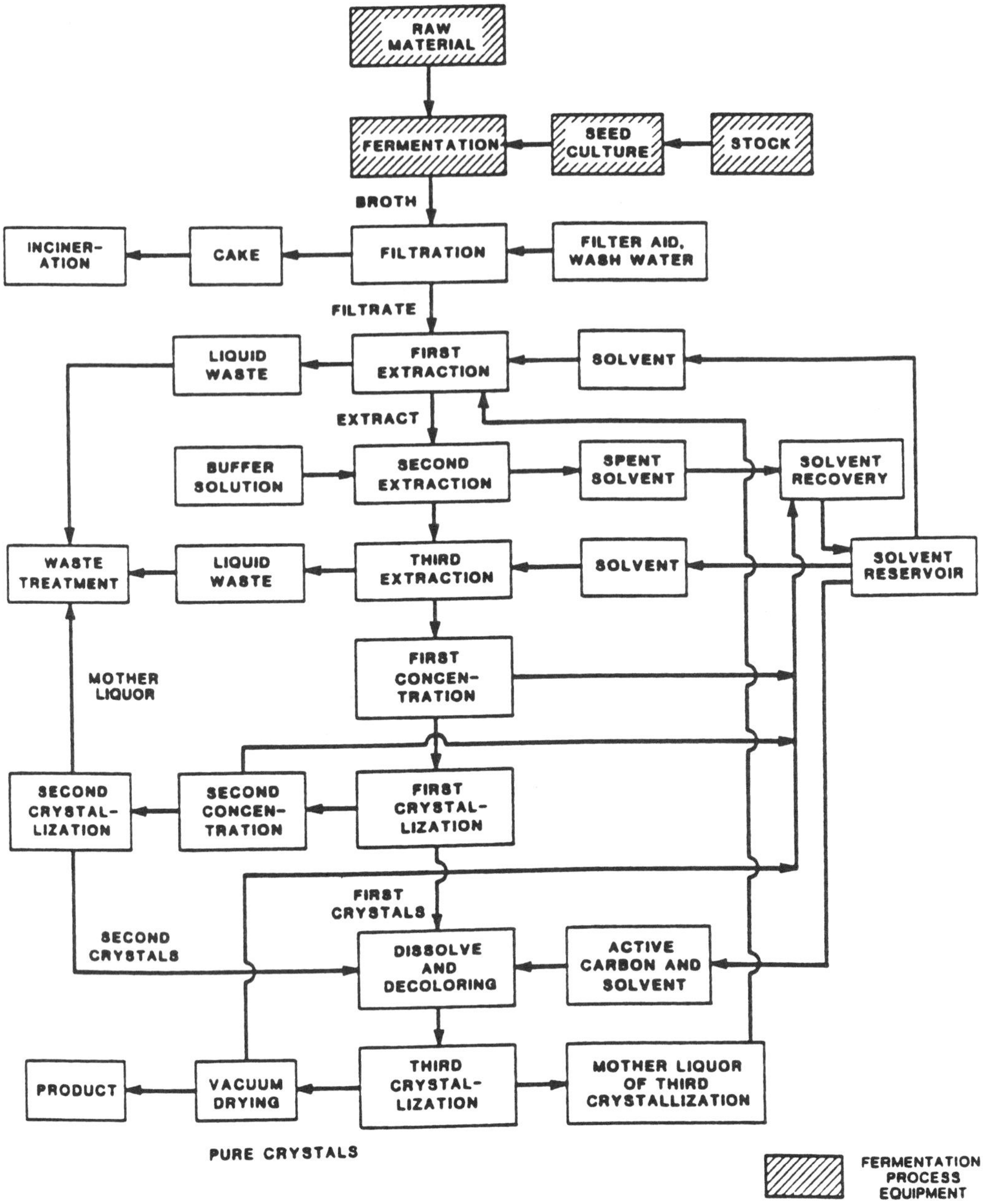

FIGURE 3.4 General process flowsheet for antibiotic production [From Bienkowski, P. R. *et al.* (1988) *Appl. Biochem. Biotechnol., 17/18,* Humana Press: Clifton, NJ, pp. 261–273].

tain up to 60 separate unit operations or bioseparation stages. Note the number of crystallization stages and the repetition of the different unit operations. It would be of interest to analyze the separation of antibiotics. Because these antibiotics are to be obtained in a highly purified form, they are crystallized out of solution in the last step.

Example 3.14

Briefly analyze the purification and crystallization of penicillin (Bienskowski *et al.*, 1988).

Solution

Bienkowski *et al.* (1988) indicated that the bioseparation protocol may be different for different antibiotics and the purity required. However, the basic procedures used still are filtration, centrifugation, liquid–liquid extraction, and crystallization. These authors emphasized the purification of antibiotics under sterile conditions to prevent or minimize their denaturation. Also, the antibiotics rapidly degrade in soluble form; thus there is a need for rapid processing. Besides, temperatures above the ambient are unfavorable because antibiotics are sensitive to temperature.

They suggested that many different protocols may be utilized for penicillin removal (Inman, 1984; Atkinson and Mavituna, 1983). Figure 3.5 shows a general purification protocol for penicillin production. The authors indicated that the process takes about 15 h. Bienkowski *et al.* (1988) suggested that extra care is taken to prevent contamination and/or degradation of the penicillin. For example, cells are carefully filtered to prevent contamination by β-lactamase-producing organisms. Also, the filtrate is cooled 0 to 4°C to minimize degradation. In the last step the penicillin is recovered from solution by adding sodium or potassium acetate and precipitating as a metal salt. The crystals may be washed and dried (by vacuum or by warm air). It would be of interest to note the extent of penicillin (or other antibiotic) loss, and the cause of this activity loss at each stage of the bioseparation protocol. This would be of assistance in controlling the losses of penicillin (or of other antibiotics) at each bioseparation stage.

Example 3.15

Briefly analyze the purification and crystallization of cephalosporin (Bienskowski *et al.*, 1988).

Solution

Cephalosporins are similar in structure to penicillins and are produced by fungi as extracellular products (Bienskowski *et al.*, 1988). These authors suggested that the following bioseparation protocol may be utilized: solvent extraction, ion-exchange resins, and salting-out procedures. For example, the fermentation broth may be filtered. The filtrate is then adsorbed on an activated carbon column. The adsorbed antibiotic is next eluted from the column by adding water and a polar organic solvent. The eluate may be contacted with an anion exchange resin. This resin is then eluted with a salt solution.

They indicated that the carbon adsorption steps may be replaced by a precipitation step as suggested by Wildfeuer (1985). The precipitation may be undertaken by crystallization of the potassium or sodium salt of the cephalosporin either by concentration or by addition of a miscible solvent. Also, the zinc (or other metal such as copper, nickel, or lead) salt may be crystallized from the purified aqueous solutions.

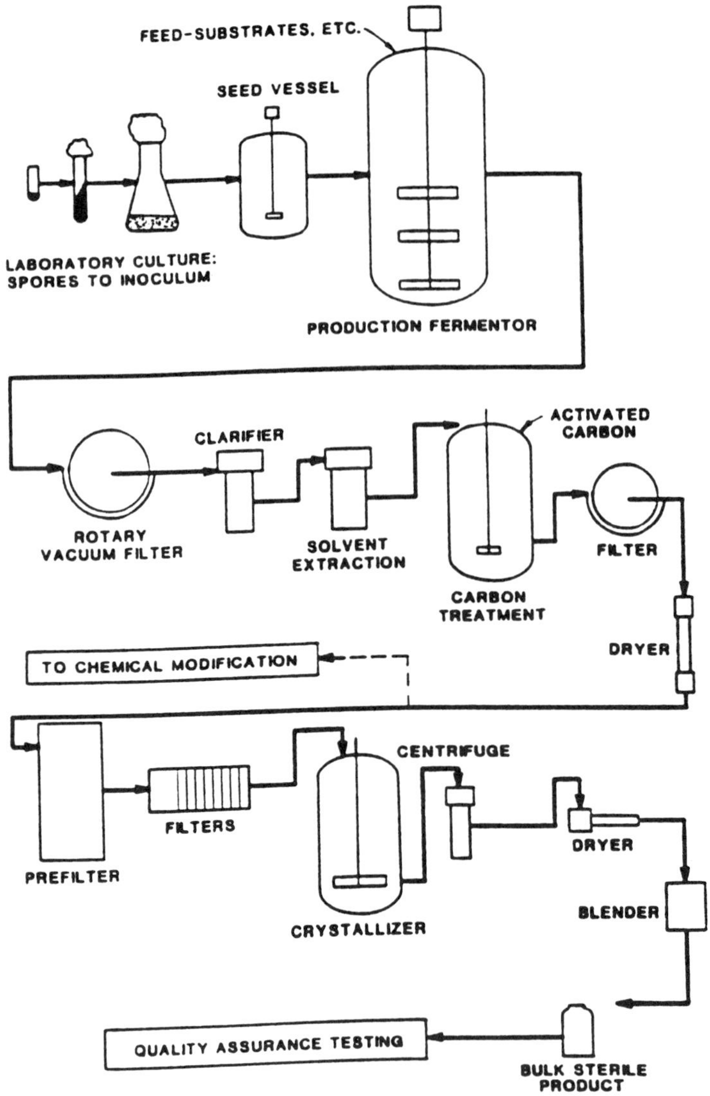

FIGURE 3.5 General process flowsheet for penicillin production [Bienkowski, P. R. *et al.* (1988) *Appl. Biochem. Biotechnol., 17/18,* Humana Press: Clifton, NJ, pp. 261–273].

These authors did not provide any numbers on cephalosporin activity losses during each of the bioseparation stages. Because these antibiotics are sensitive to temperature, the crystallization needs to be run at low temperature (for example, 2°C for about 3 h for tetracycline [Plues, 1976; Gavrilescu *et al.*, 1974]).

The preceding few examples for utilizing crystallization as a high-resolution fractionation technique for the separation of proteins and antibiotics indicate that this is effective in producing high-quality products, as required. More information is required in the open literature that clearly delineates the causes and extent of protein and antibiotic activity losses during the crystallization of different products. This would be of considerable assistance in minimizing these losses and in designing better bioseparation processes. This would then eventually lead to the development of a bioseparation protocol that may be of assistance for quite a few processes with perhaps only minor modifications.

We next present some examples of techniques that are relatively new and in the development stage, in addition to others, such as precipitation, that have been around for some time. In each of the examples presented the technique under consideration is the last step, or the high-resolution step. We start with precipitation, and then move on to techniques that seem to exhibit potential as a high-resolution fractionation step.

IV. OTHER TECHNIQUES

Some of the techniques to be analyzed here include precipitation, PAGE, and affinity-based micellar extraction and separation. In some cases, there is only partial purification of the enzyme. Then, in such a situation we have taken the last step as the high-resolution fractionation step, for example, precipitation (Gonzalez and Monsan, 1991). Further purification steps in these cases is possible, if higher purity of proteins or of other biological macromolecules are required.

Example 3.16

Briefly analyze the purification of β-galactosidase from *Aspergillus fonsecaeus* (Gonzalez and Monsan, 1991).

Solution

Gonzalez and Monsan (1991) purified β-galactosidase from *A. fonsecaeus* using ultrafiltration-diafiltration, and isopropanol precipitation. These authors indicated that this enzyme is used in the dairy industry for the hydrolysis of lactose in milk and whey (Coughlin and Charles, 1980; Burgess and Shaw, 1983). Coughlin and Charles (1980) emphasized that the main costs for this enzyme are in the production and in the separation stages.

Table 3.17 shows the bioseparation protocol; and the specific activity,

TABLE 3.17 Purification Protocol for β-Galactosidase from *Aspergillus fonsecaeus*[a]

Bioseparation step	Specific activity (U/mg of protein)	Yield (%)	Purification factor
Crude enzymatic extract	4.3	100	1.0
Ultrafiltration–diafiltration	16.0	86	3.7
Isopropanol[b]precipitation	25.1	81	5.8

[a] From Gonzalez, R. R. and Monsan, P. (1991). *Enzyme Microb. Technol.*, *13*, 349. With permission.

[b] Isopropanol–enzymatic solution: 1/1, v/v.

yield, and purification factor after each step. Gonzalez and Monsan (1991) utilized isopropanol precipitation as the high-resolution fractionation step. The enzymatic solution was concentrated by ultrafiltration using a hollow fiber cartridge Amicon HP 10. Diafiltration was utilized to remove the major low-molecular-weight contaminants. During this step these authors noted that the specific activity increased by a factor of 3.7 and the yield was 86%. The diafiltered extract was precipitated by isopropanol (1/1, v/v). These authors noted that during the precipitation step the yield dropped a little from 86 to 81%. The purification factor increased from 3.7 to 5.8. The 81% yield obtained after the isopropanol precipitation step is high with a relatively low value of the purification factor. This 81% yield is within the range of activity loss for other enzymes using the precipitation technique (Greer, 1984). These authors noted that there were no proteolytic activities in this partially purified enzyme. Detailed parameter values for the bioseparation of β-galactosidase were provided by Gonzalez and Monsan (1991) only until the isopropanol precipitation step (partial purification). Thus, this step is considered as the last step in the purification protocol.

They indicated, however, that further purification of the enzyme to homogeneity is possible using appropriate chromatographic steps. The authors recommended using hydroxylapatite chromatography (HA-Ultrogel IBF), gel chromatography on an Ultrogel AcA44, followed by anion exchange on a fast phase liquid chromatographic (FPLC) column. After the hydroxylapatite treatment the specific activity increased to 38 U/mg. After the gel filtration step on the Ultrogel AcA44 column the specific activity increased to 95 U/mg. The yield meanwhile had dropped to 76% after the gel filtration step. After the Ultrogel AcA44 column treatment the purification factor is 21.5.

The yield obtained by Gonzalez and Monsan (1991) is relatively high, but the corresponding purification factor (21.5) is relatively moderately low. Further purification on an FPLC column increased the specific activity to 314 U/mg. This corresponds to a purification factor of 73. The corresponding yield in this case was 40.5%. This yield is relatively high considering the rather high purification factor eventually obtained after the FPLC column treatment. The enzyme was purified to homogeneity after the FPLC column treatment. The analysis by these authors is of interest because they provide numbers both for partial purification (precipitation is the last step), and for a product purified to homogeneity (chromatographic steps now included).

PAGE may be used to determine the homogeneity of the product. In some cases, it may also be used to help further purify a biological product of interest. In the next example it is used as the last step in the purification of β-glucosidase.

Example 3.17

Briefly analyze the purification of β-glucosidase from the fungus *Neocallimastix frontalis* EB188 (Li and Calza, 1991).

Solution

Fungi such as *N. frontalis* (Orpin, 1975), *Sphaeromonas communis* (Orpin, 1976), and *Piormonas communis* (Orpin, 1977) play a critical role in the degradation of plant fiber in rumen. Barichievich and Calza (1990a) indicated that

β-glucosidases are present in the culture supernatants of *N. Neocallimastix*. Lowe *et al.* (1987) and Barichievich and Calza (1990b) indicated that the *Neocallimastix* spp. possess the ability to utilize a wide variety of native cellulosic materials as a carbon source. Li and Calza (1991) purified β-glucosidase to homogeneity from *N. frontalis* utilizing hydroxylapatite chromatography, gel filtration chromatography, ion-exchange chromatography, and native PAGE.

Table 3.18 shows the bioseparation protocol utilized; and the specific activity, yield, and purification factor after each bioseparation step. There was about a 56% activity loss during the initial purification steps that included the concentration by a Millipore filter and ethanol precipitation. These two steps contributed to a significant loss in enzyme activity. Thereafter, the samples were treated twice by the hydroxylapatite column. After the first treatment four peaks were obtained. The second treatment by the hydroxylapatite column yielded only two peaks. In the meanwhile, the yield had dropped to 13.41%. Li and Calza (1991) indicated that further treatment by gel filtration chromatography, DEAE ion-exchange chromatography, and preparative PAGE (last step) helped remove further contaminating proteins.

These authors demonstrated that the enzyme was purified by a factor of about 19. This was at a considerable loss in enzyme activity, because the final yield was only 0.52%. The enzyme was purified to homogeneity as confirmed by SDS-PAGE where only a single band was exhibited with a molecular weight of 125,000 Da. Initially, three β-glucosidase peaks were present in the crude enzyme extract. The yield obtained by them is relatively low (0.52%) considering the relatively low purification factor value (18.95) obtained. One possible reason could be the presence of similar enzymes as shown by three peaks obtained from the crude enzyme extract, and four peaks after the first hydroxylapatite column treatment. Furthermore, these authors suggested that these peaks may be due to distinct enzymes, degradative products of the same enzyme because of proteolysis, enzymes with different degrees of posttranslational modifications, etc. They further indicated that the aerobic fungi *Trichoderma reesei* (Knowles *et al.*, 1988) possess quite a few distinct β-glucosidases. How-

TABLE 3.18 Purification Protocol for β-Glucosidase from *Neocallimastix frontalis* EB188[a]

Bioseparation step	Specific activity (U/mg)	Yield (%)	Purification factor
Crude cell extract	0.27	100	1.0
Millipore filtration	0.36	62.54	1.33
Ethanol precipitation	0.47	43.92	1.74
Hydroxylapatite chromatography I	0.73	13.41	2.70
Hydroxylapatite chromatography II	1.32	5.27	4.88
Gel filtration chromatography	1.45	1.85	5.37
DEAE-Sephacel ion-exchange chromatography	4.36	1.51	16.15
Native PAGE	5.02	0.52	18.95

[a] From Li, X. and Calza, R. E. (1991). *Enzyme Microb. Technol.*, *13*, 622. With permission.

ever, the rather extensive bioseparation protocol was able to separate the enzyme to homogeneity but only at the cost of a low yield.

The analysis of Li and Calza (1991) is of interest even though the yield is relatively low because it shows how even similar enzymes can be separated utilizing a rather elaborate bioseparation protocol. Perhaps further improvements may be suggested in the future that provide higher yields for the separation to homogeneity of β-glucosidase and other biological products from a solution that contains similar proteins or compounds.

The reverse micellar technique has been frequently employed as a bioseparation technique. We next present a new technique that combines the principles of affinity interaction with selective extraction of an affinity complex from an aqueous phase by a reverse micelle-containing organic phase. Paradkar and Dordick (1993) emphasized that this affinity-based reverse micellar extraction and separation (ARMES) technique is different from the previous ARMES technique proposed by Woll *et al.* (1989) and by other workers because: (1) the affinity ligand is macromolecular, and (2) the ligand–ligate interaction is intraphasic.

Example 3.18

Briefly analyze the separation of peroxidase from soybean hulls by the ARMES technique (Paradkar and Dordick, 1993).

Solution

Paradkar and Dordick (1993) proposed the affinity-based reverse micellar extraction and separation (ARMES) technique to purify peroxidase from soybean hulls. This technique combines the exquisite selectivity of affinity interaction with the ease of operability and scalability of liquid–liquid extraction. These authors proposed a method where a dual selectivity (in the affinity and separation stages) is achieved. The basic mechanism utilized in ARMES is shown in Fig. 3.6.

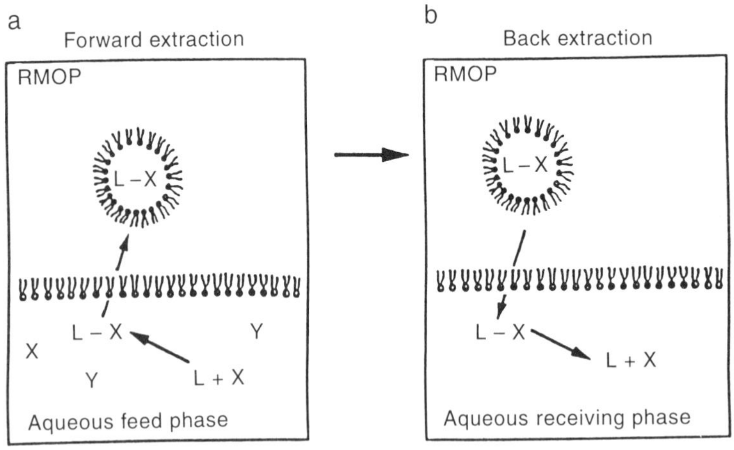

FIGURE 3.6 The ARMES mechanism. (a) The binding of the ligand (L) to the ligate (X). During the forward extraction the L-X complex is extracted from the aqueous phase (feed) to the reverse micellar phase. (b) The complex L-X is back-extracted from the reverse micellar phase into the aqueous phase [From Paradkar, V. M. and Dordick, J. S. (1993). *Biotechnol. Progr.,* 9, 199].

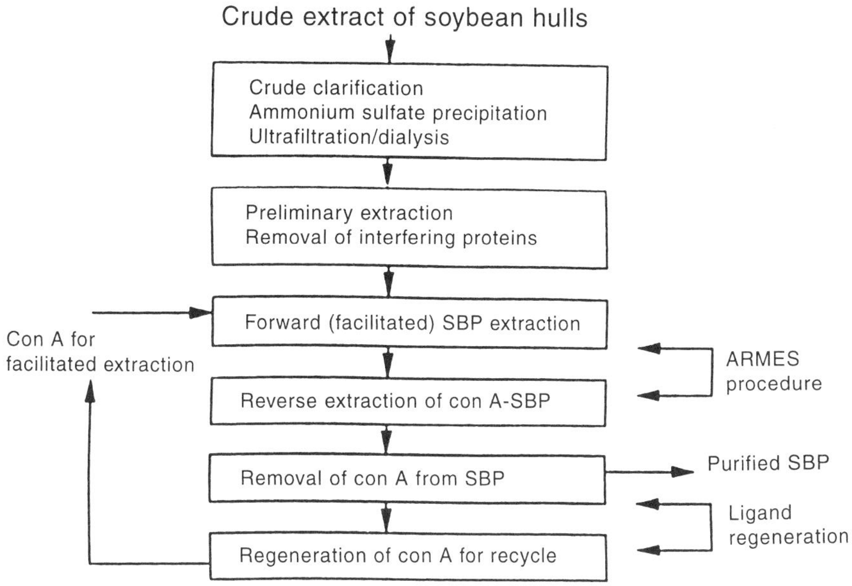

FIGURE 3.7 Bioseparation protocol for obtaining soybean peroxidase (SBP) from soybean hulls [Paradkar, V. M. and Dordick, J. S. (1993). *Biotechnol. Progr.*, *9*, 199].

The bioseparation protocol utilized for obtaining nearly pure peroxidase from soybean hulls is shown in Fig. 3.7. These authors utilized ammonium sulfate precipitation (crude clarification) and preliminary reverse micelle extraction (this step removed competing proteins such as glycosylated proteins), followed by selective ARMES extraction. The bioseparation protocol, the specific activity, the purification factor, and the yield after each step are given in Table 3.19. SDS-PAGE indicated that a nearly pure peroxidase was obtained. After the ARMES procedure, the purification factor was 29.2 and the yield was 30%. This is a relatively moderately high yield considering the not so high purification factor value obtained. Further purification by perhaps a suitable chromatographic step may yield an enzyme purified to homogeneity. This higher purity would, of course, be obtained at the cost of a decrease in yield.

The ARMES technique is a gentle technique, and the peroxidase did not undergo inactivation during the bioseparation process (Paradkar and Dordick, 1993). These authors suggested that this technique should be applied to other systems, such as the resolution of glycoform variants of therapeutic proteins.

TABLE 3.19 Purification Protocol for Soybean Peroxidase[a]

Bioseparation step	Specific activity (U/mg protein)	Yield (%)	Purification factor
Crude cell extract	21.6	100	1.0
Crude clarification (precipitation)	22.5	91	1.04
Preliminary extraction (reverse micelle)	47.4	80	2.2
ARMES	630.4	30	29.2

[a] From Paradkar, V. M. and Dordick, J. S. (1993). *Biotechnol. Prog.*, *9*, 199. With permission.

It would be advisable to apply the ARMES technique to the separation of other biological products of interest to test the broad base applicability of this technique. These studies should also provide more fundamental insights into the mechanistic details of this technique. This should considerably assist in evaluating the full potential of this technique. Besides, these studies would help minimize the inactivation of the biological macromolecules separated, and thereby help improve the quality and quantity of the products separated.

V. CONCLUSIONS

High-resolution fractionation techniques are analyzed using different examples of proteins, antibiotics, and other biological products of interest available in the literature. Quite often, as the examples selected for analysis at random suggest, the yield for obtaining a product purified to homogeneity is low (perhaps as low as 10 to 12 percent, and sometimes even lower). In some cases, where difficult separations are required, the yield can be less than a percent. Also, quite a few procedures (or steps) are required to produce a biological product purified to homogeneity. Often the procedure has to be repeated in a bioseparation protocol. All this contributes significantly to increasing the high cost of obtaining these products.

Chromatography is still the most prevalent form of high-resolution fractionation technique utilized. Different types of chromatographic procedures have been utilized to purify a wide variety of products, as the random selection of examples analyzed suggest. Gel filtration chromatography is frequently used as a high-resolution fractionation technique. Other high-resolution fractionation techniques mentioned and analyzed include hydroxylapatite (adsorption) chromatography, hydrophobic interaction chromatography, affinity chromatography, and ion-exchange chromatography. These procedures have been used not only as high resolution fractionation procedures, but also as the initial purification step (see Chap. 2). The choice and selection of which procedure to use at what stage of the bioseparation train depend primarily on the quantity and quality of the product separated, and its source.

As the tables for the different products purified suggest, the yield decreases as each chromatographic (or other bioseparation) step is utilized. In some cases, the causes for the decrease in yield are given. More detailed information is required in the open literature that provides the reasons for the activity loss of different products as they are purified by chromatographic or other bioseparation procedures used as high resolution fractionation techniques. Because, in general, there is a considerable increase in the purification factor value at this stage, there is, as expected, a concomitant decrease in the yield value.

Improvements in chromatographic as well as other bioseparation procedures used as high-resolution fractionation techniques are urgently required so as to obtain higher yields of products without a sacrifice in the quality of the product. Such improvements will only be possible if further mechanistic details of the bioseparation procedure are better understood. Because the high-resolution fractionation techniques are the last step in the bioprocessing train, product quality and quantity will also be significantly influenced by improvements upstream. For example, improvements in the initial fractionation process or even during the cell-disruption process will also significantly influence the quality and quantity of the bioproduct separated.

It is therefore recommended that considerable effort be continuously placed on improving the yield obtained from high-resolution fractionation techniques so as to maintain an economic edge. This effort is also required for the development of future biological products (particularly therapeutics) that are or will be required by mankind. Thus, there is a continual need to not only improve existing procedures but also explore, examine, and develop suitable and innovative techniques (such as the ARMES procedure briefly analyzed in the text) that exhibit the potential to serve as good candidates as high-resolution fractionation techniques.

Crystallization has been used as a high-resolution fractionation technique, especially for the purification of antibiotics. More mechanistic details are required in the open literature that give the reasons for the activity losses of antibiotics and for other biological products of interest purified using the crystallization procedure. This information will considerably assist in improving the quality and quantity of the antibiotic purified and in reducing the cost of its production.

REFERENCES

Barichievich, E. M. and Calza, R. E. (1990a). *Appl. Environ. Microbiol., 56,* 43.

Barichievich, E. M. and Calza, R. E. (1990b). *Curr. Microbiol., 20,* 265.

Barker, D. L. and Jencks, W. P. (1969). *Biochemistry, 8,* 3879.

Bienkowski, P. R., Lee, D. D., and Byers, C. H. (1988). 9th Symposium on Biotechnology for Fuels and Chemicals, Scott, C.D., Ed., Appl. Biochem. and Biotechnology, 17/18, Humana Press: Clifton, NJ, pp 261–273.

Blackwell, J. A. and Carr, P. W. (1992). *J. Chromatogr., 535,* 111.

Brookes, I. K. and Lilly, M. D. (1987). *Enzyme Microb. Technol., 9,* 217.

Brown, S. H., Sjoholm, C., and Kelley, R. M. (1993). *Biotechnol. Bioeng., 41,* 878.

Brumeanu, T. D., Zaghouani, H., and Bona, C. (1995). *J. Chromatogr., 696,* 219.

Buetler, H. O. (1984). In *Methods of Enzymatic Analysis,* 3rd ed. Vol. 6, Bergmeyer, H.U., Ed., Verlag Chemie: Weinheim, Germany, p 484.

Burgess, K. and Shaw, M. (1983). In *Industrial Enzymology,* Godfrey, T. and Reichelt, J., Eds., Nature Press: New York, pp 260–283.

Castanares, A., McCrae, S. I., and Wood, T. M. (1992). *Enzyme Microb. Technol., 14,* 875.

Connelly, I. C. and Coughlan, M. P. (1991). *Enzyme Microb. Technol., 13,* 462.

Coughlin, R. W. and Charles, M. (1980). In *Immobilized Enzymes for Food Processing,* Pitcher, W. H., Ed., CRC Press: Boca Raton, Fl., pp 153–173.

Cueille, G. and Tayot, J. L. (1985). *Biotech '85 Proceedings,* Online Publications: Pinner, Middlesex, UK, p 141.

Cussler, E. L. (1987). In Protein Purification: Micro to Macro, Burgess, R., Ed., Alan R. Liss:, New York, pp 307–314.

Datar, R. V., Cartwright, T., and Rosen, C. G. (1993). *Biotechnology, 11,* 349.

Dekker, K., Yamagata, H., Sakaguchi, K., and Udaka, S. (1991). *J. Bacteriol., 173,* 3078.

Dowd, V. and Yon, R. J. (1992). *J. Chromatogr., 627,* 145.

Dowd, V. and Yon, R. J. (1995). *J. Chromatogr. A, 693,* 15.

Evans, A. M. (1992). *Eur. J. Clin. Pharmcol., 42,* 237.

Flanagan, T. and Waites, M. J. (1992). *Enzyme and Microb. Technol., 14,* 975.

Gavrilescu, M., Pai, C., User, S., Tonescu, S., and Margineanu, N. (1974). *Process for the Isolation and Purification of Tetracycline,* Br. Patent 1368668, October 2.

Gilbert, M., Breuil, C., Yaguchi, M., and Saddler, J. N. (1992). *Appl. Biochem. Biotechnol., 34/35,* 247.

Goldstein, W. E. (1990). Enzymes in Starch Processing and Baking, In: *Enzymes in Industry,* Gerhartz, W., Ed., VCH Publishers: New York, pp 92–102.

Gonzalez, R. R. and Monsan, P. (1991). *Enzyme Microb. Technol., 13,* 349.

Greer, C. W. (1984). Ph.D. thesis, McGill University, Montreal, Canada.

He, L., Bickerstaff, G. F., Paterson, A., and Burwell, J. A. (1993). *Enzyme Microb. Technol., 15,* 13.

Hedrich, H. C., Spener, F., Menge, U., Hecht, H. J., and Schmid, R. D. (1991). *Enzyme and Microb. Technol, 13,* 840.

Heikinheimo, R., Hemila, H., Pakkanen, R., and Palva, I. (1991). *Appl. Microbiol. Biotechnol., 35,* 51.

Hethcote, H. W. and De Lisi, C. (1983). In *Affinity Chromatography and Biological Recognition,* Chaiken, I. M., Wilchek, M. and Parikh, I., Eds., Academic: London, pp 119–134.

Hirayama, C., Ihara, H., and Li, X. (1990). *J. Chromatogr., 530,* 148.

Holbrook, J. J., Liljas, A., Steindel. S. J., and Rossman, M. G. (1975). In *The Enzymes,* 3rd ed., Vol. 11, Boyer, P. D., Ed., Academic: Lomdon, pp 191–292.

Honda, F., Honda, H., and Koishi, M. (1995). *J. Chromatogr., 696,* 19.

Honda, H., Ono, K., Ishizaka, T., Matsuno, T., Katuno, T., and Koishi, M. (1987). *J. Soc. Powder Technol. Jpn., 24,* 593.

Inman, F. N. (1984). *Filtr. Sep.,* May/June, 165.

Jeroch, H., Aboud, M., Engerer, K. H., and Gerbhardt, G. (1988). *Arch. Anim. Nutr., 38,* 399.

Katayeva, I. A., Golovchenko, N. P., Chuvilskaya, N. A., and Akimenko, V. K. (1992). *Enzyme Microb. Technol., 14,* 407.

Kawashima, K., Takeshima, H., Higashi, Y., Hamaguchi, M., Sugie, H., Sugie,I., and Wada, H. (1991). *Leuk. Res., 15,* 525.

Knowles, J., Teeri, T. T., Lehtovaara, P., Penttila, M., and Saloheimo, M. (1988). In *Biochemistry and Genetics of Cellulose Degradation,* Academic: London, pp 153–169.

Knox, J. H. and Kaur, B. (1988). In *High Performance Liquid Chromatography,* Brown, P. R. and Hartwick, R. A., Eds., Wiley-Interscience: New York, p 189.

Knutsen, S. H. (1991). *Oebalia, 17*(Suppl.), 387.

Koishi, M. (1991). In *Powder Technology Handbook,* Iinoya, K., Gotoh, K., and Higashitami, K., Eds., Marcel Dekker: New York, p 453.

Krish, K. (1971). In *The Enzymes,* 3rd ed., Vol 5, Boyer, E. D., Ed., Academic: London, pp 43–69.

Lachke, A. H. and Jeffries, T. W. (1986). *Enzyme Microb. Technol., 8,* 353.

Lang, D., Erdmann, H., and Schmid, R. D. (1992). *Enzyme Microb. Technol., 14,* 479.

Le Gall, Y., Braud, J. P., and Kloareg, B. (1990). *Plant Cell Rep., 8,* 582.

Lehmacher, A. and Bisswanger, H. (1990). *J. Gen. Microb., 136,* 679.

Lewis, D. H. and Smith, D. C. (1967). *New Phytol., 66,* 143.

Li, X. and Calza, R. E. (1991). *Enzyme Microb. Technol., 13,* 622.

Liapis, A. I. (1989). *J. Biotechnol., 11,* 143.

Lisi, P. L., van Es, T., Abuchowski, A., Palczuk, N. C., and Davis, F. F. J. *Appl. Biochem., 4,* 19.

Lowe, S. E., Theodorou, M. K., and Trinci, A. P. J. (1987). *J. Appl. Environ. Microbiol., 53,* 1216.

McPherson, A. (1990). *Eur. J. Biochem., 189,* 1.

Nho, K., Glower, D., Bredehoeft, S., Shankar, H., Shorr, R., and Abuchowski, A. (1992). *Biomater. Artif. Cells Immobilization Biotechnol., 20,* 511.

Orpin, C. G. (1975). *J. Gen. Microbiol., 91,* 249.

Orpin, C. G. (1976). *J. Gen. Microbiol., 94,* 270.

Orpin, C. G. (1977). *J. Gen. Microbiol., 99,* 107.

Ostgaard, K., Wangen, B. F., Knutsen, S. H., and Aasen, I. M. (1993). *Enzyme Microb. Technol., 15,* 326.

Owusu, R. K. and Cowan, D. A. (1991). *Enzyme and Microb. Technol., 14,* 236.

Paliwal, J. K., Smith, D. E., Cox, S. R., Berardi, R. R., Dunn-Kucharski, V. A., and Elta, G. H. (1992). *J. Pharmacokinet. Biopharm., 21,* 506.

Paradkar, V. M. and Dordick, J. S. (1993). *Biotechnol. Prog., 9,* 199.

Pitkanen, K., Heikinheimo, R., and Pakkanen, R. (1992). *Enzyme and Microb. Technol., 14,* 832.
Plues, S. (1976). *Antibiotic Purification Process,* U.S. Patent 3983108, September 28.
Rivere, J. (1977). In *Industrial Applications of Microbiology,* Moss, M. O. and Smith, J. E., Eds., Surrey University Press: London, pp 230–240.
Romaniec, M. P. M., Huskisson, N., Barker, P., and Demain, A. L. (1993). *Enzyme Microb. Technol., 15,* 393.
Rombouts, F. M. and Pilnik, W. (1986). *Symbiosis, 2,* 79.
Rosenberger, R. F. (1979). In *Fungal Walls and Hyphal Growth,* Burnett, J. H. and Trinci, A. P. J., Eds., Cambridge University Press: Cambridge, p 265.
Sarwar, G., Matayoshi, S., and Oda, H. (1987). *Microbiol. Immunol., 31,* 869.
Scawen, M. D. and Hammond, P. M. (1989). *J. Chem. Tech. Biotechnol., 46,* 85.
Schneider, K. H. and Giffhorn, F. (1991). *Enzyme and Microb. Technol., 13,* 332.
Sollman, T. (1957). *A Manual of Pharmacology,* 8th ed., W. B. Saunders: London.
Stryer, L. (1981). *Biochemistry,* 2nd ed., W. H. Freeman: San Francisco, p 261.
Suekene, M., Tamura, M., and Tomimura, C. (1978). *Agric. Biol. Chem., 42,* 909.
Tombo, G. M. R., Schar, H. P., and Ghisalba, O. (1987). *Agric. Biol. Chem., 51,* 1833.
Trudinger, U., Muller, G., and Unger, K. K. (1990). *J. Chromatogr., 596,* 27.
Ulhoa, C. J. and Peberdy, J. F. (1992). *Enzyme and Microb. Technol., 14,* 236.
Usui, T., Hayashi, Y., Nanjo, F., and Sakai, K. (1987). *Biochim. Biophys. Acta, 923,* 302.
Vijayalakshmi, M. and Thomas, D. (1985). *Proceedings Biotech '85 (Europe),* Online Publications: Pinner, UK, pp 131–139.
Ward, O. P. and Moo-Young, M. (1989). *CRC Crit. Rev. Biotechnol., 8,* 237.
Wildfeuer, M. E. (1985). *Purif. Ferment. Prod.,* 155.
Woll, J. M., Hatton, T. A., and Yarmush, M. L. (1989). *Biotechnol. Progr., 5,* 57.
Wong, K. K. Y., Tan, L. U. L., and Saddler, J. N. (1988). *J. Microbiol. Res., 52,* 305.
Woodward, J. and Wiseman, A. (1982). *Enzyme and Microb. Technol., 4,* 73.
Yon, R. J. (1988). *J. Chromatogr., 457,* 13.
Zarain-Herzerg, A. and Arroyo-Begovich, A. (1983). *J. Gen. Biol., 129,* 3319.

4 INTERFACIAL PROTEIN ADSORPTION AND INACTIVATION DURING BIOSEPARATION

I. INTRODUCTION

Proteins are adsorbed at different types of interfaces and influence the processes occurring at these interfaces. Some recent examples include biocompatibility of clinical implants, mammalian and bacterial cell adhesion to surfaces, initiation of blood coagulation, solid-phase immunoassay, and protein binding to cell surfaces. An area of growing importance where protein–enzyme interfacial interactions are involved is bioseparations. In general, in bioseparation processes an interface is involved with which contact with a protein solution is likely, and this process will then be influenced by protein adsorption at the interface. In this chapter, we examine the influence of gas–liquid, liquid–liquid, and liquid–solid interfaces on protein–enzyme inactivations.

Proteins in solution diffuse to the interface. This is thermodynamically favorable because some of the conformational and hydration energy is lost at the interface (MacRitchie, 1978). Initially, at low protein concentrations, there is no barrier to adsorption; and for protein molecules that are readily adsorbed at the interface, the rate of adsorption is diffusion controlled. However, after some time, especially at high surface protein concentrations, there is an activation barrier to adsorption (Graham and Phillips, 1979), which may involve electrostatic, steric, and osmotic effects close to the interfacial and surface layers. Then the ability of protein molecules to interpenetrate and create space in the existing film and to rearrange at the surface is rate determining.

Initially, Joly (1965) suggested that enzymes adsorbed at gas–liquid interfaces are generally present in an unfolded partially active or inactive state as a more or less rigid film. Graham and Phillips (1979) then stated that the capacity of proteins to unfold at an interface depends greatly on the conformational stability of flexible segments of the protein molecule. Interfaces are primarily responsible for protein inactivation as highlighted by experiments by Virkar *et al.* (1979). Using a partially filled disk reactor these authors noted that shear-associated damage can be severe, but it arises when gas–liquid interfaces are present. Then the replenishment of the interface associated with intense shear causes interfacial denaturation. This, rather than shear alone, is the explanation for much loss of protein structure and enzyme activity in pumps (Virkar *et al.*, 1979), and in centrifuges and ultrafiltration systems (Narendarnathan and Dunnill, 1982) where air is entrained (Dunnill, 1983). Virkar *et al.* (1979) further noted that a decrease in the air–liquid interfacial area by completely filling up the reactor vessel minimized the enzyme inactivation. Similar observations may be made with liquid–liquid (e.g., present in liquid–liquid extraction systems) or liquid–solid (e.g., high-pressure liquid chromatographic separations) systems.

Proteins on adsorption at fluid interfaces undergo a change from their globular configuration in solution to an extended chain structure. This has often been referred to as surface denaturation. On energetic grounds, it is expected that the polypeptide backbone lies in the plane of the surface with the polar and nonpolar side chains directed toward and away from the aqueous phase, respectively. The surface acts as a good solvent, in this case a two-dimensional one. When a protein molecule adsorbs, an area of high surface energy is replaced by interfaces of low free energy, that is, polar side chains-water and nonpolar side chains–air. The achievement of this free energy lowering accounts for the unfolding of the molecule at the surface. Also, whereas polypeptide chains may exist exclusively in the α-helix form, proteins in their solution state generally have their chains only partially in this form with relatively larger or smaller proportions in the β-form or random structure. Because the adsorption step represents a transition from a relatively poor solvent (water) to a relatively good solvent (air–water interface), there does not appear to be a strong driving force for the retention of helical structure. The changes in the structure of the protein on adsorption at the interface would lead to activity changes.

Norde and Lyklema (1979) thermodynamically analyzed the different subprocesses that are involved in protein adsorption. They specifically wanted to assess the contribution to the characteristic thermodynamic functions of the structural arrangements. Generally, protein adsorption is a spontaneous process. Therefore,

$$\Delta G_{\text{ads}} = \Delta H_{\text{ads}} T - T\Delta S_{\text{ads}}. \tag{4.1}$$

Here ΔG_{ads}, ΔH_{ads}, and ΔS_{ads} are the overall changes in free energy, enthalpy, and entropy on adsorption. Norde and Lyklema (1979) emphasized that there is a significant redistribution–rearrangement of free and bound charges in and around the protein molecule prior to, and after, adsorption.

Structural rearrangements on protein adsorption on surfaces is central to understanding the interfacial behavior of proteins. Norde and Lyklema (1979) analyzed the contributions of the following processes. Others may be involved, too, in the different protein adsorptions at the different types of interfaces:

1. Protein molecules in aqueous solution have a hydrophobic interior and a hydrophilic exterior. On adsorption at an interface, the hydrophobic patches on the protein surface become shielded from water. This hydrophobic dehydration involves a large entropy increase. Norde and Lyklema (1979) stated that when the structure of the adsorbing molecule changes, the possible loss of internal hydrophobic interaction may be compensated for by attachment of the hydrophobic residues at the interface.
2. Structural perturbations–rearrangements inside the protein molecule involving changes in the interatomic distances on adsorption at the interface affect the intermolecular van der Waals interactions.
3. Charged groups in the apolar interior of the protein molecule tend to form ion pairs. At the aqueous interface, these protein molecules tend to be isolated and hydrated. There is thus a change in the thermodynamic quantities resulting from the transfer of ionic groups from the interior to the exterior–interface.
4. There are different possibilities for hydrogen bonding to occur in protein molecules. In the apolar phase hydrogen bonding between peptide units is favorable, whereas in aqueous medium it is not. Tanford (1970) also concluded from an analysis of helical-coil transitions of synthetic polyaminoacids that intramolecular peptide hydrogen bonds are more favorable than peptide–water interactions.
5. The variations in the hydrogen bonding, ion pairing, hydrophobic interaction, etc., on the attachment of a protein molecule at an interface affect the rotational motion (freedom) of the protein molecule (or parts of it). A decrease in the secondary structure (e.g., α-helix, β-sheet structures) would increase the rotational freedom. From classical mechanics one can estimate that an increase of one degree of rotational freedom causes an enthalpy gain of RT per mole of bonds.

In summary, protein adsorption comprises the overall effect of different subprocesses. The influence of some of these subprocesses may be more or less for the different protein adsorptions at the different interfaces. Other subprocesses not mentioned may also influence protein adsorption. In general, an increase of entropy originating from the dehydration of the interface–surface, and the dehydration of and structural changes in the protein molecule are the driving force in the adsorption process.

pH and shaking will significantly affect the rate of transport of the protein to the interface. In the absence of shaking, a film forms that creates a barrier to diffusion of more protein molecules to the air–liquid interface. On shaking, precipitated film material is removed from the interface leaving it clean for more proteins to be adsorbed and precipitated. Surfactants are adsorbed and spread at the interface, but their energy barrier to their adsorption is much less than that for the protein. Thus, when surfactants are added, they are preferentially adsorbed, occupy the surface, and by so doing prevent the unfolding and denaturation of the protein.

Proteins adsorbed at the interface may exist in more than one state. The adsorption of a protein molecule at a site is likely to be influenced by the existence of molecules already adsorbed at nearby sites, by either geometric reduction of the area available for adsorption as the surface sites become occupied, or by repulsive forces that are expected whenever the molecules are close enough. These factors may cause the adsorbed population to be intrinsically heterogeneous with respect to the protein–surface interactions. Also, multiple binding modes may also exist due to the mixed site nature of a real interface. A perfectly homogeneous, perfectly clean interface is difficult if not impossible to produce, and in any case is probably not representative of actual materials used in adsorption studies. Evidence supporting multiple states of protein adsorption include observations indicating the presence of weakly and tightly bound proteins. Removal of adsorbed proteins occurs in different stages or fractions, and not all of the adsorbed protein is removed under one or a specified set of conditions. The multiple adsorption states exhibited by proteins would yield, in general, a plethora of different structures at the interface exhibiting slightly different functionalities. This multiple state of protein adsorption at different sites of the interface should exhibit heterogeneous deactivation behavior at the interface. In any realistic model for protein–enzyme inactivation at interfaces this heterogeneity of adsorption and the subsequent heterogeneity in deactivation should be taken into account. In general, a heterogeneity in enzyme deactivation leads to enhanced enzyme stabilization when compared with a homogeneous enzyme (Sadana and Malhotra, 1987a,b).

Information on protein adsorption–inactivation at different interfaces is present, in general, in different articles and restricted to either one or a few proteins and to a single type of an interface. We present later a wide variety of examples of protein–enzyme adsorption–inactivation at gas–liquid, liquid–liquid, and solid–liquid interfaces. It is a formidable task to seek general principles or guidelines that provide physical insights into protein adsorption at different interfaces, and the subsequent generalizations that may result. These generalizations should be of considerable assistance in the understanding of subsequent protein adsorptions at different interfaces. This chapter attempts to do this.

We initially analyze, by presenting examples, the influence of liquid–liquid interfaces on protein adsorption–inactivation. This is followed by protein adsorption–inactivation at gas–liquid and solid–liquid interfaces.

II. REACTION AND INACTIVATION AT LIQUID–LIQUID INTERFACES

A. Lipid–Water Interfaces

Gargouri *et al.* (1986) studied the inactivation of *Rhizopus delmar* lipase at the lipid–water interface. These authors studied the inhibitory effect of different proteins on lipase activity at the interface. Lipases catalyze hydrolysis or esterification of ester bonds at the interface between an insoluble substrate phase and an aqueous phase in which the enzyme is dissolved (Verger, 1989; MacRae, 1984). The rate of hydrolysis of oil by a nonoil soluble lipase is a direct function of the surface area of the oil–water interface (Benzonna and Desnuelle, 1965).

The inhibitory protein prevents the lipase from binding to the mixed dicaprin monolayers, or the lipase is desorbed from the interface when the protein is added later. When no inhibitory protein is added, there is no loss of lipase activity. Serum albumin and β-lactoglobulin result in lipase activity losses of 93 and 92%, respectively. A large proportion of the lipid film remained accessible to the enzyme in the presence of the inhibitory protein. It may be concluded that the observed decrease in the lipase binding to the interface is due to a variation of the physicochemical properties of the lipid–water interface following binding of the inhibitory protein.

Wilson and Dahlquist (1985) detected two conformational states of the coat protein of the filamentous bacteriophage M13 in detergent solution by magnetic resonance techniques. The different conformational states exhibit different activities and thus the activities of this protein could be regulated. There is a nonspecific lipid solvation of the membrane proteins. The structure of the lipid environment can easily regulate protein conformational equilibria. By providing certain important membrane proteins with conformational equilibria coupled to membrane structure, the activities of these proteins can be regulated by variations in the fatty acid chain length of the head group. These factors in turn could be controlled by the cell in response to environmental signals.

B. Aqueous Two-Phase Systems

Kula (1987) utilized phase partitioning in aqueous two-phase systems using polyethylene glycol–salt or polyethylene–dextran to separate alcohol dehydrogenase from *Saccharomyces cerevisiae* and from *Candida boidinii*. In all the cases reported high yields between 91 and 96% were obtained. The purification factor value was 1.8. The method exploits the behavior of hydrophobic polymers in solution and provides a fast and gentle separation technique well suited for large-scale operation.

Partition-based bioseparations are likely to find increasing use in biotechnology product development due to the system's biocompatibility and amenability to scale-up. Kula (1987) also analyzed the influence of scale-up on extracting formate dehydrogenase from *C. boidinii*. The author utilized an aqueous two-phase system using polyethylene glycol–salt. A yield of 95% was obtained on a 10-ml scale. The yield dropped insignificantly from 95% to 94% as the process was scaled-up from 10-ml to a 250-liter process.

Hariri *et al.* (1989) indicated that for industrial-scale partitioning multistage separations would rely on a liquid–liquid partitioning column. A pump would feed the heavy polymer phase into the top of the column and a lighter phase into the bottom of the column. These two immiscible phases move in opposite directions through alternating mixing and separation stages. The mixing steps facilitate mass transfer, whereas the settling stages allow for phase separation. The feed sample may be introduced either with one of the two polymer phases, or directly into the mixing-settling stages. Mansoori *et al.* (1987) emphasized a particular advantage of the system is that it permits strict product quality control.

Hughes and Lowe (1988) purified human serum albumin (HSA) by aqueous two-phase partitioning in novel acrylic copolymer systems. The authors were able to recover 95% of HSA and obtained a purification factor of 2.2, with 5% of the albumin being located at the interface. Carboxypeptidase G2 also exhibited similar partitioning behavior in these systems. Protein precipitation and loss of enzyme activity were observed under acid conditions at high salt concentrations, although at higher pH values the effect was reduced. At higher pH values there was a comparatively minor loss in enzyme activity due to interfacial adsorption. There is thus a need to analyze the reasons for the interfacial binding of proteins, and how this may be effectively minimized. The preceding results suggest that electrostatic interactions play a significant part in protein adsorption and interactions at the interface. It would be of interest to develop mechanistic descriptions of protein unfolding-folding and deformations of the protein-laden interface. Factors such as diffusion and electrostatic interactions that influence the forward and backward transport kinetics of the protein with regard to the interface are important.

Considerable improvements have been made in understanding liquid-liquid protein systems so that they may be effectively utilized as effective bioseparation systems. Improvements in process performance need to be made continuously, and should be aggressively researched. Two examples presented later highlight the importance of process improvements and better physical insights into the process.

Example 4.1

A two-phase system that exhibits potential for bioseparation other than the classical polyethylene glycol (PEG)-dextran system is described (Pathak *et al.*, 1991).

Solution

Pathak *et al.* (1991) presented phase diagrams for two aqueous-phase systems based on polyethylene-glycol-$FeSO_4$-water and polyethylene-glycol-Na_2SO_4-water over a wide range of temperatures. Their main intention was to improve partitioning of the biomolecules between the phases and to present an economically better system. The PEG-salt-water systems are attractive compared with the PEG-dextran-water systems due to the lower cost of the salts.

These authors noted that the concentration levels of PEG in the salt-rich phase are significantly lower than those found in the PEG-dextran-water or PEG-phosphate-water systems. This is of significance because the recovery of biomolecules is easier from the salt-rich phase using ultrafiltration in the absence of PEG. Additionally, low levels of the PEG concentration in the salt-rich phase lead to limited loss of PEG. However, no numbers were provided by these authors to demonstrate this fact. It would be of interest to note quantitatively the increase in recovery of the product, as well as the quality of the product recovered.

They also noted that an increase in temperature shifts the bimodal curve toward the origin. Thus, the concentration of the salt in the PEG phase and the concentration of PEG in the salt-rich phase decrease with an increase in temperature. This is contrary to observations with the PEG-dextran-water systems. For example, the phase diagram for the PEG-$FeSO_4$-water system reveals that at 10% of $FeSO_4$, an increase in the temperature from 20 to 40°C decreases the PEG weight percent from 22 to 16. This represents about a 37% decrease in PEG for a 20°C increase in temperature. These types of systems should be carefully investigated to examine the enhancements in the yield and the quality of the product that may be obtained.

Liquid–liquid extraction provides an attractive alternative to conventional procedures for the recovery and purification of proteins and other biopolymers, because it combines moderate selectivities with high volumetric capacities. Reversed micelles are surfactant aggregates in organic solvents capable of solubilizing significant quantities of proteins in the bulk organic phase. The protein can be induced to move from a bulk aqueous phase into the micelle-containing organic solvent, and vice versa by manipulation of the pH, ionic strength, and surfactant concentration (Goklen and Hatton, 1985, 1987). Woll *et al.* (1987) utilized reversed micelles to extract proteins selectively from aqueous solutions. There is a trade-off of the recovery of protein activity as the pH of the solution is varied. Lower pHs increase the driving force for protein transfer because of increasingly favorable electrostatic interactions between the protein and the surfactants, but below a pH of 5 there is a dramatic falloff in the protein activity. For example, at pH 5.5 there is a 23% recovery of the detergent enzyme, an extracellular alkaline protease from *Bacilus* sp. ATCC 21536. The authors emphasized that the transfer of proteins from an aqueous solution to a reversed micellar organic phase appears to be dominated by electrostatic interactions between the charged protein and the surfactant layer forming the walls of the micellar polar core.

C. Reversed Micelles

Different protein molecules exhibit different affinities for water-in-oil microemulsions. This property may be utilized to achieve a selective separation of proteins and other valued products from other material in an aqueous broth (Luisi and Laane, 1986; Hatton, 1989). These proteins can be recovered from the reversed micelles by contacting the organic phase with fresh aqueous solution leading to a back transfer of the newly purified product (Dekker *et al.*, 1987; Rahaman *et al.*, 1988). Little information is available, however, about the rate at which the microemulsion system approaches equilibrium when the two phases are contacted. Therefore, scant knowledge is available about the interfacial kinetics and mechanistic details associated with the transfer of proteins into and out of reversed micellar systems. It would be of significant interest to develop mechanistic descriptions-kinetics of proteins and other valued product interactions at liquid-liquid interfaces.

Example 4.2

Present a brief analysis of interfacial transport processes in reversed micellar extraction of proteins (Dungan *et al.*, 1991).

Solution

Dungan *et al.* (1991) measured the interfacial mass transfer coefficients for the transfer of α-chymotrypsin and cytochrome c between a bulk aqueous and a reversed micellar phase using a stirred diffusion cell. The authors obtained the forward and the back transfer rates for both of these proteins under a variety of pH and ionic strength conditions spanning the ranges over which significant protein solubilization is possible. The authors emphasized that because changes in the aqueous pH and salt concentration influence both the forward and the back transfer rates of the protein, electrostatics plays a significant role in the mechanism associated with protein solubilization. A Derjaguin-type analysis was utilized by Dungan *et al.* (1991) to elucidate the electrostatic interactions between the protein and the interface as the protein-filled reversed micelle is being formed during the forward transfer process.

As the protein approaches the proximity of the surfactant-laden organic phase-aqueous phase interface a modified Derjaguin analysis indicates that charge interactions will produce a significant electrostatic force at the interface. As the protein is enveloped at the interface by the surfactant molecules it undergoes deformation. No comments were provided by the authors on the nature-degree of deformation experienced by the proteins that would affect the quality of the product separated. The extent of deformation decreases as the pH or ionic strength increases. It would be of interest to theoretically analyze the extent of deformation experienced by the proteins as the electrostatic interactions change at the interface. Reasonably, the larger extents of deformation would lead to deleterious effects on the quality of the protein separated.

The authors emphasized that once close to the bulk interface the protein particle participates cooperatively in the mechanism for its solubilization by a reversed micellar droplet. This is brought about by the fact that the protein forward transfer into the reversed micellar phase occurs up to three orders of magnitude faster than does forward transfer of small solutes. This indicates a facilitated solubilization process. It is of interest to know the relative protein deformations-subsequent denaturations in the cooperative solubilization process compared with a regular solubilization process. Also, the reverse process of back transfer of proteins out of the reversed micellar phase into an aqueous phase occurs at rates two or three orders of magnitude slower than the rates reported in the literature for the back transfer of small solutes. These measurements suggested to Dungan *et al.* (1991) that the presence of protein hinders the desolubilizaton process.

The preceding study by Dungan *et al.* (1991) is useful because it sheds physical insights into the mechanistic descriptions of protein interfacial interactions. More such studies are required, particularly those that clearly delineate the deformations (reversible-irreversible) undergone by proteins at interfaces during reversed micellar extraction, and the subsequent alterations in the quality-yield of protein product separated.

Bausch *et al.* (1992) indicated that the kinetics of extraction and reextraction of hydrophilic solutes in and out of reversed micelles is of interest for primarily two reasons:

1. The kinetic measurements provide better physical insights into the mass transfer process itself. Besides, the location of the process is delineated. Is the rate determining step in the bulk or at the interface?
2. The rational design of extraction apparatus is facilitated.

These authors utilized a two-phase stirred cell to measure the kinetics of reextraction of amino acids or chymotrypsin from reversed micelles of the anionic surfactant sodium salt of bis-(2-ethylhexyl) sulfonate (AOT) in aliphatic hydrocarbons. The reversed micellar solutions were contacted with aqueous buffers with high ionic strength, thus facilitating the reextraction of the hydrophilic solute. Bausch *et al.* (1992) noted that the reextraction of all the solutes was controlled by an interfacial process. Also, the reextraction rate of the solutes was nearly independent of the solute.

Extraction kinetics in reversed micelles is controlled by convective transport in the aqueous phase. This was observed by Plucinski and Nitsch (1989) in the extraction kinetics of lysozyme in AOT-reversed micelles. Dekker *et al.* (1989, 1990) noted that the extraction of α-amylase was controlled by the diffusion of the enzyme in the aqueous phase boundary layer. For the reextraction (of proteins) step, authors (Dekker *et al.*, 1990; Dungan *et al.*, 1991; Bausch, 1989) are in agreement that the coalescence of reversed micelles with the macroscopic interface is the critical step. Later, Bausch *et al.* (1992) extended this to other hydrophilic substances such as amino acids and salts.

The mechanism suggested by Bausch *et al.* (1992) includes these steps:

1. The reversed micelle is atteached to the interface. This includes a "rupture" of the organic solvent film between the reversed micelle and the interface. This is a rapid step and is not rate determining.
2. The reversed micelle may discharge its contents to the aqueous phase by two methods. For water, the osmotic pressure difference (between the micelle and the aqueous phase) facilitates the exchange of water between the two phases (permeation). The other method is coalescence where the reversed micellar AOT exterior has to fuse or merge with the AOT monolayer at the interface. Then there is a minimum time where the reversed micelle exchanges or discharges its contents to the bulk aqueous phase. The reversed micelle may or may not keep its old surfactant layer. This depends on the deformations-reformations occurring locally at the interface. Nevertheless, these authors emphasized the *merging* of the two surfactant layers is the rate determining step in the reextraction process.

The analysis by Bausch *et al.* (1992) is of particular interest because it sheds novel physical insights into the interfacial reactions. It is hoped that a better understanding of interfacial reactions will lead the way to a better control of not only liquid–liquid interfacial reactions but also gas–liquid and solid–liquid interfacial reactions.

III. REACTION AND INACTIVATION AT GAS–LIQUID INTERFACES

Adsorption of proteins and other valued products is of importance both theoretically and from a practical standpoint. Adsorbed protein films assist in the stabilization of foams and emulsions (Viesturs *et al.*, 1982; Halling, 1981), and they may affect the transport of dissolved gases in fermentation broths (Yagi and Yoshida, 1974; Atkinson and Mavituna, 1983). Foam fractionation and froth flotation rely on adsorption at gas-liquid interfaces to facilitate biomolecule separation (Grieves, 1982). Stirring can also introduce air into a protein-enzyme solution.

A. Shear and Agitation

Initially, Joly (1965) suggested that enzymes adsorbed at gas-liquid interfaces are generally present in an unfolded partially active or inactive state as a more or less rigid film. Graham and Phillips (1979) then stated that the capacity of proteins to unfold at an interface depends greatly on the conformational stability of flexible segments of the protein molecule. Interfaces are primarily responsible for protein inactivation as highlighted by the experiments by Virkar *et al.* (1979). By using a partially filled disk reactor these authors noted that shear-associated damage can be severe, but it arises when gas-liquid interfaces are present. Then the replenishment of the interface associated with intense shear causes denaturation. Dunnill (1983) suggested that this, rather than shear alone, is the explanation for much loss of protein structure and enzyme activity in pumps (Virkar *et al.*, 1979), and in centrifuges and ultrafiltration systems (Narendarnathan and Dunnill, 1982) where air is entrained. Virkar *et al.* (1979) further noted that a decrease in the air-liquid interfacial area by completely filling up the reactor vessel minimized the enzyme inactivation. Similar observations may be made with liquid-liquid, for example, present in liquid-liquid extraction systems; or liquid-solid, for example, high-pressure liquid chromatographic separation (HPLC) systems. Stirring can also introduce air into a protein/enzyme solution. Enzymes with sulfhydryl (SH) groups are known to be easily inactivated by the oxidation of such groups of disulfides. Two examples of these are yeast alcohol dehydrogenase (Swartz, 1985) and glyceraldehyde-3-phosphate-dehydrogenase (Harris and Perham, 1965).

Elbaum *et al.* (1976) analyzed the inactivation of human hemoglobin A and S at the air-water interface. Hemoglobin A is present in normal humans and hemoglobin S is predominant in the red cells of humans with the sickle cell disease. These authors noted that hemoglobin S solutions tend to form precipitates when shaken, unlike solutions of the normal hemoglobin A. Because shaking of protein solutions induces bubble formation and because agitation without bubble formation (by slow stirring) causes a much slower rate of precipitation, the enhanced precipitation rate of hemoglobin S solutions was attributed to an enhanced rate of surface denaturation at the air-water liquid interface. A sharp increase in the resistance to further compression at the interface (attributed to a monolayer formation) occurred at an area of 8000 Å^2 per molecule for hemoglobin S compared to 5000 Å^2 per molecule for hemoglobin A. The greater area per molecule suggests a greater degree of unfolding of the hemoglobin S molecule compared with that of the hemoglobin A.

Reese and Robbins (1981) studied the effect of agitation on the inactivation of β-lactoglobulin. According to these authors shear is a factor but a minor one. Air-liquid surface effects play the major role. Protein molecules are adsorbed at the air-liquid interface where they are unfolded, form aggregates, and are later precipitated. In the absence of shaking, a film forms that creates a barrier to diffusion of more protein molecules to the interface. On shaking, precipitated film material is removed from the interface, leaving it clean for more protein to be adsorbed and precipitated.

Shear rate plays an important role in the inactivation of proteins at the air-water interface in aqueous solution. Shear rate is known to influence the inactivation of catalase, rennet, and carboxypeptidase (Charm and Wong, 1970); urea (Tirrell and Middleman, 1975); and cellulase from *Trichoderma reesei* (Reese and Ryu, 1980). These authors concluded that the enzymes lost their activity due to the denaturation of their catalytic sites by mechanical shear. The next example analyzes the mechanism of shear inactivation of lipase from *Candida cylindracea.*

Example 4.3

Briefly describe the kinetics and mechanism of shear inactivation of lipase from *C. cylindracea* (Lee and Choo, 1989).

Solution

Lee and Choo (1989) conducted shearing experiments in a stirred tank reactor with 0.1% lipase solutions of *C. cylindracea.* Lipases catalyzed the hydrolysis or esterification of ester bonds at the interface between an insoluble substrate phase and an aqueous phase in which the enzyme is dissolved (Verger, 1989). Lee and Choo (1989) indicated that the rate of hydrolysis of oil by a nonoil soluble lipase directly depends on the surface area of the oil-water interface. Good mixing is therefore essential in these systems; however, mixing produces shear, and therefore it is appropriate to analyze the influence of shear on lipase inactivation in reaction systems.

These authors emphasized that the inactivation of lipase is a shear-induced interface effect and involves the following steps:

1. Shear increases the rate of adsorption of the lipase at the air-water interface when compared with diffusion alone (Blank, 1969; MacRitchie and Owen, 1985).
2. At the air-water interface lipase splits up like hemoglobin (MacDonnell *et al.,* 1950). It may also change its conformation, unfold, or coagulate.
3. Shear may also assist in the interface inactivation by rupturing the molecular network formed and by replacing fresh molecules at the interface, which in turn are inactivated by the interface tension (Virkar *et al.,* 1981).

When 0.05 ml of the antifoam agent, polypropylene glycol (PPG), was added to the shearing experiment, the rate of lipase denaturation decreased significantly by 93% compared with the case when PPG was not added (Lee and Choo, 1989). These authors needed to distinguish whether the antifoam agent: (1) binds directly to lipase to protect its structure from inactivation, or (2) it decreases the surface tension and thereby protects lipase from inactivating due to interface tension. They filled the lipase to the brim of the reactor during the shearing experiments and noted that the inactivation decreased by 97%. Thus, the loss in lipase activity during shearing in the reactor is due to inactivation by the air-water interface tension. Also, the addition of the antifoam agent reduced the interface tension.

Reese and Robbins (1981) indicated that the rate of denaturation can be reduced to very low levels by the addition of small amounts of surfactants or large polymers like PEG or methylcellulose. A high percentage of the shake denatured β-lactoglobulin can be renatured by dissolving it in dilute acid. Surfactants may also be utilized to minimize protein adsorption at the interface. The surfactants are adsorbed and spread at the interface. Nevertheless, the energy barrier to their adsorption is much less than that for protein. As a result, when surfactants are added, they are preferentially adsorbed, occupy the surface, and by doing so prevent the unfolding and the denaturation of the protein.

B. Techniques for Protein Adsorption

The various techniques that have been utilized to study the adsorption of proteins at interfaces include film balances, pendant drop methods, surface tension reduction by the drop-volume technique, ellipsometry, and various types of ellipsometry (Andrade, 1985). Hunter *et al.* (1990) emphasized that the only method that has been used to directly quantify protein concentration at gas-liquid interfaces is the radiotracer method (Muramatsu, 1973). In the radiotracer technique the radiation emitted by a weak β-emitting radionuclide is attenuated much more by water than by air. Therefore, the surface-adsorbed radiolabeled species contributes proportionately more to the radiation measured by a detector positioned above a solution containing radiolabeled surface-active molecules than do the same species in the bulk solution. This technique is highly specific and sensitive. The technique also does not disturb the adsorbing protein solution as a measurement is made. Therefore, both adsorption and desorption rates are possible. Chen *et al.* (1982) emphasized the only disadvantage of the technique is that chemical changes introduced by the radiolabeling procedure could alter the conformation or hydrophobicity of the molecule.

Studies of protein adsorption at gas-liquid interfaces by the radiotracer technique have used ^{14}C labeling (Graham *et al.*, 1975; LeCompte *et al.*, 1983; Davies *et al.*, 1983). Hunter *et al.* (1990) indicated that the most common method of radiolabeling is acetylation of the terminal amine and lysine residues. This reaction removes the charge from the derivatized residues and hence increases the hydrophobicity of the protein. Frequently, radiation detection in these studies has been by gas flow proportional counters. Table 4.1 shows some of the different protein adsorptions at gas-liquid (air-water) interfaces that have been analyzed by the radiotracer technique.

The radiotracer technique seems to be well suited to quantitatively assess the adsorption of proteins at gas–liquid interfaces when they are present either alone in a solution, or as a mixture of two or possibly more proteins. More studies using radiotracer techniques or modifications thereof are required that both quantitatively and qualitatively assess protein adsorption at gas–liquid or gas–water interfaces. The next example presents in a little more detail an analysis of protein adsorption at the air–water interface using the radiotracer technique.

Example 4.4

Describe an example where protein adsorption at an air–water interface has been studied by the radiotracer technique. Briefly describe the information that is made available (Hunter *et al.*, 1990).

TABLE 4.1 Protein Adsorption at the Air–Water Interface Analyzed by the Radiotracer Technique

Protein	Comments	References
Ribonuclease A	This protein had two distinct adsorption plateaus indicating monolayer and second layer adsorption	Khaiat and Miller (1966)
^{14}C-acetylated lysozyme, β-casein, and bovine serum albumin (BSA)	A definite monolayer plateau observed only for β-casein; all three proteins exhibited multilayer adsorption at high concentration	Graham and Phillips (1979a,b, 1980a,b)
Reductively methylated chicken egg-white lysozyme	Monolayer saturation at low concentrations of bulk protein (below 2×10^{-5} wt%); multilayer adsorption at high concentrations (above 10^{-3} wt%); at intermediate concentrations, an abrupt increase in surface concentration with increasing bulk concentration indicates a change in the orientation of the adsorbed protein molecules	Hunter *et al.* (1990)
Reductively methylated lysozyme–β-casein mixtures	Coadsorption and exchange of lysozyme and β-casein studied; the amount of protein adsorbed from mixtures containing 1.1 to 1.5 μg/ml of each protein exceeded the amount expected if the proteins adsorbed in a simple competitive fashion; the flexible, rodlike structure of β-casein allows it to exchange with itself, while the rigid, globular structure of lysozyme prevents it from exchanging	Hunter *et al.* (1991)

Solution

Hunter *et al.* (1990) described the adsorption of reductively methylated chicken egg-white lysozyme to the air-water interface. The protein was radiolabeled by reductive methylation of the terminal amine and lysine residues. These authors utilized improved calibration techniques to reduce the errors associated with the conversion of the measured count rate to surface concentration. The improvements included by the authors permitted them to obtain an adsorption isotherm for lysozyme measured over a range spanning six orders of magnitude of bulk protein concentration.

Based on their observations and results the authors proposed the following adsorption mechanism that included four steps:

1. Below a critical concentration lysozyme molecules adsorb at the air-water interface in the side on conformation/orientation, until a saturated monolayer (of side-on) of molecules is present.
2. At the critical concentration all the side-on adsorbed molecules change to the end-on orientation. This represents a higher packing density at the interface.
3. As the concentration increases above the critical concentration a saturated monolayer of end-on molecules is attached at the interface. At

the same time, orientation-independent adsorption of a diffuse second layer or multilayers occurs.

4. At high concentrations a saturated monolayer of end-on molecules and the diffuse second layer are both present.

These authors indicated that the air-water lysozyme adsorption isotherm exhibits monolayer saturation at low bulk protein concentrations (below 2×10^{-5} wt%), and nonsaturating multilayer adsorption at high concentrations (above 10^{-3} wt%). The authors were able to explain an abrupt increase in surface concentration with increasing bulk concentration by a change in the adsorbed protein orientation. They were also able to show that lysozyme molecules adsorbed in the first layer do not exchange significantly with lysozyme in the bulk solution.

A kinetic model that incorporated their experimental observations was developed by these authors. Two of their basic steps (equations) are presented. Below the critical concentration of the protein in the bulk solution, protein adsorption follows classical Langmuirian kinetics. The rate of increase of side-on surface concentration of lysozyme is given by

$$\frac{d\Gamma_1}{dt} = k_a e^{-(E_a/RT)} C_p(1 - b\Gamma_1) - k_d e^{-(E_d/RT)} \Gamma_1. \tag{4.2}$$

Here Γ_1 is the concentration of the side-on protein adsorbed at the interface, C_P is the bulk concentration of the protein, b is the average area occupied by the molecule at the interface. k_a and k_d are the preexponential factors for the adsorption and the desorption constants, respectively. E_a and E_d are the activation energies for adsorption and desorption, respectively.

Guzman *et al.* (1986) initially proposed that the activation energies are dependent on surface coverage. Similarly, Hunter *et al.* (1990) suggested that

$$E_a = E_a^0 + \beta(\Gamma_1)^\delta \tag{4.3a}$$

and

$$E_d = E_d^0 - \gamma(\Gamma_1)^\delta. \tag{4.3b}$$

δ is an integer. β and γ are constants. Equations (4.3a and 4.3b) imply that as the surface coverage increases, the activation energies for adsorption and desorption increase and decrease, respectively. By utilizing the fact that the time rate of change of Γ_1 is zero at equilibrium, the authors were able to obtain an expression for Γ_1. A value of 4 for δ by the authors indicated to them that there is some cooperativity among adsorbing molecules and molecules already present at the interface.

Heterogeneity in protein adsorption is a fact and is a more realistic picture of the events occurring at the interface. In that sense the model presented by Hunter *et al.* (1990) is intriguing as well as partially, if not completely, satisfying. The model presented includes orientation-dependent adsorption until monolayer saturation occurs; then molecular rearrangement from a side-on to an end-on orientation occurs, followed by orientation-independent adsorption in the multilayer region. More such models should be developed to include the heterogeneity aspects which are closer to the real picture of the reactions occurring at the interface.

C. Models for Protein Adsorption

It would be of interest to develop models of protein adsorption at the air–water interface. In general, models of protein adsorption at the air–water interface avoid the complications of a heterogeneous surface (interface) for adsorption. Andrade and Hlady (1986) emphasized that these models should at least include the rate of adsorption, the rate of desorption, and the rate of surface reorientation-reconformation. Beissinger and Leonard (1982) proposed a model to include these effects; however, their model required five parameters. Because most experimental data generated are not extensive enough to do justice to a five-parameter model, most protein adsorption studies involve a simple model.

Ward and Tordai (1952) initially introduced a simple model for protein adsorption at the air–water interface that became quite popular due to its simplicity. The rate of protein adsorption at the air–water interface was given by

$$\frac{d\Gamma_1}{dt} = (k_{\text{ads}}c_{\text{b}} - k_{\text{des}}\Gamma_1)\exp\left(-P_{\text{surf}}\frac{\Delta A}{kT}\right). \tag{4.4}$$

Here c_{b} is the bulk concentration of the protein. Γ_1 is the concentration of the protein at the air–water interface. k_{ads} and k_{des} are the adsorption and desorption rate constants, respectively. P_{surf} is the surface pressure, and the parameter A is the surface area required to "anchor" the protein to the interface. The exponential type term in Eq. (4.3) has the same form as the exponential term in a classical rate constant. Therefore, Ward and Tordai (1952) equated it to an energy barrier for adsorption. In their formulation the energy barrier for adsorption is the same for adsorption as well as for desorption. MacRitchie (1978) referred to $P_{\text{surf}}\,\Delta A$ as the additional work that has to be done against the surface pressure, P_{surf} to create the hole of area ΔA.

It is perhaps appropriate to indicate the simplifications inherently present in Eq. (4.4). k_{ads} and k_{des} need not necessarily be constant, but may be considered as rate coefficients that may exhibit a temporal dependence. Also, if there are surface conformational changes of the protein at the interface (from, for example, side-on to end-on, or otherwise), then the A (anchor dependence term) will also exhibit a temporal dependence. All these factors point to include a heterogeneity factor in the models. This would increase the parameters in the model requiring much more extensive data for an appropriate analysis. This is, however, required if one wants to deal with the true picture of the reactions occurring at the interface.

In any case, let us go back to the Ward-Tordai model (1952). MacRitchie and Alexander (1963) simplified this model by assuming a negligible rate of desorption. Then Eq. (4.4) simplifies to yield

$$\frac{d\Gamma_1}{dt} = k_{\text{ads}}c_{\text{b}}\exp\left(-P_{\text{surf}}\frac{\Delta A}{kT}\right). \tag{4.5a}$$

If P_{surf} is a linear function of Γ_1 in an interval, then

$$\frac{dP_{\text{surf}}}{dt} = k'_{\text{ads}}c_{\text{b}}\exp\left(-P_{\text{surf}}\frac{\Delta A}{kT}\right). \tag{4.5b}$$

A plot of ln (dP_{surf}/dt) versus P_{surf} should be linear. Equation (4.5b) has been widely used for the adsorption of proteins at the air–water interface (Graham and Phillips, 1979; Tornberg, 1978; Ward and Regan, 1980). Paulsson and Dejmek (1992) were unsuccessful in applying this equation to the adsorption of β-lactoglobulin, α-lactalbumin, and bovine serum albumin (BSA) at the air–water interface over a surface pressure range of 0 to 25 mN/m. Some modifications to the preceding model have been suggested. De Feijter and Benjamins (1987) also raised doubts about the interpretation of the A parameter as the area necessary for anchoring a protein.

As an initial model, the model proposed by Ward and Tordai (1952) is satisfying due to its simplicity. Nevertheless, additional parameters such as heterogeneity are required to account for the real-life situation. This would in turn force experimenters to obtain precise and more extensive data for protein adsorption at the air–water interface.

More complex models of protein adsorption are also available. Douillard and Lefebvre (1990) proposed a kinetic and statistical mechanics model of protein adsorption at the gas–liquid interface based on the kinetic approach used previously by Guzman *et al.* (1986). These authors emphasized that the protein adsorption at the gas–liquid interface should include two adsorption layers. The first layer is actually located at the interface, and the second one results from adsorption of protein onto the second layer. These authors allowed for the dependence of the activation energies for adsorption and desorption on the adsorbed protein concentration. This model gave a good fit for the adsorption of β-casein, BSA, and lysozyme.

This model included two flaws as indicated by Douillard and Lefebvre (1990). It is realistic to assume that each layer is in equilibrium with the protein solution just below the surface. Besides, it did not account for known conformational changes that occur during protein adsorption at the gas–liquid interface (MacRitchie, 1978, 1986). Douillard and Lefebvre (1990) made the two adjustments in the model to correct these flaws. The two layers at the interface were assumed to be in equilibrium with each other. Also, protein in the first layer may occur in two configurations (Graham and Phillips, 1979). Note that in the first layer the protein can assume two conformations–configurations. This layer allows for saturation. In the second layer proteins are adsorbed less specifically and without saturation. This second layer permits the calculation of the concentration isotherm.

More realistic models like the Guzman *et al.* (1986) and the Douillard and Lefebvre (1990) models for protein adsorption at the gas–liquid interface need to be developed to provide better physical insights into the protein conformational changes and adsorption at the different gas–liquid interfaces.

The adsorption of proteins at air–water interfaces has been studied extensively. Only a few studies are available in the literature that analyze the adsorption of peptides at the air–water interface. Peptide adsorption at the liquid–solid and air–water–liquid interfaces is important in peptide-hormone drugs, where low doses make the influence of adsorption significant. Also, Arnebrant and Ericsson (1992) emphasized that the use of peptides as model substances for adsorption studies is important, because they provide the opportunity to analyze the influence of small structural units on the adsorption behavior by substituting amino acids within the peptide chain. Arginine vasopressin (AVP) is a peptide hormone involved in kidney function and blood pressure regulation. Desamino 8-D-arginine vasopressin (dAVP) is a commercial analogue. These compounds were chosen for the study due to their medical relevance and their availability in sufficient quantities. Both peptides consist of nine amino acids and contain one internal disulfide bridge. The molar masses of AVP and dAVP are 1084 and 1069, respectively.

The adsorption at the air–water interface was monitored by the surface tension reduction method (Arnebrant and Ericsson, 1992). The surface tension was measured using the drop-volume apparatus of Tornberg (1977, 1978) and automated by Arnebrant and Nylander (1985). Arnebrant and Ericsson (1992) measured the time-dependence of the surface tension decay and determined the plateau values after 10 min.

Only small reductions of the surface tension for the peptide solutions were noted (Arnebrant and Ericsson, 1992). For solutions of dAVP in water (pure) no effect of surface tension was noticeable. The peptides reduced the surface tension at high concentrations in a 0.15 M sodium chloride solution. In general, the kinetics of surface tension reduction was rapid, and the main part of the reaction took place within a few seconds. The authors noted that at a pH of 7.5 the kinetics for surface tension reduction appeared to be faster for AVP than for dAVP. Higher amounts of adsorption were obtained for AVP compared with those of dAVP. From this, Arnebrant and Ericsson (1992) concluded that a different structure in the adsorbed state exists for AVP compared with that of dAVP at the air-water interface. The analysis of Arnebrant and Ericsson (1992) is of significant interest because it presents data on peptide-hormone adsorption at the air–water interface. More such studies are urgently required due to their medical relevance in the treatment of common and serious ailments.

D. Foam Fractionation

Example 4.5

Briefly describe protein separation by differential drainage from foam (Mohan and Lyddiatt, 1994).

Solution

Mohan and Lyddiatt (1994) analyzed protein separation by differential drainage from foam. These authors stated that foams are colloids made up of gas bubbles dispersed in liquid. The foams are stabilized when surface-active agents present in a liquid adsorb at the gas-liquid interface. Proteins contain such surface-active agents. Thus, when a gas is bubbled through a protein solution, proteins adsorb at the gas-liquid interface. In a column, as the gas bubble rises, the liquid hold up decreases due to drainage, and the protein concentrates. At the top of the column, the gas bubble collapses and the protein concentrates. This is the basis of foam fractionation.

These authors utilized commercially available beer as a model system to demonstrate foam fractionation beyond the primary foaming stage. They wanted to determine if preferential drainage from foam obtained from dilute feedstocks might provide a general method for further fractionation. Initially, it had been shown that the surface-active proteins present in commercial beer were recovered in a continuous foam tower (Mohan *et al.*, 1992). Furthermore, most of the protein components present in the initial beer were also present in the foam.

On using drained foam preparations Mohan and Lydiatt (1994) were able to obtain purification factors ranging from 2.4 to 3.8 for the protein fractions. Using fast phase liquid chromatography (FPLC) on different fractions, these authors demonstrated that the various components present in the foam drained differentially. This indicated the possibility of further fractionating the proteins concentrated in the liquid films. Different parameters such as pH; temperature; and presence of salts, sugars, and lipids affect the foaming behavior of proteins (Bhattacharya *et al.*, 1991; Montero *et al.*, 1993; Vellissariou, 1992).

The height of the column and the column diameter also affect the foaming behavior. Mohan and Lyddiatt (1994) indicated that a manipulation of these parameters could help maximize the concentration of the protein in the initial form. Further purification of particular components is possible by subsequent drainage. These authors emphasized that the method exhibits potential as a primary processing step during the downstream separation of proteins, especially if there is a low concentration in the initial feedstock. Furthermore, because the conditions to produce and collapse the foam are relatively mild, the integrity of the product is conserved and protein inactivation is minimized.

The column height and the column diameter also affects the foaming behavior. Mohan and Lyddiatt (1994) indicated that a manipulation of these parameters could help maximize the concentration of the protein in the initial foam. Further purification of particular components is possible by subsequent drainage. These authors emphasized that the method exhibits potential as a primary processing step during the downstream separation of proteins, especially if there is a low concentration in the initial feedstock. Furthermore, because the conditions to produce and collapse the foam are relatively mild, the integrity of the product is conserved and protein inactivation is minimized.

In general, the adsorption of proteins, hormones, peptides, and other valued products at solid-liquid interfaces is of significant concern, especially because of its impact in the medical and related fields and other areas. The next section examines the adsorption of proteins and other valued products at solid–liquid interfaces.

IV. REACTION AND INACTIVATION AT LIQUID–SOLID INTERFACES

The adsorption of proteins at solid–liquid interfaces plays a major role in the following processes including biofouling (Fletcher *et al.*, 1980), thrombosis arising from medical prostheses (Brash, 1987), immunologic reactions on solid supports (Smith *et al.*, 1978; Graves, 1988), chromatographic separation (Scopes, 1982; Lesins and Ruckenstein, 1988), biotechnology applications requiring mass culturing of cells on surfaces with an intervening protein "glue," and new delivery methods for protein drugs (e.g., insulin, human growth factor, tissue plasminogen inactivator) where proteins are in contact with polymeric materials.

A. Quantitative Aspects and Conformational Change

A great deal of attention has been paid to the determination of the quantitative aspects of protein adsorption to surfaces. However, in addition to the amount of protein adsorbed, the biological consequences of proteins at solid–liquid surfaces often depend on the nature and the state of protein layer. In particular, information about the protein conformation and orientation in the adsorbed layer is required. The lack of information about how the proteins are organized has hindered the delineation of the role of the interface in protein adsorption studies in spite of three decades of research. Elgersma *et al.* (1990) emphasized a shortcoming of several studies is that an insufficient number of variables is studied.

More detailed protein adsorption followed by possible subsequent denaturation studies are required to delineate the mechanisms involved at the solid–liquid interface. Ideally, what one is looking for is the time-dependent composition and conformational changes occurring in the protein adsorbed at the interface. This is of primary importance because the initial protein layer or layers mediate and control further interactions at the interface. For example, the adhesion of blood platelets to glass surfaces is well known in the field of clinical chemistry, and blood platelets adhere more on surfaces coated with fibrinogen (Zucker and Broman, 1969). The amounts of proteins adsorbed on the surface of glassware have a significant effect on the quantitative analysis of very small amounts of protein in the case of radioimmunoassay or enzyme immunoassay (Rosselin *et al.*, 1966).

Adsorption of proteins on a glass surface is not a specific phenomenon but instead a general phenomenon that is usually neglected because of the low surface area of typical glassware. Though adsorption of proteins to glass is important, especially in clinical studies, one really needs to know more about the adsorption of proteins, in general, and blood proteins, in particular, to different polymeric surfaces that have significant biomedical usage.

In the past investigators did not pay much attention to the possibility that a protein can be displaced by other proteins with different characteristics. However, this line of thinking has changed, because Vroman (1980) clearly demonstrated that fibrinogen is displaced by high-molecular-weight kininogen (HMWK). Grinell and Feld (1982) also showed that the Vroman effect is a common phenomenon. These authors carried out studies on the adsorption properties of plasma fibronectin on hydrophobic and hydrophilic surfaces. At low serum concentrations (up to 0.1%) there was increased adsorption of plasma fibronectin with increasing serum concentrations. Above 1% serum there was a marked decrease in fibronectin adsorption, and at 10% very little adsorption occurred. This indicated that at high serum concentrations other serum proteins were able to compete with and possibly displace fibronectin from surface adsorption sites.

The surface-induced coagulation of blood is a critical factor in the design and application of most devices for use with the cardiovascular system. The clotting time of blood is dependent on the material with which it is in contact. It is important to study the adsorption behavior of proteins that are a major component of plasma such as albumin, γ-globulin, and fibrinogen in relation to the antithrombogenicity of polymer materials. The possible effects of a given surface on a protein (mixture) would include, among others, permanent or reversible adsorption with or without concomitant denaturation or conformational changes, preferential adsorption of specific proteins, and changes in the microenvironment of enzymes. Because the adsorption of proteins on a surface depends on both the protein and the surface, it is important to characterize both the nature of the protein sample and the surface. How homogeneous or heterogeneous is each of these? Does heterogeneity affect the adsorption and further properties; and if it does, by how much?

Adsorption of proteins at solid–liquid interfaces has been reviewed in the literature (MacRitchie, 1978; Norde, 1986; Andrade and Hlady, 1986; Lundstrom *et al.*, 1987). Not much information is presented in a concise manner concerning the heterogeneity of either the protein adsorbate or the surface, and the subsequent effects of this heterogeneity on denaturation of the adsorbed protein on the surface. In fact, the previous reviews either seem to neglect (or ignore) or perhaps treat lightly the subsequent denaturation of the adsorbed protein on the surface. Lok *et al.* (1983) correctly pointed out the factors that influence protein adsorption onto surfaces include intrinsic protein adsorption kinetics, chemical equilibrium between surface-adsorbed protein and free solution proteins, and flow of the solution past the adsorbing surface. They also hinted that the conformation of proteins in the adsorbed layer may be an important factor. Also, though much is known about the quantitative nature of protein adsorption on different surfaces, there are still apparently no rigorous mathematical theories that describe protein adsorption on different surfaces. This is not surprising because this is a complex problem.

Quiquampox and Radcliffe (1992) emphasized that the absence of techniques for determining the fine three-dimensional structure of adsorbed proteins is a major limitation. This is true because structure determines the activity of the protein; and from a thermodynamic point of view, changes in entropy arising from a modification of the structure may determine the degree of adsorption (Norde and Lyklema, 1978, 1979; Norde *et al.*, 1986). Quiquampox and Radcliffe (1992) described a nuclear magnetic resonance (NMR) method that allows conformational changes in adsorbed proteins to be followed on the basis of a measurement of the protein–solid interfacial area occupied by a single protein macromolecule.

Tan and Martic (1990) emphasized that protein adsorption is intimately associated with the intrinsic molecular structures and dynamics of protein molecules. Because of their multifunctionalities, protein molecules can exist in several conformational states. The free energy required in going from one structure to another is relatively small (that is, several kcal/gmole, which is equivalent to the dissociation of a few hydrogen bonds). Tan and Martic (1990) stressed that because of the flexible and dynamic structure of the protein molecules, conformational change of proteins on, for example, solid–liquid interfaces is expected to be a natural response of these molecules to adapt to their microenvironment at the expense of some intramolecular bonding. As a consequence, the driving force for adsorption (Norde and Lyklema, 1978) is mainly attributed to entropic gains resulting from the dehydration resulting from the hydrophobic interaction between proteins and surfaces, as well as the unfolding of the adsorbed proteins to accommodate their new environment.

Furthermore, Tan and Martic (1990) indicated that even though protein adsorption behavior may appear to be inherently related to their specific and unique structures and functions of protein molecules (Fleer and Lyklema, 1988), synthetic manipulation of surfaces (e.g., polymers) can provide opportunities to modulate such interaction. For example, polymers can be tailor-made with specific surface characteristics to render them protein resistance. Recognize that the orientation and conformational states of the adsorbed proteins can and do have significant influence on their biological function. For example, the biological activities of immobilized enzymes and antibodies, or the biological pathways for coagulation or complement activation, are expected to be affected by surface adsorption (7th International Symposium on Molecular Interactions, 1988).

It would be of interest to: (1) minimize protein adsorption by modifying polymer particles with surface-adsorbed polymeric surfactants, and (2) characterize the conformations of proteins adsorbed-desorbed from polymer particles.

Example 4.6

Present an example where proteins are adsorbed on small particles. Also, describe the conformational changes (Tan and Martic, 1990).

Solution

It is well known that adsorption of proteins and their conformational changes on adsorption on artificial surfaces are expected to play a significant role in determining the subsequent biological processes-interfacial reactions when a polymeric material is brought in contact with biological fluids. Lee *et al.* (1989) initially showed that protein adsorption on polystyrene particles can be modulated by coating the particles with a series of Pluronic copolymer surfactants. These authors noted that the adsorption of albumin or fibrinogen, as well as whole plasma, on Pluronic F108- or Tetronic 908-coating was negligibly small. This was demonstrated utilizing techniques such as photon correlation spectroscopy, sedimentation, and fluorescence spectroscopy.

Tan and Martic (1990) analyzed the conformational changes experienced by the three major plasma proteins-albumin, fibrinogen, and immunoglobulin on Pluronic F108. These authors analyzed the conformational changes of the adsorbed protein *in situ* and those desorbed from the particles. The technique of time-resolved anisotropy decay was used utilizing pyrene-labeled protein.

They measured the hydrodynamic diameters of the bare monodisperse polystyrene and the surfactant-coated polystyrene particles on addition of the proteins by a Brookhaven light-scattering instrument. The fluorescent lifetime, τ, and the rotational correlation time, Φ, for a given fluorphore in a rigid sphere is given by:

$$I(t) = I_0 \exp(-t/\tau) \tag{4.6}$$

and

$$r(t) = r_0 \exp(-t/\phi). \tag{4.7}$$

Here $I(t)$ and I_0 are the transient and initial intensities. $r(t)$ and r_0 are the anisotropies. Tan and Martic (1990) indicated that if the fluorophore has more than one kind of environment during its excited-state lifetime or if there are several modes of molecular motions, then

$$I(t) = \sum_i A_i \exp(-t/\tau_i) \tag{4.8}$$

and

$$r(t) = \sum_i B_i \exp(-t/\phi_i). \tag{4.9}$$

An SIM 4800 spectrofluorimeter was utilized to scan the fluorescence spectra. A photocounting PRA system 3000 lifetime instrument equipped with polarizers was used to make the transient measurements.

Photon correlation spectroscopy was utilized by these authors to probe the size increase of the polystyrene particles subsequent to protein adsorption. Because the technique is accurate to $\pm 2\%$, it is particularly suited to size increases for particles in the 0.01 to 0.03 μm range. The important feature of the anisotropy decay technique is its ability to detect changes in the local structure of the probe molecule that reflects the three-dimensional structural rearrangement of the host proteins. On adsorption on the polystyrene particle, the freedom of rotational correlation time, Φ, of the protein is reduced; this leads to an increase in the rotational correlation time, Φ, of the probe in the host protein. An observed decrease in Φ for an adsorbed protein molecule may be due to structural changes of this protein molecule. Also, an increase of the local segmental mobility resulting from the unfolding or denaturation of an adsorbed molecule would decrease the rotational correlation time of the probe molecule.

These authors noted minimum protein adsorption on polystyrene particles if they were precoated with Pluronic F108 or Tetronic 908. The rotational correlation times of the three plasma proteins on adsorption on F108-coated polystyrene particles were unchanged. This is consistent with the absence of protein-surface interactions. Tan and Martic (1990) indicated that evidence was available for protein unfolding subsequent to adsorption on untreated polystyrene particles. The zero-rotational correlation time of the adsorbed albumin particles while remaining adsorbed on polystyrene particles provided the evidence of unfolding. This behavior was also observed for fibrinogen and immunoglobulin after being desorbed by F108 from the polystyrene particles.

More studies like the Tan and Martic (1990) analysis are required that provides (1) direct evidence for conformational changes experienced by proteins subsequent to adsorption; and (2) methods by which protein adsorption resistant materials may be developed, by precoating or otherwise. It is well known that protein interactions at solid-liquid interfaces increase with surface hydrophobicity (Lyklema, 1984; Andrade and Hlady, 1986). An increase in the protein–surface interactions should increase the conformational changes exhibited by the protein at the solid–liquid interface.

Example 4.7

Present an analysis of the influence of surface hydrophobicity on the conformational changes of adsorbed fibrinogen (Lu and Park, 1991).

Solution

The kinetics of fibrinogen adsorption and the relative amount of adsorption on solid–liquid interfaces have been studied using Fourier transform infrared (FTIR) spectroscopy coupled with attenuated total reflection (ATR) optics (FTIR-ATR) (Jakobsen *et al.*, 1983; Kellner and Gotzinger, 1984; Pitt *et al.*, 1986; Chittur *et al.*, 1986). Lu and Park (1991) analyzed the conformational changes of adsorbed fibrinogen on germanium, poly-(hydroxylmethacrylate) and polystyrene surfaces using FTIR-ATR. The extent of protein conformational changes was analyzed by calculating the sum of the shift of the individual peaks using a weighted-peak shift method. Lu and Park (1991) noted that the degree or extent of conformational changes experienced by fibrinogen is related to the surface hydrophobicity. Iwamoto *et al.* (1985) by utilizing total internal reflection fluorescence (TIRF) also found that fibronectin experienced greater conformational changes on a more hydrophobic silica surface. Lu and Park (1991) indicated that when a protein adsorbs on a solid surface with high hydrophobicity, the hydrophobic core will become exposed to the surface due to hydrophobic interaction. This will lead to large conformational changes on more hydrophobic surfaces.

Lu and Park (1991) noted several peak shifts on protein adsorption. For all four spectra, both peaks near 1682 and 1666 cm^{-1} (structures and bends) were shifted to higher wave numbers (1688 and 1672 cm^{-1}, respectively). Turns and bends occur at 1688 and 1670 cm^{-1}. Therefore, these authors suggested that adsorption induces conformatonal changes between the *different modes* of the turns and bends. These authors also indicated that on protein adsorption, some of the α-helical structure was also broken and changed into an unordered structure. The peak contribution by this structure was of the highest intensity. A shift of this peak indicates a strong contribution to the conformational changes. They noted that the extent of the shift of this peak correlated well with the surface hydrophobicity. Because the sum of the weighted-peak shifts reflects all the information in amide I and II regions, it provides an overall picture of the conformational changes and physical insights into the protein and solid–liquid interfacial reactions.

B. Adsorption–Desorption Kinetics Theory

Adsorption as well as desorption significantly affects the dynamic behavior of molecules from solution adsorbed at a solid–liquid interface. Bychuk and O'Shaughnessy (1994) have recently analyzed the adsorption–desorption kinetics of molecules in solution from a liquid–solid interface. These types of kinetics control the dynamic behavior of the molecules at the interface. A detailed analysis of these types of kinetics is important in the adsorption–desorption of proteins (Vroman and Adams, 1986), in adsorption of synthetic polymers on colloidal particles (Liang *et al.*, 1992), and in manufacturing self-assembly monolayers and multilayers. Also, these adsorption–desorption kinetics are important in the relaxation of foams (composed of surfactant–stabilized bubbles) (Clint, 1992). These authors emphasized that in spite of the significant amount of studies done on protein adsorption, quite a few questions still remain unanswered, or are, at best, only partially answered. For example, Bychuk and O'Shaughnessy (1994) wanted to know how changes in the molecular weight of the molecules in solution affect the kinetics of adsorption–desorption. Furthermore, if surface inhomogeneities develop, how long does it take the surface to relax, or recover from these types of inhomogeneities?

Two quantities were defined to help characterize the bulk-interface system (Bychuk and O'Shaughnessy, 1994). The term *attractiveness* measures how quickly molecules in solution are adsorbed on the surface, and furthermore how far into the solution the penetration depth or adsorbing power extends from the surface. The term *retentiveness* defines the duration the surface holds onto the adsorbed molecule prior to desorption. Figure 4.1 shows the free-energy profiles for strongly and weakly adsorbing systems. These authors indicated that an attractive surface exhibits a low bulk-surface barrier or activation energy (F_a) that the adsorbing molecule in solution must surmount or "climb over." A retentive surface, on the other hand, exhibits a high barrier to desorption (F_d). By using these definitions, they indicated that strong systems not only are attractive, but also are retentive. Weak systems not only are unattractive but also exhibit low retentiveness. These authors emphasized that weak and strong systems do exhibit different dynamics and kinetics for adsorption-desorption of the molecules at the interface.

Furthermore, they indicated that weak systems are characterized by sur-

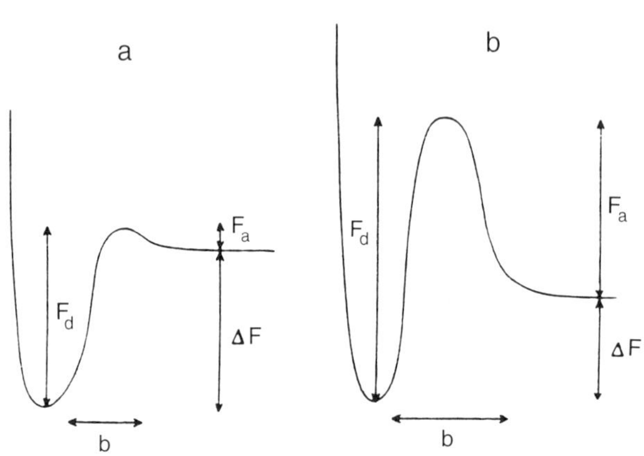

FIGURE 4.1 Free energy profiles experienced by a solute particle at the surface: (a) strongly adsorbing system, (b) weakly adsorbing system. [From Bychuk, O. V. and O'Shaughnessy, B. (1994). *J. Colloid Interface Sci., 167,* 193.]

face-bulk exchange kinetics that are rapid. The interface releases the adsorbed molecules back into the solution before diffusion can affect the bulk density profile close to the surface. In this case, the surface coverage relaxes exponentially. Strong systems significantly slow down the exchange kinetics, so much so that the bulk solution close to the surface is affected and diffusion-controlled effects are more noticeable.

The analysis of Bychuk and O'Shaughnessy (1994) is of interest for protein adsorption systems, as well as for protein purification. It would be of interest to optimize a surface-bulk system in terms of "weak" or "strong" systems that help minimize or alleviate the denaturation of proteins during their purification. Apparently, weak systems should help minimize protein denaturation, but how weak a system should be so that it effectively purifies the protein from solution while simultaneously minimizing the extent of denaturation needs to be determined. More theoretical analysis of real-life protein adsorption studies along the lines mentioned by Bychuk and O'Shaughnessy (1994) should prove extremely beneficial.

Different properties of proteins affect their adsorption at solid–liquid interfaces. Some of these include molecular size, charge, and hydrophobicity of the protein. Kondo and Higashitani (1992) indicated that the flexibility of proteins is also a crucial factor, because the magnitude of structural changes in protein molecules on adsorption on silica increases with their adiabatic compressibility, β_s (Gekko and Hasegawa, 1986), or their flexibility (Kondo *et al.*, 1991). Thus, the structural adaptability of proteins would naturally affect their adsorption behavior.

Example 4.8

Describe adsorption behavior of different proteins with wide variations in their molecular properties (Kondo and Hagashitani, 1992).

Solution

Kondo and Hagashitani (1992) analyzed the adsorption isotherms of ribonuclease A (RNase A), cytochrome c, lysozyme, α-lactalbumin, and BSA on colloidal particles of polystyrene, styrene-2-hydroxyethyl methacrylate, and silica as a function of pH and ionic strength. Care was taken to select those proteins with a wide range of molecular flexibility (β_s ranges from 0 to 11). These proteins also covered a wide range of hydrophobicities and isoelectric points. These authors noted that the adsorption behavior of the proteins could be elucidated by the protein-surface interactions (electrostatic and hydrophobic interactions), and the lateral interaction between the adsorbed molecules.

They noted that all the proteins utilized in their study exhibited a high affinity for the polystyrene particles above their isoelectric point, even though both the proteins and the particles were negatively charged. Thus, the hydrophobic interactions for the protein-surface interaction are stronger than the electrostatic repulsion. In contrast, the affinities between the proteins and the hydrophilic surface decreased significantly with increasing pH irrespective of the proteins. In this case, electrostatic interactions between the protein-surface dominate. According to their calculations, Kondo and Hagashitani (1992) indicated that the maximum plateau adsorption ranged between the adsorbed amounts of closely packed side-on and end-on protein adsorption at the interface.

Furthermore, these authors noted that the magnitude of structural changes for BSA on adsorption increased with decreasing pH. This is in contrast to the hypothesis that the maximum adsorption at the isoelectric point is a consequence of a minimum of structural changes. They also noted that based on their results the difference between small and large proteins was not attributable to the structural changes. Kondo and Hagashitani (1994) proposed that lateral interaction between the larger protein molecules is stronger because of the thicker adsorption layers at the solid–liquid interfaces. This reduces the plateau adsorption to a greater extent at pHs distant from the isoelectric points. Thus, the larger protein molecules show maximum adsorption around the isoelectric points. This trend is less pronounced for the smaller proteins due to the smaller effect of lateral interactions. The analysis by these authors emphasizes the added importance of lateral interactions between proteins at the solid–liquid interface. One needs to consider a significant number of factors while analyzing protein interactions at solid–liquid interfaces.

We next analyze the driving forces involved in protein adsorption at liquid–solid interfaces. Thereafter, we will examine the structural perturbations involved during the adsorption of two similar-sized globular proteins at solid–liquid interfaces. Then we will analyze another example where these same enzymes are adsorbed on a different surface. Finally, we will consider some of the more recent kinetic models for protein adsorption at solid–liquid interfaces.

Example 4.9

Briefly describe the driving forces involved in the adsorption of the enzyme savinase at solid–liquid interfaces. Also, determine the major driving forces (Duinhoven *et al.*, 1995).

Solution

Duinhoven *et al.* (1995) analyzed the driving forces involved in the adsorption of savinase at solid–liquid interfaces. These authors indicated that the formation of an enzyme-substrate complex in heterogeneous reactions may be considered to be an adsorption process, and one needs to understand the forces involved to optimize the reaction. Furthermore, because the enzyme is a protein, enzyme adsorption is anticipated to be similar to protein adsorption. Also, they indicated that the following four sets of interactions are involved in the adsorption of proteins:

1. Protein-surface interactions—these could be of an electrostatic or van der Waals nature. The Coulombic interaction due to the *net* charge of the surface and the protein is the major interaction.
2. Dehydration of interfaces on the exterior protein surface and on the solid surface-dehydration of hydrophobic interfaces promotes protein adsorption whereas dehydration of hydrophilic interfaces opposes it.
3. Structural changes in the protein molecule on adsorption—these authors indicated that the densely packed protein structure arises because of the

hydrophobic interactions in the interior of the molecule that more than offset the intramolecular electrostatic repulsive interactions along with the loss of entropy on folding. A hydrophobic surface promotes protein unfolding. Also, electrostatic repulsion between the charges on the adsorbed protein promotes unfolding. Furthermore, as expected, protein unfolding leads to higher surface area contact of the protein.

4. Lateral interactions on the surface due to accumulation of proteins at the interface-monolayer coverage is resisted due to electrostatic repulsion between similar charged protein molecules on the surface.

These authors analyzed the relative importance of these driving forces during the adsorption of savinase from the alkalophilic bacterium *Bacillus lentus* on polystyrene and on glass. They noted that electrostatic interactions are the major driving forces for savinase adsorption at the solid-water interface. There is some small contribution due to adsorption by dehydration at the hydrophobic interfaces and lateral interactions between the adsorbed savinase molecules. The authors indicated that the enzyme adsorbs in its native state, that is, without unfolding.

Haynes and Norde (1995) analyzed the adsorption of hen egg-white lysozyme and bovine milk α-lactalbumin (LAC) to negatively charged polystyrene (PS^-) latex and variably charged hematite (α-Fe_2O_3). These authors used the preceding proteins as model compounds to analyze the forces and subprocesses that are involved in the adsorption of globular proteins to non-porous solid surfaces. Because protein adsorption is a complex process, they were careful to use a well-characterized protein or proteins, a well-characterized sorbent surface, and an aqueous solution containing nonbuffering ions. Furthermore, the authors believed that the conformations of adsorbed proteins on interfaces are different from their native structure in solution. In fact, they emphasized this is one aspect of protein adsorption that is poorly understood.

These authors indicated that subprocesses leading to an increase in entropy provide a major driving force for protein adsorption. Furthermore, significant entropic contributions to adsorption arise due to sorbent and protein surface dehydration. Both lysozyme and α-lactalbumin exhibited a significant amount of denaturation on the hydrophobic polystyrene surface. On the hydrophilic α-Fe_2O_3 surface α-lactalbumin is almost completely denatured, whereas lysozyme loses only a fraction of its activity. This is largely due to the high structural stability of lysozyme in its native state as indicated by microdifferential scanning microcalorimetry (micro(DSC)). Furthermore, even on adsorption, lysozyme maintains a high level of internal cohesion.

The analysis of Haynes and Norde (1995) is particularly useful because these authors took care to select their protein-sorbent system, and were able to isolate (to a large extent) the reasons for structural rearrangements exhibited by the proteins on adsorption at the solid–liquid interface. More such studies are needed that provide the physical insights required to better understand protein adsorption at interfaces.

Example 4.10

Briefly describe the adsorption of the fungal lipase lipolase at solid–liquid interfaces (Duinhoven *et al.,* 1995).

Solution

Duinhoven *et al.* (1995) analyzed the adsorption of the fungal lipase lipolase at solid-liquid interfaces. Lipolase is the extracellular lipase from the thermophilic fungus *Humicola lanuginosa* S-38 (Liu *et al.*, 1973), expressed in *Aspergillus oryzae.* Aaslyng *et al.* (1991) indicated that lipolase is utilized in detergents to help remove fatty soil from clothes. Duinhoven *et al.* (1995) indicated that lipases (glycerol ester hydrolases, EC 3.1.1.3) catalyze the hydroloysis and the formation of ester bonds, for example, in triacyl glycerol molecules. There is a significant increase of this activity at an interface (Sarda and Desnuelle, 1958; Brockman *et al.*, 1973). At the interface there is a high concentration of the substrate (Brockman *et al.*, 1973), and a better orientation and conformational change in the enzyme (Entressangles and Desnuelle, 1974).

Duinhoven *et al.* (1995) indicated that the functional viability of lipase is strongly dependent on its adsorption on the water-surface interface. These authors also pointed out that the (surface active) reaction products formed at the interface may also influence subsequent binding. This could lead to a negative feedback mechanism, where the reaction products formed are deleterious to subsequent binding. The adsorption of lipase lipolase on polystyrene lattices and on glass beads was analyzed by these authors.

These authors further indicated that the adsorption of lipase on solid-liquid interfaces is determined by a delicate balance between surface dehydration and electrostatic interactions between the solid-enzyme and enzyme-enzyme. This lateral enzyme-enzyme repulsion reaction influences the plateau value in the adsorption process. They emphasized that their analysis helps provide insights into the actual reaction taking place at the natural substrate, triacylglycerol-water (liquid-liquid) interface. The relative importance of the hydrophobic interaction will change as the lipolytic hydrolysis proceeds. This will result in a change in the surface characteristics. These authors suggested that the delicate balance of forces that determine lipolase adsorption on negatively charged surfaces delineates the fact that lipolase adsorption is sensitive to the presence of other surface-active components (Aaslyng *et al.*, 1991). Finally, Duinhoven *et al.* (1995) mentioned that lipolase adsorbs as a hard protein; in other words, it exhibits hardly any unfolding on adsorption.

Example 4.11

Briefly compare the adsorption of hen lysozyme (LS2) and milk LAC on colloidal AgI (Galisteo and Norde, 1995).

Solution

Galisteo and Norde (1995) indicated that protein adsorption is a complex process, and a successful study requires the use of well-defined systems. These authors analyzed the adsorption of LS2 and bovine milk LAC onto an inorganic colloidal silver iodide surface. They emphasized that their structure is simple, they have common aspects in their structure, and they exhibit similarity in shape and size. Some aspects where they differ are in isoelectric point, structural stability, and number of charged groups. Furthermore, they point out that silver iodide has been extensively analyzed as a model to examine and test electrical, double-layer, and colloid stability theories (Bijsterbosch and Lyklema, 1978).

Both LS2 and LAC exhibit high-affinity adsorption on the AgI surface (Galisteo and Norde, 1995). LAC is adsorbed in a more irreversible way compared with LS2, because LAC does not desorb after plateau adsorption, whereas about 15% of the LS2 can desorb after plateau adsorption. For both proteins analyzed, their results indicate that maximum amounts are apparently controlled by electrostatic interaction between the adsorbed protein molecules. Both proteins exhibit a stronger dependence on protein charge than on sorbent surface charge. Finally, microcalorimetry measurements indicate that the mode of adsorption for LS2 is dependent on surface coverage, whereas this is not true for LAC.

Cornelius *et al.* (1992) indicated that most experimental methods of studying adsorption involve contacting the protein solution with the solid-liquid interfaces for a predetermined length of time. After adsorption, the protein is then rinsed-off the solid-liquid interface, and the amount of protein is determined often by radiolabeling. These authors emphasized that this *ex situ* method of protein determination has distinct advantages: time resolution is low and the time required to rinse the protein may result in significant losses of protein. They proposed an *in situ* experimental approach that determines the amount of protein adsorbed. Their method is based on the serum replacement method (Ahmed *et al.*, 1987). In essence, the serum replacement method is a solution depletion method wherein the decrease in protein concentration in a solution due to adsorption is measured on-line continuously.

Cornelius *et al.* (1992) included the effects of diffusion and adsorption during the initial stages of fibrinogen adsorption by coupling the diffusion equation to the Langmuir adsorption kinetic terms. These authors noted that the rate-determining steps in protein adsorption were due to the protein supply and the kinetics of surface binding. In a well-stirred system, as expected, the diffusion effects were negligible. They noted that fibrinogen adsorption in their system exhibited the classical Vroman effect. For example, adsorption phenomena occurred at times that depended on the concentration of plasma fed to the cell and decreased with increasing concentration. The adsorption of immunoglobulin (IgG) from plasma also exhibited the Vroman effect. The Vroman effect is basically the manifestation of a maximum in adsorbed protein (e.g., fibrinogen) versus bulk concentration.

Example 4.12

Describe by appropriate modeling: (1) the principle of the replacement method, and (2) the simulation of adsorption in a well-mixed particle suspension (Cornelius *et al.*, 1992).

Solution

Case One. Cornelius *et al.* (1992) indicated that the serum replacement method is a kinetic solution depletion method in which the decrease in concentration of a protein solution due to adsorption is measured continuously (on-line). The protein solution is fed to a cell filled with spherical beads in buffered solution. A simple mass balance permits the determination of the amount of protein adsorbed.

These authors obtained the following expression for the mass of protein adsorbed per unit area as

$$\Gamma(t) = \frac{C_{p0}}{S_A}\left[Ft - \frac{F}{\alpha_0}\int_0^t \alpha dt - \frac{V_c\alpha}{\alpha_0} \right]. \tag{4.10}$$

Thus, $\Gamma(t)$ may be obtained *in situ.*

Case Two. The amount of protein adsorbed may be obtained from a classical partial differential diffusion equation in, for example, spherical coordinates with appropriate boundary conditions. Rather than describe the boundary conditions mathematically, they are written down to describe the physics of the process. The partial differential equation for adsorption on spherical beads is given by

$$\frac{\partial C_p(r,t)}{dt} = \frac{D}{r^2}\frac{\partial}{\partial r} r^2 \frac{\partial C_p(r,t)}{\partial r}. \tag{4.11}$$

D, the diffusivity, is assumed constant; and r is the radial position with $r = 0$ at the center of the bead.

The three boundary conditions may be formulated from (Cornelius *et al.*, 1992):

1. The initial protein concentration of zero throughout the flow cell, including the unstirred region of thickness outside the beads of radius, R
2. An overall mass balance throughout the well-stirred free volume of liquid outside the unstirred layer
3. Changes in protein concentration at the solution-glass-bead interface.

These authors utilized their model to simulate the effects of diffusion and adsorption kinetics on the initial stages of fibrinogen adsorption. They noted that the principal adsorption rate limitations were due to protein supply and the kinetics of surface binding. In a well-stirred system, the diffusion effects were negligible.

Two comments are perhaps appropriate at this time:

1. More emphasis needs to be made on the boundary conditions to more adequately describe or consider the various possibilities that may be occurring at the solid-liquid interface. For example, Giroux and Cooper (1990) chose five kinetic rate expressions to be used as boundary conditions. In them, the flux is related to the rate expression by different forms. This accounts nicely for the different possibilities.

2. Conformational changes on the interface, and aggregation on the interface or in solution should be borne in mind, especially at higher protein concentrations, as suggested by Brynda *et al.* (1990).

V. CONCLUSIONS

A better understanding of protein and other valued product adsorption at different types of interfaces (gas–liquid, liquid–liquid, liquid–solid) is essential to provide better physical insights into, and the control of, the subsequent reactions occurring at the interface. Proteins and other valued products have a (thermodynamic) tendency to diffuse toward and react at the interface. These compounds are, in general, delicate and complex structures, and one may reasonably expect significant as well as subtle rearrangements in the compounds of interest as they adsorb and subsequently react at the interface. The adsorbed molecule may continuously change its configuration–conformation at the interface. This alters the interface properties that thereby affect the subsequent adsorption and reaction of further molecules from solution. All this indicates the importance of the heterogeneity of the adsorbed protein or other valued product at an interface.

Most current models for adsorption at an interface include the basic steps for diffusion and reaction-adsorption at an interface. Though models, in general, qualitatively indicate the presence of heterogeneity in the adsorbed state at the interface, these models do not provide a measure of heterogeneity of the adsorbed states at the interface. This is not an easy task, because precise techniques to estimate or evaluate the different adsorbed states at an interface are not available. Most measurement techniques provide quantitative estimates of the adsorbed states (of the protein or other valued products) without providing some estimate of the qualitative nature or quality of the adsorbed states. In a situation, one would like to know not only the amount adsorbed but also the nature of the adsorbed states.

In the future as newer techniques develop or the present-day techniques are further refined, it is estimated that the qualitative and the quantitative nature of the adsorbed states will be made available. This will definitely assist in the control of the adsorption-reaction of different compounds occurring at different types of interfaces. Once these techniques become available, reliable measures of heterogeneity of adsorbed states will need to be developed, incorporated in appropriate models, and tested or validated by applying the models to experimental data for protein or other valued product adsorption at different types of interfaces.

These types of analyses including refining of models and experimental data collection should lead to the development of more realistic and appropriate models for protein adsorption and subsequent reaction at different types of interfaces. Eventually, a reasonable framework should become available that permits different people working with compounds and interfaces to assist in comparing their own findings with the general data available, and the general principles generated in building up this framework. The development of such a framework is essential and should be done quickly because reactions and adsorptions of compounds at interfaces occur in so many different and critical areas such as biomedical, immunologic, biotechnological, environmental, and other applications.

REFERENCES

Aaslyng, D., Gormsen, E., and Malmos, H. (1991). *J. Chem. Technol. Biotechnol., 50,* 321.

Ahmed, S. M., El Aasser, M. S., Pauli, G. H., Pohlein, G. W., and Vanderhoff, J. W. (1980). *J. Colloid Interface Sci., 73,* 388.

Andrade, J. D. (1985). Ed., Interfacial Aspects of Biomedical Polymers, Vol. 2, Plenum: New York.

Aptel, J. D., Carroy, A., Dejardin, P., Pefferkorn, E., Schaaf, P., Schmitt, R., Varoqui, R., and Voegel, J. C. (1987). In "Proteins At Interfaces: Physicochemical And Biochemical Studies," Brash, J. L. and Horbett, T. A., Eds., ACS Symposium Series, No. 343, American Chemical Society: Washington, DC, p 222.

Arnebrant, T. and Nylander, T. (1985) *J. Dispersion Sci. Technol., 6,* 209.

Arnebrant, T. and Ericsson, B. (1992). *J. Colloid Interface Sci., 150*(2), 428.

Atkinson, B. and Mavituna, F. (1983). In Biochemical Engineering and Biotechnology Handbook, Nature Press: New York, Chapter 9.

Bausch, T. E. (1989). Diploma Thesis, Technical University, Munich, Germany.

Bausch, T. E., Plucinski, P. K., and Nitsch, W. (1992). *J. Colloid Interface Sci., 150,* 226.

Bausch, T. (1989) Diploma thesis, Technical University, Munich.

Beissenger, R. L. and Leonard, E. F. (1982). *J. Colloid Interface Sci., 85,* 521.

Benzonna, G. and Desneulle, P. (1965). *Biochim. Biophys. Acta, 105,* 123.

Bhattacharya, P., Ghosal, S. K., and Sen, K. (1991). *Sep. Sci. Technol., 26,* 1279.

Bijsterbosch, B. H. and Lyklema, J. (1978). *Adv. Colloid Interface Sci., 9,* 147.

Blank, J. L. (1985). *Macromol. Chem. Suppl., 9,* 69.

Brash, J. L. (1987). In: Proteins at Interfaces. Physicochemical and Biochemical Studies, Brash, J. L. and Horbett, T. A., Eds., ACS Symposium Series 343, American Chemical Society: Washington, DC, Chapter 30.

Brockman, H. L., Law, J. H., and Kzedy, F. J. (1973). *J. Biol. Chem., 248,* 4965.

Brynda, E., Hlady, V., and Andrade, J. D. (1990). *J. Colloid Interface Sci., 139*(2), 374.

Bychuk, O. V. and O'Shaughnessy, B. (1994). *J. Colloid Interface Sci., 167,* 193.

Charm, S. E. and Wong, B. L. (1970). *Science, 170,* 466.

Chen, J., Dong, D. E., and Andrade, J. D. (1982). *J. Colloid Interface Sci., 89, 577.*

Chittur, K. K., Fink, D. J., Leininger, R. I., and Hutson, T. B. (1986). *J. Colloid Interface Sci., 11,* 419.

Clint, J. H. (1992). *Surfactant Aggregation,* Blackie: Glasgow.

Cornelius, R. M., Wojciechowski, P. W., and Brash, J. L. (1992). *J. Colloid Interface Sci., 150*(1), 121.

Davies, R. J., Goodwin, G. C., Lyle, I. G., and Jones, M. N. (1983). *Colloids Surf., 8,* 29.

De Feijter, J. A. and Benjamins, J. (1987). In "Food Emulsions And Foams," Dickinson, E., Ed., Royal Society of Chemistry: London.

Dekker, M., Van't Riet, K., Wignans, J. M. G. M., and Baltussen, J. W. (1987). In Proceedings International Congress on Membranes, Tokyo, Japan, p. 793.

Dekker, M., Hilhorst, R., and Laane, C. (1989). *Anal. Chem., 178,* 217.

Dekker, M., Van't Riet, K., Bijsterbosch, B. H., Fijneman, P., and Hilhorst, R. (1990). *Chem. Engg. Sci., 45,* 2949.

Douillard, R. and Lefebvre, J. (1990). *J. Colloid Interface Sci., 139*(2), 488.

Duinhoven, S., Poort, R., van der Voet, G., Agterof, W. G. M., Norde, W., and Lyklema, J. (1995). *J. Colloid Interface Sci., 170,* 340.

Duinhoven, S., Poort, R., van der Voet, G., Agterof, W. G. M., Norde, W., and Lyklema, J. (1995). *J. Colloid Interface Sci., 170,* 351.

Dungan, S. R., Bausch, T., Hatton, T. A., Plucinski, P., and Nitsch, W. (1991). *J. Colloid Interface Sci., 145*(1), 33.

Dunnill, P. (1983). *Process Biochem., 18*(10), 9.

Elbaum, D., Harrington, J., Roth, Jr., E. F., and Nugel, R. L. (1976). *Biochim. Biophys. Acta, 427,* 57.

Elgersma, A. V., Zsom, R. L. J., Norde, W., and Lyklema, J. (1990). *J. Colloid Interface Sci., 138,* 145.

Entressangles, B. and Desnuelle, P. (1974). *Biochim. Biophys. Acta, 341,* 437.

Fleer, G. J. and Lyklema, J. (1978). *J. Colloid Interface Sci., 66,* 257.

Fletcher, M., Latham, M. J., Lynch, J. M., Rutter, P. R. (1980). In Microbial Adhesion to Surfaces, Berkeley, R. C. W., Lynch, J. M., Melling, J., Rutter, P. R., and Vincent, B., Eds., Chapter 3, Ellis Horwood: Chichester, UK.

Galisteo, F. and Norde, W. (1995). *J. Colloid Interface Sci., 172,* 502.

Gargouri, Y., Pieroni, G., Riviere, C., Sarada, L., and Verger, R. (1986). *Biochem., 25,* 1733.
Gekko, K. and Hasegawa, H. (1986). *Biochemistry,* 6563.
Giroux, T. A. and Cooper, S. L. (1990). *J. Colloid Interface Sci., 139*(2), 374.
Goklen, K. and Hatton, T. A. (1985). *Biotechnol. Prog., 1,* 69.
Goklen, K. and Hatton, T. A. (1987). *Sep. Sci. Technol., 22,* 831.
Graham, D. E. (1976). Structure of Adsorbed Protein Films. Ph.D. thesis, CNAA, London.
Graham, D. E. and Phillips, M. C. (1979). *J. Colloid Interface Sci., 70,* 427.
Graves, H. C. B. (1988). *J. Immunol. Methods, 111,* 157.
Grieves, R. B. (1982). Adsorptive Bubble Separation Methods, Reprints from Treatise on Analytical Chemistry, Part 1, Vol. 5, Wiley, New York, p. 372.
Grinell, F. and Feld, M. K. (1982). *J. Biol. Chem., 257,* 4888.
Guzmann, R. Z., Carbonell, R. G., and Kilpatrick, P. K. (1986). *J. Colloid Interface Sci., 70,* 427.
Halling, P. J. (1981). In CRC Critical Reviews In Food Science and Nutrition, *155.*
Hariri, M. H., Ely, J. F., and Mansoori, G. A. (1989). *Biotechnology, 7,* 686.
Harris, J. I. and Perham, R. N. (1965). *J. Mol. Biol., 13,* 876.
Hatton, T. A. (1989). In Surfactant-Based Separation Processes," Scamehorn, J. F. and Harwell, J. H., Eds., Dekker: New York, p. 55.
Haynes, C. A. and Norde, W. (1995). *J. Colloid Interface Sci., 169,* 313.
Hughes, P. and Lowe, C. R. (1988). *Enzyme Microb. Technol., 10,* 115.
Hunter, J. R., Kilpatrick, P. K., and Carbonell, R. G. (1990) *J. Colloid Interface Sci., 137*(2), 462.
Jakobsen, R. J., Brown, L. L., Winter, S., and Gendreau, R. M. (1983). *J. Biomed. Mater. Res., 16,* 199.
Joly, M. (1965). A Physicochemical Approach to the Denaturation Of Proteins, Academic: London, p. 15.
Kellner, R. and Gotzinger, G. (1984). *Makrochimica Acta, 2,* 61.
Khaiat, A. and Miller, I. R. (1966). *Biochim. Biophys. Acta, 183,* 309.
Kondo, A. and Higashitani, K. (1992). *J. Colloid Interface Sicence, 150,* 344.
Kondo, A. Oku, S., and Higashitani, K. (1991). *J. Colloid Interface Sci., 143*(1), 214.
Kula, M. R. (1987). In *Protein Purification: Micro to Macro,* Burgess, R., Ed., Proceedings of Cetus-UCLA Symposium, Frisco, CO, Alan R. Liss, Inc.: New York, p. 99.
Le Compte, M. F., Rubinstein, I., and Miller, I. R. (1983). *J. Colloid Interface Sci., 91,* 12.
Lee, H. J., Kopeckova, P., Zhang, J., Kopecek, J., and Andrade, J. D. (1988). *Polym. Mater. Sci. Eng., 54,* 381.
Lee, Y. K. and Choo, C. L. (1989). *Biotechnol. Bioengg., 33,* 183.
Lesins, V. and Rickenstein, E. (1988). *Colloid Polym. Sci., 266,* 1187.
Liang, W., Tadros, T. F., and Luckham, P. F. (1992). *J. Colloid Interface Sci., 153,* 131.
Liu, W. H., Beppu, T., and Arima, K. (1973). *Agric. Biol. Chem., 37,* 2493.
Lok, B. K., Cheng, Y. L., and Robertson, C. R. (1983). *J. Colloid Interface Sci., 91,* 104.
Lu, D. R. and Park, K. (1991). *J. Colloid Interface Sci., 144,* 271.
Luisi, P. L. and Laane, C. (1986). *TIBTECH,* June.
Lundstrom, J., Ivarsson, B., Johnson, B., and Elwing, H. (1987). In: *Polymers Surfaces and Interfaces,* Feast, W. D. and Munro, H. S., Eds., J. Wiley & Sons, New York, p. 201.
Lyklema, J. (1984). *Colloids Surf., 22,* 273.
MacDonnell, L. R., Hanson, H. L., Silva, R. B., Lineweaver, H., and Feeney, R. E. (1950). *Food Ind., 22,* 273.
MacRae, A. R. (1984). In *Biotechnology of the Oils and Fats Industry,* Ratledge, C., Dawson, P., and Rattray, J., Eds., American Chemical Society: Champaign, IL, p. 340.
Macritchie, F. and Alexander, A. E. (1963). *J. Colloid Sci., 18,* 453.
MacRitchie, F. and Owen, N. F. (1969). *J. Colloid Interface Sci., 29,* 66.
MacRitchie, F. (1978). *Adv. Protein Chem., 32,* 283.
MacRitchie, F. (1986) *Adv. Colloid Interface Sci., 25,* 341.
Mansoori, G. A. and Ely, F. G. (1987). *Partitioning of Monodisperse/Polydisperse Polymers and Biological Macromolecules in Aqueous Two-Phase Systems,* National Bureau Of Standards, Technical Note.
Mohan, S. B., Smith, L., Kemp, W., and Lyddiatt, A. (1992). *J. Inst. Brew., 98,* 187.
Mohan, S. B. and Lyddiatt, A. (1994). *Biotechnol. Bioengg., 44,* 1261.
Monterro, G. A., Kirchner, T. F., and Tanner, R. D. (1993). *Appl. Biochem. Biotechnol., 39/40,* 467.

Morrissey, B. W. and Stromberg, R. R. (1974). *J. Colloid Interface Sci.*, *46*, 152.
Muramatsu, M. (1973). *Surf. Colloid Sci.*, *6*, 101.
Narendarnathan, T. J. and Dunnill, P. (1982). *Biotechnol. Bioeng.*, *24*, 2103.
Norde, W. (1986). *Adv. Colloid Interface Sci.*, *25*, 267.
Norde, W. and Lyklema, J. (1978) *J. Colloid Interface Sci.*, *66*, 295.
Norde, W. and Lyklema, J. (1979). *J. Colloid Interface Sci.*, *71*(2), 350.
Norde, W., MacRitchie, F., Nowicka, G., and Lyklema, J. (1986). *J. Colloid Interface Sci.*, *112*, 447.
Pathak, S. B., Sudha, S., Sawant, S. B., and Joshi, J. B. (1991). *Chem. Eng. J.*, *46*, 831.
Paulsson, M. and Dejmek, P. (1992). *J. Colloid Interface Sci.*, *150*(2), 394.
Pitt, W. G., Park, K., and Cooper, S. L. (1986). *J. Colloid Interface Sci.*, *111*, 343.
Plucinski, P. and Nitsch, W. (1989) *Ber. Bunsenges. Phys. Chem.*, *93*, 994.
Quiquampox, H. and Ratcliffe, R. G. (1992). *J. Colloid Interface Sci.*, *148*, 343.
Rahaman, R. S., Chee, J. Y., Cabral, J. M. S., and Hatton, T. A. (1988). *Biotechnol. Progr.*, *4*, 218.
Reese, E. T. and Robbins, F. M. (1981) *J. Colloid Interface Sci.*, *83*(2), 393.
Rosselin, G., Assan, R., Yallow, R. S., and Berson, S. A. (1986). *Nature* (London), *212*, 355.
Sadana, A. and Malhotra, A. (1987a). *Biotechnol. Bioeng.*, *30*, 1041.
Sadana, A. and Malhotra, A. (1987b). *Biotechnol. Bioeng.*, *38*, 108.
Sarda, L. and Desnuelle, P. (1958). *Biochim. Biophys. Acta*, *30*, 513.
Scopes, R.K. (1982). *Protein Purification. Principles and Practice*, Springer-Verlag: New York.
7th International Symposium on Affinity Chromatography and Interfacial Macromolecular Interactions (1988). *Makromol. Chem. Macromol. Symp.*
Smith, J. A., Hurrell, J. G. R., and Leach, S. J. (1978). *Anal. Biochem.*, *87*, 299.
Swartz, J. R. (1985). In Comprehensive Biotechnology, Vol. 2, Cooney, C. L., and Humphrey, A. E., Eds., Pergamon: New York, p 299.
Tan, J. S. and Martic, P. A. (1990). *J. Colloid Interface Sci.*, *136*, 415.
Tanford, C. (1970). *Adv. Protein Chem.*, *24*, 1.
Tirrell, M. and Middleman, S. (1975). *Biotechnol. Bioengg.*, *17*, 299.
Tornberg, E. (1977). *J. Colloid Interface Sci.*, *60*, 50.
Tornberg, E. (1978). *J. Colloid Interface Sci.*, *64*, 391.
Vellisariou, M. (1992). Foam Fractionation of Biopolymers: Study of Protein Behavior in Analytical and Preparative Systems, Ph.D. Thesis, University of Birmingham, UK.
Verger, R. (1989). In *Methods in Enzymology*, Vol. 64, Purich, D. L., Ed., Academic: New York.
Viesturs, V. E., Kristapsons, M. Z., and Levitans, E. S. (1982). *Adv. Biochem. Eng.*, *21*, 169.
Virkar, P. D., Narendarnathan, T. J., Hoare, M., and Dunnill, P. (1981). *Biotechnol. Bioeng.*, *23*, 425.
Virkar, P. D., Hoare, M., Chan, M. Y. Y., Dunnill, P., Humphrey, A. E., and Lilly, M. D. (1979). *Enzyme Fermentation Technology*, J. Wiley and Sons: New York, Chapter 12.
Vroman, L. and Adams, A. (1986). *J. Colloid Interface Sci.*, *111*, 391.
Vroman, L., Adams, A. L., Fisher, G. C., and Munoz, P. C. (1980). *Blood*, *55*, 156.
Ward, A. F. H. and Tordai, L. (1952). *Recueil*, *71*, 572.
Ward, A, J. I. and Regan, L. H. (1980). *J. Colloid Interface Sci.*, *78*, 389.
Wilson, M. L. and Dahlquist, F. W. (1985). *Biochem.*, *24*, 1420.
Woll, J. M., Dillon, A. S., Rahaman, R. S., and Hatton, T. A. (1987). In Protein Purification: Micro to Macro, Burgess, R., Ed., Proceedings Cetus-UCLA Symposium, Frisco, Alan R. Liss, Inc.: New York, p 117.
Yagi, H. and Yoshida, F. (1974). *J. Ferment. Technol.*, *52*, 905.
Zucker, M. B. and Broman, N. (1969). *Proc. Soc. Exp. Biol. Med.*, *131*, 318.

5

PROTEIN INACTIVATIONS DURING CHROMATOGRAPHIC METHODS OF SEPARATION

I. INTRODUCTION

The application of high-performance liquid chromatography (HPLC) to the separation of proteins has increased considerably over the last 10 to 15 years leading to significant benefits to biochemists and others (Ackland *et al.*, 1991). Wilson (1989) indicates that stationary phases useful for protein separations have been combined with HPLC instrumentation leading to increased sensitivity, reproducibility, and a considerable time saving for analysis. It is imperative to develop appropriate industrial-scale methods for the separation and purification of proteins-enzymes and other biological methods of interest from plant, animal, or microbiological sources. There is tremendous interest in the separation of biological molecules by chromatographic techniques for medical diagnostics and pharmaceuticals (Parikh and Cuatrecasas, 1985; Clonis, 1987; McCormick, 1988). Dunnill (1983) feels that chromatographic techniques exhibit the potential and are close to being applied usefully on an industrial scale. A major problem in the effective utilization of chromatographic techniques is the extent of protein denaturation encountered on different columns. Though individuals have mentioned this aspect in their studies, there has been, to the best of this author's knowledge, no extensive and comprehensive analysis of causes and mechanisms of the denaturation of proteins and other biological macromolecules of interest on different columns.

Regnier (1987) indicated that interfacial phenomena and surface–surface interactions play a key role in the organization and control of biological systems. The highly specific interactions apparently occur when there is sufficient geometric complementarity between a reasonable number of groups on the surface of two macromolecules to initiate "intermolecular docking." This author further emphasized that there is considerable similarity between the variables that direct intermolecular docking in biological macromolecules and those that determine the chromatographic behavior of proteins. Chromatographic retention is largely determined by a relatively small number of amino acids located in a chromatographic contact region on the polypeptide surface. Structural changes that alter the chromatographic contact region will alter the chromatographic properties. Also, proteins generally have a preferred conformational state under physiological conditions. More often than not, nonphysiological conditions may be used in the chromatographic separation and purification of proteins. Thus, the protein of interest may be denatured and require refolding to regain the native structure. Refolding can be achieved relatively easily for small proteins but this may be a formidable task with large proteins.

Basically, the column is competing in the contact region with the rest of the protein structure for functional groups at the interface. When competing hydrophobic and coulombic forces on the column are greater than those maintaining protein structure, denaturation can occur. Mobile phases may also initiate denaturation. Also, strongly retained solutes often require mobile phase additives that disrupt intramolecular hydrophobic and coulombic forces beyond the chromatographic contact region. Regnier (1987) emphasized that protein denaturation can increase sample complexity to the point that it is impossible to differentiate between sample components and artifacts. Besides, chromatographic retention increases with protein denaturation in hydrophobic interaction chromatography (HIC) and reversed-phase chromatography (RPC). Thus, there is more than just a single reason to minimize protein denaturation on chromatographic columns.

Ion-exchange chromatography continues to play a major role in the recovery and purification of biomolecules in the biotechnology industry. James and Do (1991) analyzed the equilibria of biomolecules on ion-exchange adsorbents. They indicated that considerable work is being done to characterize the fundamental adsorption processes occurring in bioseparation systems. Optimization of chromatographic operations such as adsorption, selective elution, and regeneration become increasingly important as systems are scaled up.

Huang and Horvath (1987) utilized the Langmuir approach to model the equilibrium isotherms of proteins on cation exchangers. Later on, Huang *et al.* (1990) noted that the Langmuir isotherm did not adequately represent the adsorption of some proteins due to the initial very steep slope of the adsorption isotherm. These isotherms are better approximated with a Langmuir equation that incorporated two types of noncooperative independent adsorption sites. The Freundlich isotherm that is derived from the Langmuir isotherm, and uses a distribution of adsorption energies for the surface sites, has been used to model adsorption data.

James and Do (1991) utilized an extended Langmuir-Freundlich isotherm equation to fit the equilibrium adsorption isotherms of bovine serum albumin (BSA) on a (diethylamino) ethyl (DEAE)-Sepharose Fast Flow ion exchanger. These authors noted that the model correctly approximates the adsorption of various proteins and amino acids at different salt concentrations using a single set of parameters. The adsorption characteristics of the proteins on the columns would significantly affect the conformational dynamics of the proteins on the surfaces, thereby affecting the denaturing tendencies of the proteins.

It is of considerable interest to collate and understand the different mechanisms of protein denaturation on different columns. Such an analysis is timely and appropriate. Various chromatographic techniques are now available in the literature. At the outset we recognize that this is a wide area, and perhaps justice may not be possible for all the possible aspects. Nevertheless, we have made a conscious attempt to include most techniques and have attempted to treat them fairly, at least those where an extent of protein conformational changes and inactivation during separation was available. The examples presented should serve just as model examples suggesting the extent of information available in the literature.

II. CHROMATOGRAPHIC TECHNIQUES

Ion exchange, gel permeation, affinity, and hydrophobic interaction chromatography are among the popular methods for the separation of proteins and other bioploymers. HPLC of proteins–enzymes or bioploymers, in general, is an area of rapid development. The availability of rigid chromatographic supports that allow high flow rates has enabled the separation of proteins by HPLC based on the same criteria as those used in conventional liquid chromatography. These criteria are differences in isoelectric point, in net charge at different pH values, or in apparent molecular weight. In the past, separations of proteins-enzymes were typically possible within hours; now these same separations are possible within minutes. Chromatographic techniques have also been applied for the separation of other biological macromolecules of interest such as antibiotics, vitamins, steroids, peptides, and others of interest in the pharmaceutical-drug industry.

A. Ion-Exchange Chromatography

Ion-exchange chromatography (IEC) is a method of separating proteins based on differences in their electric charge. When introduced into an ion-exchange column, the adsorbate, due to attraction of opposite charges, will attach itself to the oppositely charged adsorbent on the ion-exchange column. The adsorbate is eluted from the column by: (1) continued elution with the initial mobile phase (adsorbate bound mildly to the column); (2) pH gradient; or (3) ionic strength gradient. The latter two procedures reduce the electrical attraction between the adsorbate and the adsorbent, thereby reducing elution.

Lesins and Ruckenstein (1989) indicated that the ion-exchange adsorption process is not always clearly understood. Often this technique involves the *net charge* concept. That is, the attractive electrostatic interactions between the oppositely charged adsorbate and adsorbent can be represented by Coulomb's law. Retention of the solute on the column is a result of the strength of these interactions. The basic concepts involved in the net charge theory include (1) at their isoelectric point (pI) values proteins will not be retained beause the net charge is zero; (2) proteins will be retained on positively charged anion exchangers when the pH of the mobile phase is greater than their pI because they have a net negative charge; (3) on negatively charged cation columns the proteins will be retained below their pI values bercause they carry a net positive charge; and (4) there is a functional relationship between net charge and retention time of proteins. Lesins and Ruckenstein (1989) indicated, along with others (Kopaciewicz *et al.*, 1983; Kopaciewicz and Regnier, 1983; Rounds and Regnier, 1984) that the net charge theory is an over-simplification of protein adsorption on charged sorbents. This is due to two reasons: significant retention in IEC may occur at pI, and correlation between net charge and retention is poor.

Lesins and Ruckenstein (1989) emphasized that charge localization on the protein surface that is quite different from the net charge of the molecule may occur. Regnier (1987) emphasized the role of heterogeneity on the protein surface and its relation to protein adsorption. Lesins and Ruckenstein (1989) emphasized that if there is a nonuniform charge distribution on the net surface, it is not necessary for the net charge of the protein to be opposite to that of the adsorbent for adsorption due to electrostatic interactions to occur; instead the occurrence of an oppositely charged patch is all that is necessary. These authors emphasized that the patch of the molecule that contacts the adsorbent can sometimes be identified by examining protein structure and ionization characteristics of its individual amino acids. This allows for the determination of the protein area that controls adsorptive behavior. These charged patches and the subsequent heterogeneity in the adsorbate–adsorbent contact would significantly affect the conformational changes encountered by proteins and other biological macromolecules of interest during the chromatographic separation process.

Work by Kopaciewicz *et al.* (1983) suggested that the protein retention depends on both the distribution of charge within a molecule and the number of charged sites interacting with the support surface. Rounds and Regnier (1984) proposed a model for the retention of proteins on high performance ion-exchange supports. The model was based on the equilibrium:

$$P \cdot C_i + ZD_b \rightleftarrows P_b + Z \cdot aD_o + Z \cdot bC_i. \tag{5.1}$$

Here $P \cdot C_i$ refers to protein in solution with accompanying counter-ion, C_i, P_b is the protein bound on the ion-exchange column, and D_b and D_0 refer to displacing ions associated with the ion-exchange surface and in the mobile phase, respectively. The displacing power of an ion is proportional to its ionic strength, and the constants a and b are needed to adjust for valence, activity coefficient, and relative displacing power differences between ions. For this, an expression was derived that relates the retention of a solute (expressed as a capacity factor, k') to the concentration of displacing agent in the mobile phase (D_0), and the number of charged groups involved in the adsorption–desorption process (Z). Z could be defined more precisely as the number of electrostatic interactions between the protein and the ion-exchange support. Retention time increases directly with k' and Z. It should be noted that the physicochemical properties of the displacing salt would contribute to the Z values. Different variables affect the extent of the electrostatic interactions to different degrees. Thus, the *Z* factor is quite complex and further studies are required to elucidate this *Z* factor. This will then considerably assist in understanding HPIEC, in general.

Geoffery *et al.* (1984) used HPIEC to purify catalytically active enzymes present in minor quantities in plant material. The three o-methyl-transferases (*S*-adenosyl-*L*-methionine: catechol o-methyl-transferases, EC 2.1.1.6) of tobacco leaves were separated. Excellent recovery of enzyme activity (70 to 100 percent) was obtained. Purification achieved by HPLC was evidenced by electrophoretical analysis of the active fractions on sodium dodecyl sulfate (SDS)-polyacrylamide gels.

Displacement mode chromatography has been used to efficiently purify amino acids (Horvath *et al.*, 1983), peptides (Subramanian *et al.*, 1988; Vigh *et al.*, 1987), steroids (Subramanian *et al.*, 1988; Kalasz and Horvath, 1982), antibiotics (Kalasz and Horvath, 1981; Valko *et al.*, 1987), and proteins (Cramer *et al.*, 1987). The sample mixture in a carrier solvent that has a low affinity for the stationary phase is loaded and the bound components are displaced by a solution of the displacer that has a greater affinity for the stationary phase than any of the sample components (Horvath *et al.*, 1981). Hodges *et al.* (1988) extended the preceding idea to sample displacement mode chromatography (SDM), where during loading there is competition among the sample components for the hydrophobic adsorption sites of the stationary phase. The hydrophobic components compete more successfully for these sites than the hydrophillic components, which are displaced and eluted from the column. Finally, the adsorbed components are eluted with an aqueous organic eluent.

Veeragavan *et al.* (1991) utilized SDM chromatography to purify ovalbumin and soybean trypsin inhibitor (STI) by the use of a high-performance anion-exchange column. Conditions such as the Mono Q ion-exchange column dimensions, load flow rate, buffer type and temperature were optimized to maximize the yield and purity of the target protein. These authors applied the crude samples of ovalbumin and STI to one- and two-column systems, and the sequence and purity of both the displaced and the eluted components were determined by off-line analysis. These authors initially utilized ovalbumin samples to prove the working principle of their model-system. They then studied the second set of examples, which was a crude mixture of a trypsin inhibitor. The purity and yield of this inhibitor compound was calculated from activity determination. Table 5.1 shows and compares the purification of STI by SDM chromatography.

In the one-column process, Veeragavan *et al.* (1991) expected and observed competition between the various proteins present in the mixtures for the limited number of binding sites available in the column, which resulted in a sample displacement process. The authors also indicated that the impurities that bind more strongly than the products would be caught in the first column. The authors also expected the second column to be saturated with pure product only. The authors admitted that they had not optimized the size of the columns to capture impurities and products and therefore were able to collect a small percentage of the applied amount as pure products. The initial experiments were planned just to test the feasibility of SDM ion-exchange chromatography.

TABLE 5.1 Purification of Soybean Trypsin Inhibitor (STI) by Sample Displacement Mode (SDM) Chromatography[a]

Fraction	Total protein (mg)	Specific activity (ImU/mg protein)	Purification (fold)	Total ImU	Recovery (%)
Crude	500	1940	1.0	970,000	100
Column 1	26	3233	1.7	84,058	8.7
Column 2	22	3798	2.0	83,556	8.8

[a] From Veeragavan, S. *et al.* (1991). *J. Chromatogr.*, *541*, 207 with permission.

It is anticipated that optimization of the process parameters would lead to higher product recoveries, along with increasing the specific activity [or purification (fold)] of the STI recovered. These studies, if successful, could then also be applied to the recovery of other proteins from crude mixtures. No attempt has apparently been made to utilize SDM chromatography for the recovery of antibiotics, steroids, amino acids, and peptides from crude mixtures. Perhaps, as the SDM technique develops, the recovery of these chemicals will also be attempted.

Steindl *et al.* (1987) described a preparative scale purification procedure of monoclonal IgM from hybridoma culture supernatant with high protein content. Hybridoma culture supernatants were concentrated by ultrafiltration, precipitated, and further purified by DEAE ion-exchange chromatography (Steindl *et al.*, 1987). In cases where more highly purified antibodies were required a second column chromatographic step was added. Monoclonal IgM can further be purified by Sephacryl S300 gel chromatography. The procedure leads to product recovery of 40% and purity of 99% related to the total protein. The high purification level obtained by this procedure suggests an insignificant amount of inactivation in the antibody that is recovered. More attention needs to be paid to increase the low recovery (40%) of the antibody. It may not be surprising to observe lower values in the purity when significantly higher recovery levels of antibody are attained.

Ueda and Ishida (1987) examined the fundamental characteristics of a new series of silica-based ion-exchange columns and the use of these columns for the separation of a complex mixture of biopolymers. These authors noted that these columns are quite stable and that it is important to select the appropriate mobile phase conditions for the optimization of practical separations. Torres *et al.* (1984) showed that carboxymethyldextrans can be used to displace proteins from high-performance ion-exchange columns. The high resolving power of the method is demonstrated by the total separation of the A and B genetic variants of the β-lactoglobulins, which have only 0.1 pH difference in their isoelectric points.

Only a few typical examples of ion-exchange chromatography for a separation of proteins are provided where protein recovery and extent of inactivation are given. Surely many more examples are available in the literature where IEC has been utilized for protein separation, but few examples provide the extent of protein inactivation and possible reasons or causes as to why this may be occurring. IEC is a widely used chromatographic technique in industry for the separation of proteins. In fact, IEC on agarose media is probably the most widely used column in downstream processing. More articles like the work by Regnier and coworkers (Kopaciewicz *et al.*, 1983; Rounds and Regnier, 1984) are required to shed further physical insights into IEC.

Alpert (1983) developed a poly(aspartic acid)-silica column. Figure 5.1 shows the steps that were involved in the preparation of the column. A microparticulate silica gel was selected having pores wide enough to give most proteins free access to the pore interior. The silica was given a covalently bonded coating of aminopropyl groups. Poly(succinimide) was allowed to react with the amino groups. This produced a poly(succinimide)-silica in which the polymer was immobilized through amide bonds to the surface. Subsequent treatment with base catalyzed the hydrolysis of unreacted succinimide rings, producing the poly(aspartic acid) coating.

This author indicated that the poly(aspartic acid) coating is quite stable.

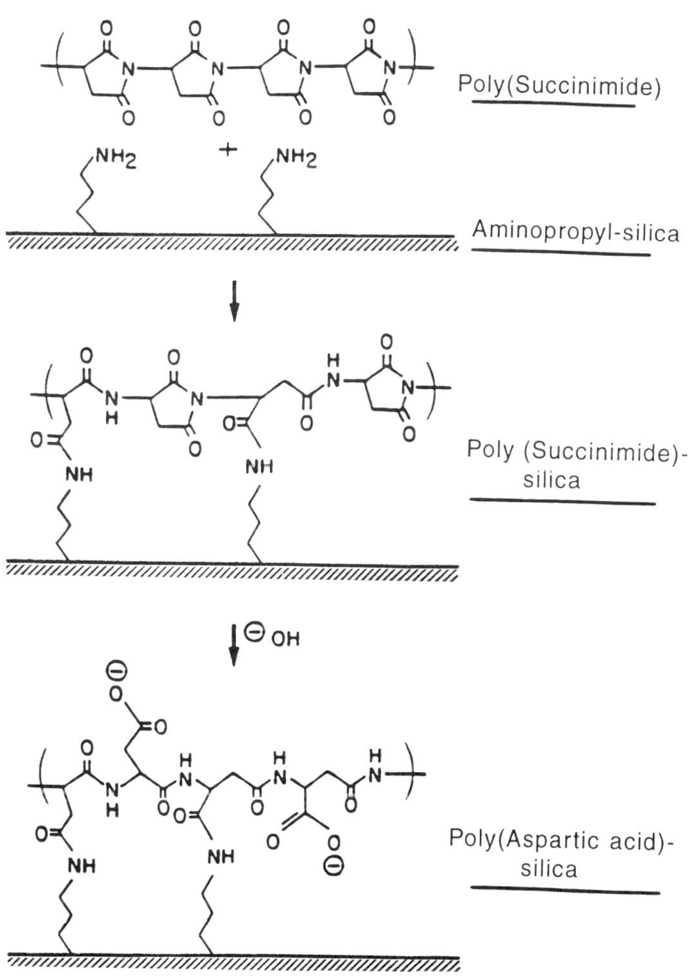

FIGURE 5.1 Steps involved in the preparation of the poly(aspartic acid)–silica column. [From Alpert, A. J. (1983). *J. Chromatogr.*, *266*, 23.]

These columns lasted for hundreds of hours when eluted with 0.05 M potassium phosphate, pH 6.0, without a decrease in efficiency or capacity. Adenylosuccinate synthetase, thiolase I, and β-hydroxyacyl-CoA dehydrogenase were eluted with 100% recovery of activity introduced into the column. Furthermore, the polypeptide nature of the coating makes it suitable for interaction with enzymes without causing denaturation. This leads to the high recovery of this enzyme. This technique seems to more than satisfy both the aspects of the separation, recovery and purity. The separation of other proteins-enzymes and other biological macromolecules may be attempted by this technique. This would enhance the flexibility and applicability of this process. This is especially attractive because both recovery and purity is high for the enzymes selected. What needs to be seen is how effective this technique is for other enzymes-proteins.

Finally, Vanacek and Regnier (1980) showed that clinical hemoglobin samples are frequently analyzed with cation-exchange columns. Alpert (1983) indicated that poly(aspartic acid)-silica functions especially well in this application. The hemoglobins A,F,S, and C were separated on a poly-Vyadac (10 m), 20 $\times$ 0.46 cm column. The elution was a 140-min linear gradient of 25 to 100% 0.04 M Bis-tris-Cl + 0.004 M potassium cyanide + 0.2 M sodium chloride, pH 6.8. No information was available about either the recovery or the purity of the hemoglobins separated.

Example 5.1

Describe a procedure for the HPIEC separation of biopolymers especially suited for applications at high pH and to high-molecular weight samples (Kato *et al.,* 1984).

Solution

Kato *et al.* (1983) utilized a weak anion exchanger TSK-GEL IEX-645 DEAE (Toya Soda, Tokyo, Japan) derived from TSK-GEL G500PW. This is a hydrophilic-polymer-based material of large pore size and is utilized for high-performance gel filtration.

Tables 5.2a and 5.2b show the recovery of protein and enzymatic activity, respectively. The recovery of proteins is greater than 90% for the proteins applied, and greater than 80% enzyme activity for the enzymes applied. No information was provided concerning the purity of any of the enzymes separated.

In Table 5.2b for lipoxidase (i) the conditions for 60-min linear elution gradient from 0.02 M ethanolamine-HCl buffer of pH 9.0 to 0.02 M ethanolamine-HCl buffer of pH 9.0 containing 0.5 M sodium chloride; flow rate is 1 ml/min. The same conditions in (i) are found in (ii). The conditions for (iii) include a 60-min linear elution gradient from 0.02 M 1,3-diaminopropane-HCl buffer of pH 9.8 to 0.02 M 1,3-diaminopropane-HCl buffer containing 0.5 M sodium chloride; flow rate is 1 ml/min.

Example 5.2

A method for the separation of mRNAs (van der Mast *et al.,* 1991).

TABLE 5.2a Protein Recovery on IEX-645 DEAE[a]

Protein	Recovery (%)
β-Lactoglobulin	98
Bovine serum albumin (BSA)	99
Ferritin	98
γ-Globulin	98
Hemoglobin	94
Myoglobin	97
Ovalbumin	99
Thyroglobulin	91
Trypsin inhibitor	101

[a] Buffer: 0.05 M Tris-HCl, pH 8.3 containing 0.5 M sodium chloride at 25°C.

Solution

van der Mast *et al.* (1991) separated mRNAs by reversed phase ion-pair HPLC. The separation parameters of mRNAs are only the amount of negative charge, that is, the length or molecular mass of these molecules; and the differences in hydrophobicity due to differing base compositions. These authors indicated that the separation of mRNAs is entirely empirical at present. They suggested that reversed-phase ion-pair (RPIP) chromatography has been used in the separation of nucleotides (Hoffman and Liao, 1977; Perrone and Brown, 1984), oligonucleotides (Schott *et al.*, 1987; Makino *et al.*, 1987), and RNAs (Nguyen *et al.*, 1982). The positive alkylammonium ions have been successfully used as ion-pairing agents. The separation is dependent on the amount of the ion-pairing compound bound to the molecule, which is a fraction of the molecule length. Elution is accomplished with a buffered gradient of organic solutions.

These authors utilized two types of Nucleosil C_4 columns (particle size 5 μm, pore diameter 30 nm; and particle size 7 μm, pore diameter 100 nm) which were obtained from Macherey and Nagel (Duren, Germany). The column di-

TABLE 5.2b Enzyme Recovery on IEX-645 DEAE

Enzyme	Recovery (%)
(i) Lipoxidase	95
(ii) Catalase	80
(iii) α-Chymotrypsin	93

mensions were 250 × 4 mm I.D. They noted a direct relation between the length of the mRNA and the percentage of the organic phase needed to release the molecules from the column material. van der Mast *et al.* (1991) emphasized that the organic phase leads to some unfolding of the secondary structures of the molecules. This accounts for the separation according to length, in agreement with the suggestion of Nguyen *et al.* (1982). van der Mast *et al.* (1991) also noted that the addition of methylmercuric hydroxide leads to complete unfolding with a subsequent increase in the resolution on the columns utilized. Methylmercuric hydroxide is a powerful inhibitor of mRNAs, and not utilizing it leads to significant breakdown of the mRNAs.

Habibi-Najafi and Lee (1994) purified X-prolyl dipeptidyl peptidase to homgeneity from crude cell-free extracts of *Lactobacillus casei* subsp. *casei* LLG using fast protein liquid chromatography (FPLC) equipped with preparative and analytic anion exchange columns. These authors utilized ammonium sulfate fractionation followed by two ion-exchange chromatographic steps. Using their procedure these authors were able to obtain a 34% enzyme recovery, and obtained a purification factor of 274. X-prolyl dipeptidyl peptidase is considered to play a significant role during cheese ripening. Habibi-Najafi and Lee (1993) analyzed its role as a debittering agent in dairy and in protein-based products.

The crude cell-free extracts were fractionated by salting out with solid ammonium sulfate. The precipitate formed was collected by centrifugation. The dialyzed fraction was then applied successively to two ion-exchange chromatography columns. The first was a preparative Mono Q column equilibrated with 0.02 M Tris-HCl buffer, pH 8.0. The second ion-exchange column was analytic Mono Q equilibrated with 0.02 Tris-HCl buffer, pH 7.0.

The crude cell-free extract contained 312.5 units of total enzyme activity and the total protein was 1217.5 mg. The specific activity (U/mg), recovery or yield (%), and purification (fold) for the different stages (in order) are: cell extract 0.26, 100, 1; ammonium sulfate 0.34, 97, 1.3; preparative ion exchange 10.0, 70, 38.5; and analytic ion exchange (II) 71.12, 34, 273.5.

Finally, Habibi-Najafi and Lee (1994) indicated it is now well established that many proline-containing peptides are hydrophobic and may be responsible for the bitter flavor development in cheese. Thus, the X-prolyl dipeptidyl peptidase, along with other proline-specific peptide hydrolases, could play an important role in the debittering process during cheese ripening.

Arora and Lee (1994) purified an aminopeptidase of broad specificity from *Lactobacillus casei* subsp. *rhamnosus* S93. The cell-free extract was precipitated with $(NH_4)_2SO_4$ (70% saturation). The precipitate was desalted on a Sephadex G-25 gel filtration column (Pharmacia). They used a Pharmacia FPLC system to chromatographically separate the crude enzyme. These authors indicated that the two ion-exchange chromatographic steps during the initial stages of the purification help remove greater than 99% of crude extract protein with good enzyme recovery. They emphasized that enzyme activity loss was minimized due to the speed of the FPLC system and the subsequent pH adjustment of the eluate from the chromatofocusing column.

Four chromatographic steps were utilized after the $(NH_4)_2SO_4$ precipitation step (Arora and Lee, 1994). The enzyme was purified by a factor of approximately 195, though the recovery was only 7%. The low yield was due to denaturation of the enzyme during storage and the purification step. The final specific activity of the enzyme was 1584 U/mg.

The specific activity (U/mg), yield (%), and purification factor obtained after each purification step are as follows: crude extract (8, 100, 1.0); $(NH_4)_2SO_4$ precipitation (12, 100, 1.4); ion-exchange step I (188, 47. 23.0); ion-exchange step II (482, 16, 59.2); chromatofocusing (1051, 9, 129.1); and gel filtration (1584, 7, 194.5). These authors were able to purify the enzyme to homogeneity because the active fraction after gel-filtration chromatography yielded a single protein bound either by SDS-polyacrylamide gel electrophoresis (PAGE) or by isoelectric focusing.

Furthermore, these authors emphasized that the aminopeptidase purified from *L. casei* subsp. *rhamnosus* was active in the pH range 6 to 9. This is of assistance in the cheese industry, because there is very little pH control during the processing and maturation of cheese.

Example 5.3

Briefly describe the separation of lipase from *Pichia burtonii* (Sugihara *et al.*, 1995).

Solution

Lipases catalyze the hydrolysis of triglycerides to glycerol and fatty acids (Brockerhoff and Jensen (1974). Sugihara *et al.* (1995) indicated that lipases exhibit higher activities toward water-insoluble esters than soluble ones due to interfacial activation. It has been indicated that some of the industrial chemicals from oils and fats could be produced faster with more specificity under milder conditions on utilizing lipases (Jones, 1986; Klibanov, 1990; Davis *et al.*, 1990; Vulfson, 1994). Thus, there is considerable effort being undertaken to utilize lipases for industrial application.

The yeast *P. burtonii* produces an extracellular, heat-labile lipase, which should be useful for food processing, for example, flavor development in dairy products. The culture supernatant was desalted using a Sephadex G-25 column and put on a DEAE-Sephadex A-50 column. A 10-ppm Nonidel-P40 (buffer) was added to all of the purification steps to prevent or minimize protein inactivation. After the DEAE-Sephadex A50 column, the solution was sent to a Sephadex G-100 column followed by recycling isoelectric focusing. Thereafter, SDS-PAGE indicated that the preparation was homogeneous. The preceding steps increased the purity by a factor of 310 and the yield was 6%. The specific activity (U/mg) and the yield (%), respectively, for the different steps were: culture supernatant (0.4, 100); DEAE-Sephadex A-50 (12.4, 32); Sephadex (48.1, 19); and isoelectric focusing (124; 6). These authors emphasized that the low yields are due to a combination of irreversible adsorption to the chromatographic columns or concomitant denaturation.

They indicated that a rapid loss of enzyme activity occurred above 30 °C in the absence of olive oil. The addition of olive oil or trimethylolpropane significantly stabilized the lipase. Other researchers also indicated that the addition of olive oil or its related substances is essential for the production of lipases (Suzuki *et al.*, 1988; Omar *et al.*, 1987; Sugihara *et al.*, 1991).

Finally, Sugihara *et al.* (1995) indicated that the purified lipase was rendered much more heat stable in the presence of substrate. Lipase adsorption at the substrate–water interface followed by activation has been indicated (Benzonana and Desnuelle, 1965). These authors analyzed the effect of a substrate analogue, trimethylolpropane diallyl ether on the heat stability of lipase. Sugihara *et al.* (1995) noted that *P. burtonii* lipase enzyme-substrate (or substrate analogue) interaction or conformational changes increased heat stabilization.

Human placenta (an available hospital waste) is an abundant source of different biological compounds of significant biotechnological interest. Some of these compounds include hormones, activators, inhibitors, proenzymes, enzymes, immunoregulatory factors, receptors, etc. Liautaud (1986) indicated that placental blood may be utilized to economically extract serum albumin, immunoglobulins, and other proteins.

Example 5.4

Briefly describe a process to separate basic fibroblast growth factor (bFGF) and alkaline phosphatase (PALP) from human placenta (Costa *et al.*, 1993).

Solution

Costa *et al.* (1993) described a coupled process to purify bFGF and PALP from placenta. Their bFGF purification process involves three steps: (1) extraction, (2) S-Sepharose chromatography, and (3) heparin-Sepharose chromatography. bFGF is a multifunctional heparin-binding growth factor with mitogenic, angiogenic, and neurotrophic properties (Burgess and Maciag, 1989; Baird and Klagsburn, 1991). Costa *et al.* (1993) indicated that bFGF can be used as a component of cell culture media, as a wound-healing component (Dijke and Iwata, 1989), and in ulcer treatment (Folkman *et al.*, 1991).

Figure 5.2 shows the flow diagram for bFGF and PALP purification. Costa *et al.* (1993) indicated that the crude extract was adsorbed on an S-Sepharose column. The eluate was then loaded onto a heparin-Sepharose column and eluted step wise with 1.1 M and 2.0 M NaCl. These authors indicated that the purified bFGF yielded about 1 μg/kg of placenta and it was purified by a factor of four orders of magnitude. The specifc activity (U/mg) and activity recovery (%) of bFGF in the three fractions are crude extract (444, 100), S-0.6 (2702, 22), and H-2.0 (8×10^{6}, 3.7). On assuming that the purification factor (PF) is 1 for the crude extract, yields are PF values of 6 and 1.8×10^{4} for the S-0.6 and H-2.0 fractions, respectively.

The purification of PALP is a bit more complex. The second cellular mass (see Fig. 5.2) was obtained from the first cellular mass after bFGF extraction. This was homogenized and extracted with butan-1-ol at pH 5.5. Thereafter, the PALP was sequentially purified by Q-Sepharose, Con A-Sepharose, and Q-Sepharose chromatography. The specific activity (U/mg) and activity recovery (%) of PALP in the four fractions are aqueous phase (25.4, 100), Q-0.2 (60.0, 50), Con A-0.4 (221.3, 28.7), and Q-200 (803.9, 22.7). On assuming that the purification factor is 1 for the aqueous phase, yields are PF values of 2.4, 8.7, and 31.7 for the Q-0.2, Con A-0.4, and Q-200 fractions, respectively.

There is keen interest in producing enzymes that are capable of hydrolyzing efficiently bound aroma from fruit juice or wine. Gueguen *et al.* (1994) indi-

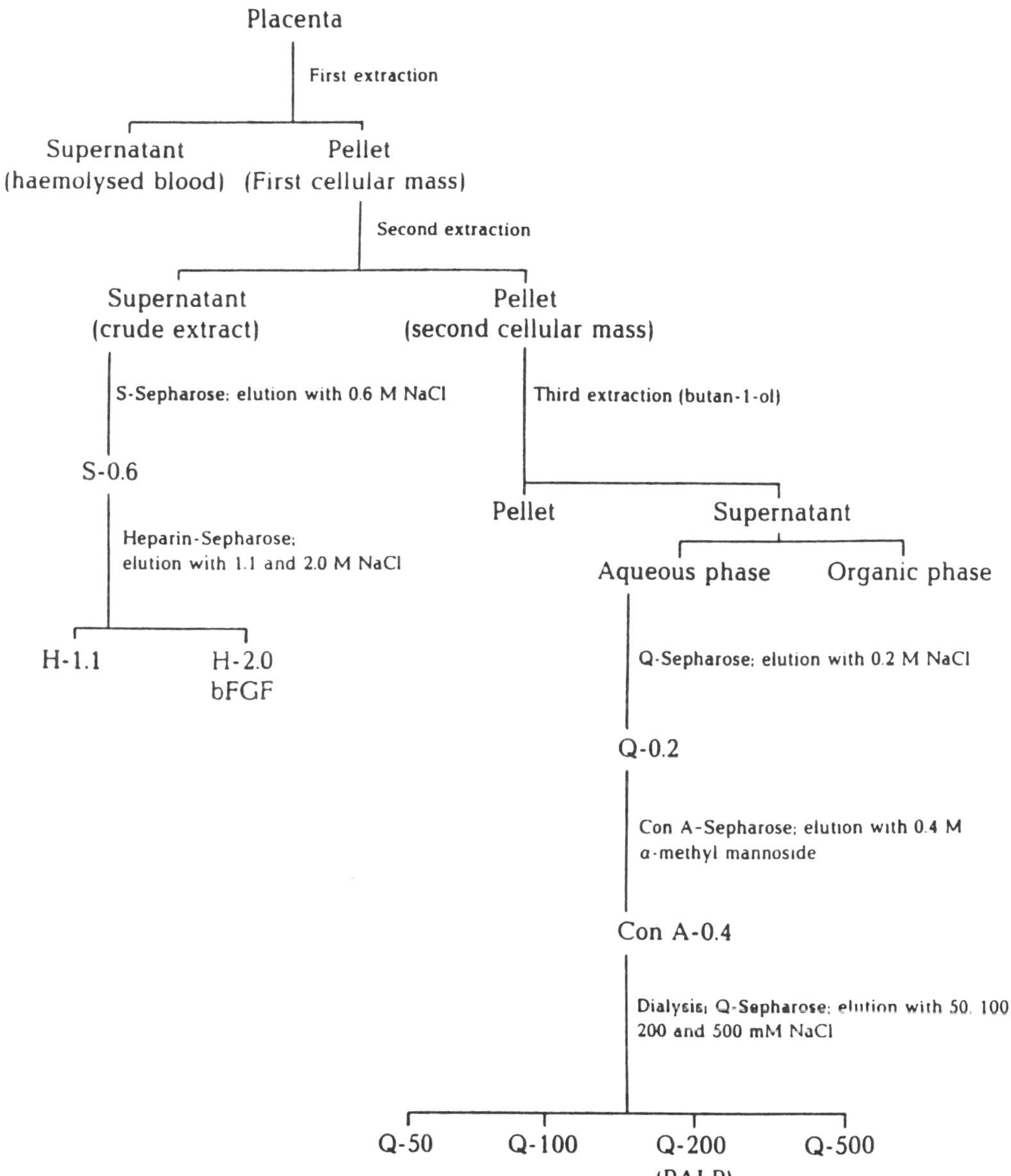

FIGURE 5.2 Flow diagram for bFGF and PALP purification. [From Costa, M. H. B. *et al.* (1993). *Biotechnol. Appl. Biochem.*, *17*, 155.]

cated the requirement of an enzyme that possesses this activity at (1) low pH (2.5 to 4) for most wine and fruit juices; and (2) high glucose concentration (some fruit juices), and high ethanol concentration (for wines).

These authors purified β-glucosidase from the yeast *Candida entomophila*. These authors purified β-glucosidase from yeast by ion-exchange chromatography and gel filtration. The purification steps, along with the specific activity, the purification factor, and the yield for each of the purification steps are: (1)

the S1 supernatant fluid (0.166 U/mg, 1.0, 100%) was centrifuged at 180,000 g for 90 min to yield the S2 supernatant; and (2) the S2 supernatant fluid (0.22 Units/mg, 1.3, 83%) was fractionated on a Q-Sepharose column. After the Q-Sepharose fractionation the specific activity was 9.4 U/mg, the purification factor was 169, and the yield was 25.9%. Thereafter, the β-glucosidase was chromatographed on a Sephacryl S-300 column. The final enzyme had a specific activity of 27.99 U/mg, a purification factor of 169, and a yield of 12.9%.

Finally, these authors emphasized that the β-glucosidase purified by them exhibited a broad specificity to hydrolyze a diversity of bonds between terpinols and aglycones. This would increase the liberation of bound aromatic compounds. However, the enzyme is inhibited by glucose. Thus it would have limited use in fruit juices, but, it could be used for the flavor enrichment of wine.

Antibiotics are widely used subtherapeutically to prevent disease and promote growth (Nouws, 1981; Jackson, 1980), and to treat a variety of infections (Agarwal, 1989). Badar and Edward (1985) indicated that gentamycin, streptomycin, dihydrostreptomycin, and neomycin are the aminoglycoside antibiotics that have been approved for veterinary use (only for oral administration) in food-producing animals.

Gentamycin has a broad spectrum of activity against Gram-positive and Gram-negative bacteria; and consists of three major components C1, C1a, and C2. All these resemble each other in chemical nature and have similar antibacterial activities.

Agarwal (1989) described an HPLC method for the determination of gentamycin in animal tissue. This antibiotic is quantitatively extracted from animal tissue with potassium phosphate buffer and the extract is deproteinated. The deproteinated extract is acidified to a pH of 6.4 to 6.5 with sulfuric acid and is purified by ion-exchange and gel chromatography (CM-Sephadex). A gradient elution of 20% methanol at zero time to 60% methanol in 15 min was used while maintaining a flow rate of 1.6 ml/minute. After elution from the ion exchange column with alkaline buffer, the eluent is further purified using a silica Sep-pak cartridge and derivatized. Ethanol was then used to elute the derivatized gentamycin from the Sep-pak cartridge, and analyzed by liquid chromatography with fluorometric detection.

This author noted that an acidic buffer did not extract gentamycin from the ground beef tissue. Also, the addition of sodium sulfate to the buffers improved recoveries. The recoveries, calculated based on the sum of the peak heights C1a and C2 ranged from 85.9% to 107.6% in samples fortified between 0.5- to 10-ppm levels. There were apparently no structural changes in the gentamycin on recovery after the separation procedure.

Because the recovery of gentamycin is high (above 85.9%) and there are apparently no structural changes in the gentamycin recovered, the technique may be used for the recovery of higher concentrations of gentamycin and for other antibiotics of similar nature.

Example 5.5

Provide an example for the HPLC separation of an enzyme exhibiting microheterogeneity (Wong *et al.,* 1988).

Solution

Colvin *et al.* (1954) defined microheterogeneity as follows: "A protein preparation will be said to be microheterogeneous if there is experimental evidence for one or more minor differences between individual protein molecules of the preparation, over a period of time which is long compared to the duration of the experiment(s)." Many glycoproteins exhibit microheterogeneity that arises due to the type and amount of carbohydrate attached to the polypeptide chain (Albert *et al.*, 1972; Beckman and Beckman, 1967; Robinson and Stirling, 1968). Wong *et al.* (1988) indicated that the carbohydrate residues may play an important role in the regulation of the various biological activities of the proteins. Mouse α-fetoprotein (MAFP) is a serum glycoprotein known to be heterogeneous and exists in many different molecular variations (Smith and Kelleher, 1980). Zimmerman *et al.* (1977) indicated that MAFP possesses an immunosuppressive activity and that the carbohydrate residues of the protein are responsible for this effect.

Wong *et al.* (1985, 1988) developed a novel methodology using anion-exchange HPLC to purify the various isoforms of MAFP from biological specimens. Wong *et al.* (1988) were able to detect six isoforms for MAFP in fetal development. These six isoforms were present at different times in the development period. Iso-1 and Iso-2 accumulated during the early gestation period. Iso-3 and Iso-4 accumulated during the midgestation period, and Iso-5 and Iso-6 accumulated during the late gestation period. Wong *et al.* (1988) indicated their results suggest that the higher isoforms are derived gradually and at the expense of the lower isoforms. It is of interest to note that Zimmermann and Madappally (1973) detected a similar trend in conversion for their six MAFP isoforms.

Zimmermann *et al.* (1977) identified the Iso-5 form to be immunosuppressive as determined in mouse splenic lympocyte cultures by measuring the primary immune response to sheep erythrocytes. Wong *et al.* (1988) emphasized that the corroboration of Zimmerman *et al.* (1977) studies by identifying the Iso-5 form as the specific immunoactive component together with the knowledge of its concentration during the fetal development of the mouse will provide better physical insights into the immune system development in the mouse.

HPLC can be of considerable assistance in the detection of microheterogeneity and also of the possible changes of microheterogeneity of proteins-enzymes with time. This is of significant interest because microheterogeneity may be utilized as an effective measure or index of protein quality. Microheterogeneity is especially of interest with reference to the consistency of the antibiotics-drugs to be utilized for human consumption.

Not much information is available concerning the mechanisms of conformational changes encountered by proteins, antibiotics, peptides, and other biological macromolecules of interest during their chromatographic separation. This type of analysis is absolutely necessary so that one may obtain appropriate physical insights into the conformational changes of the biological macromolecules occurring at the adsorbent–adsorbate interface.

B. Mechanistic Considerations

Kopaciewicz *et al.* (1983), and Rounds and Regnier (1984) have derived a model for HPIEC that has been shown to be useful for a number of proteins. These models are nonmechanistic because neither the nature of interaction of protein with the surface nor the implications of the protein structure in determining retention are fully known. The complex structure of the protein molecule further exacerbates the problem of protein retention.

Parente and Wetlaufer (1984) examined the retention characteristics of both native and denatured forms of α-chymotrypsinogen-A. According to these authors a sharp drop in retention can be attributed to the conformational changes in the protein that occur due to urea-thermal denaturation. Because the binding strength of the denatured protein is lower than the native protein, there is a drop in HPIEC retention of the denatured protein. The authors presumed a reversible two-state denaturation model and proposed the scheme for the binding of native and denatured protein in the ion-exchange stationary phase

$$\begin{array}{ccc} E_{\text{free}} & \rightleftarrows & E_{\text{d,free}} \\ \downarrow\uparrow & & \downarrow\uparrow \\ E_{\text{bound}} & \rightleftarrows & E_{\text{d,bound}}. \end{array} \tag{5.2}$$

Here E and E_{d} are the native and the denatured forms, respectively. The authors stated that when the interconversion between native and denatured states is slow compared with the chromatographic time scale, two peaks could be expected to be observed corresponding to the native and denatured forms. For a fast equilibration rate, one peak corresponding to the average retention time could be predicted. In an intermediate kinetic region, asymmetrical bands may result.

More mechanistic models like the Parente-Wetlaufer (1984) model are required that shed physical insights into the denaturation of proteins during chromatographic separations. The Parente-Wetlaufer (1984) model is based on a simple two-state enzyme transition $E \rightarrow E_{\text{d}}$. Intermediates have been utilized to model complex enzyme inactivations (Henley and Sadana, 1986, 1987). A primary knowledge of modeling complex enzyme inactivations by a series-type mechanism involving first-order steps, is that it provides one possible avenue to visualize specific activity changes of the different conformational states ($E \rightarrow E_1 \rightarrow E_2$). Furthermore, it is perhaps the only mechanism that satisfactorily describes an increasing specific activity of the intermediate state with respect to the initial state. This leads to an initial increase in the enzyme activity with time. It is thus very probable that a series model involving an intermediate ($E \rightarrow E_1 \rightarrow E_2$) may be involved during chromatographic separation–denaturation of different proteins–enzymes. These intermediates, if present, will be difficult to detect because they may be present in very small concentrations and also because there are complexities involved in the chromatographic separation procedure. Complexities may also arise due to heterogeneity of proteins (Malhotra and Sadana, 1987) on chromatographic columns caused by the heterogeneity of the adsorption or other processes. These factors and others should be taken into account in further models of protein–enzyme denaturations during chromatographic separations.

C. Reversed-Phase High-Performance Liquid Chromatography

In reversed-phase chromatography the most commonly used solvent systems for the separation of proteins involve linear gradients, starting with water and increasing strengths of organic solvent (methanol, acetonitrile, or propanol). These solvent systems usually employ low concentrations of perfluorinated organic acids at a concentration of 0.05 to 0.1% (v/v) in both water and the organic solvent.

Denaturation of the proteins may occur due to the presence of the alkyl silane residues of the column, the organic solvents in the mobile phase, and the very acidic conditions. Regnier (1987) stated that multimeric proteins with dissimilar subunits pose a problem. This is because denaturation of this protein causes dissociation and separation of the subunits. This makes it difficult to locate the subunits and to reassemble them correctly to the native protein.

Most reversed-phase matrices have relatively high hydrophobic loadings, which cause a strong binding of the protein to the matrix; and the concentration of the organic solvent is usually greater than 10% for protein elution. It is reasonable to expect that the hydrophobicity of the matrix or nonpolarity of the solvent could denature a protein on binding to, or elution from, the column. The hydrophobicity of a protein in its native conformation is dramatically different from its unfolded state. This is true because side chains are exposed during denaturation (Cohen *et al.*, 1984a,b; Hearn *et al.*, 1985). It is thus advisable to avoid a situation where the folding state of the protein is changing during the separation procedure.

Regnier (1987) emphasized that even though protein structure is altered in reversed-phase high-performance chromatography (RP-HPLC), substantial secondary and tertiary structure can be retained. This is due to the fact that (1) more than one peak is observed for a single protein, implying that different conformational states exist (Hearn *et al.*, 1985; Ingraham *et al.*,1985); and (2) spectral techniques show the retention of protein helical content on the sorbent surface (Sadler *et al.*, 1984). Furthermore, the addition of organic solvents during RP-HPLC under acidic conditions produces a much greater disruption of tertiary and quaternary structure than of secondary structure (Sadler *et al.*, 1984; Lau *et al.*, 1984). Regnier (1987) indicated that this is true because hydrophobic forces are major contributors to tertiary and quaternary structure and hydrogen bonds are more stable in organic solvents than in water.

Conformational changes in proteins in either the stationary phase or mobile phase play an important part in the RP-HPLC separation of proteins. Such changes, if reversible during chromatographic elution, can lead to distorted and broadened peaks. However, if the kinetic processes of conformational change are slow or irreversible during elution, multiple peaks may be observed. Melander *et al.* (1984) dealt with the influence of this behavior on chromatographic performance. Lu *et al.* (1986) analyzed the reversible conformational effects of ribonuclease A in RP-HPLC. By using absorbance ratio measurements

Lu *et al.* (1986) proposed a two-state model in which an early eluted broad band is associated with the folded or native state and a late eluted sharp band is associated with an unfolded state. By using the results of this model system the authors were able to predict the optimum conditions for elution of species with sharp elution peaks when reversible unfolding takes place in the column.

Example 5.6

Provide an example of protein separation using conformational differences (Regnier, 1987).

Solution

The contribution of disulfide binding to chromatographic behavior in RP-HPLC was studied for interleukin-2 (IL-2) variants that were substituted at positions 1, 58, 104, 105, and 125 (Kunitani *et al.*, 1986). Natural IL-2 has a single disulfide bridge between the cysteines at positions 58 and 105. The molecule can be forced to form disulfide bonds at positions 58 and 125 or 105 and 125. The different IL-2 species had approximately the same molecular weight (15,000) and relative hydrophobicity. Nevertheless, their Z numbers varied by a factor of 25. This is primarily due to significant conformational changes in the different molecules resulting in different chromatographic contact regions. Because the "unnatural" IL-2 species gave a much larger Z than the native species, this implied that the external hydrophobicity of the unnatural species was substantially greater than that of the natural species.

Low protein recoveries have been a long-standing problem in RP-HPLC (van Enckevart *et al.*, 1984). Sadek *et al.* (1985) indicated that the mechanism or mechanisms responsible for the irreversible adsorption (on the chromatographic time scale) appear to be quite complex. These authors further stated that 100% mass recovery of proteins is difficult, especially when the total injected amount decreases to 100 μg. Stainless steel frets have specifically been found to be a significant contributor to irreversible protein loss, particularly when protein sample sizes are on the order of 1 μg or less. The authors emphasized that the major contact occurs at the entrances and at the exits of the columns. Sadek *et al.* (1985) indicated that significant amounts of proteins are adsorbed on the frets. These frets have a Brunauer-Emmett-Teller (BET) determined surface area of 190 cm^2, which is much greater than the total available surface area of the column tubing and the connecting tubing.

The separation of complex samples (e.g., peptides, proteins, oligonucleotides, and nucleic acids) is affected by a large number of experimental variables whose role according to Snyder *et al.* (1991) is often poorly understood. These authors utilized computer simulations based on the Craig distribution model to examine how separation varies with experimental conditions for the case of heavily overloaded (overlapping bands) gradient elution in RP-HPLC. They showed that the production rate in isocratic and gradient elution is essentially the same for the separation of a two-component mixture. Both separation procedures show that the production rate increases steeply as the separation factor, α, of the two bands (product and impurity) increases. The close similarity noted by these authors between gradient separations and corresponding isocratic runs occurred when isocratic capacity factors, k' are equal to gradient capacity factors, k. These authors indicated that the present treatment is based on an idealized model (the Langmuir isotherm) that is known to be inapplicable in some respects for the HPLC separation of macromolecular samples. Nevertheless, their study does provide some physical insights and overall practical guidelines. Finally, these authors emphasized that even though their conclusions apply to RP-HPLC, similar conclusions would also be applicable to other HPLC methods, for example, ion exchange.

RPC of proteins was, in general, popular until the mid-1980s (Melander *et al.*, 1984; Hearn *et al.*, 1985; Cohen *et al.*, 1985). Other than the purification of insulin on Zorbax C8 columns and a few other small peptides, RPC is not used in downstream processing. Other workers soon realized that the denaturation problems on RPC columns are so serious with proteins that they are, in general, no longer used. Nevertheless, we have discussed RPC in detail and now present some mechanistic considerations observed during RPC. Even though this aspect of chromatographic separation is, in general, not now under active consideration, chronologically in the evolution of separation techniques RPC did play an important role in understanding these processes.

D. Mechanistic Considerations

Based on their results, Lu *et al.* (1986) proposed the following model of conformational changes of RNase A during RP-HPLC

$$\begin{array}{ccc} F_m & \underset{k_{21}}{\overset{k_{12}}{\rightleftharpoons}} & U_m \\ k_{31} \upharpoonleft\downharpoonright k_{13} & & k_{42} \upharpoonleft\downharpoonright k_{24} \\ F_s & \underset{k_{43}}{\overset{k_{34}}{\rightleftharpoons}} & U_s. \end{array} \tag{5.3}$$

Here F represents the folded (native) state, and U, an unfolded state. The k_{ij} are the respective first-order rate constants for folding or unfolding. The subscripts m and s represent the mobile and stationary phases, respectively. It is important to realize that conformational behavior can, in principle, be used to probe chromatographic distribution or adsorption. Thus, examination of such protein behavior is of importance in HPLC.

STI was observed to follow similar irreversible denaturation as papain on the alkyl bonded-phase surface (Cohen *et al.*, 1983). These authors emphasized that the denaturation observed in their work can be viewed in terms of a classical two-state model (native, denatured). They state that the chromatographic procedure is not sufficiently fast to observe intermediates in the deactivation reaction.

Cohen *et al.* (1984) also analyzed the RP-HPLC behavior of papain, a proteolytic enzyme, under acidic condition with 10 mM phosphoric acid. An n-butyl bonded phase and 1-propanol as the modifier give two major peaks widely separated from one another in gradient elution at pH 2.2 and 5°C. As the column temperature was increased, the later eluting peak increased at the expense of the earlier peak. The earlier peak was the active fraction and the later peak was the inactive fraction. The authors interpreted their results in terms of the irreversible denaturation of papain using a two-state (active, inactive) model. The irreversibility of the model was suggested due to reinjection of the second peak. No native (first) peak was obtained. These authors noted a large difference in the retention times of the native and the denatured state. This is expected because a denatured species would possess a different conformation that would be more unfolded than the globular shape. The size of such a species adsorbing to the bonded phase surface would thus be expected to be larger than in the native state. Therefore, a much stronger binding of the denatured state to the surface may be anticipated.

Example 5.7

Provide an example of kinetics of denaturation of an enzyme or enzymes on a surface used in RP-HPLC (Benedek *et al.,* 1984)

Solution

Benedek *et al.* (1984) presented measurements on the kinetics of denaturation of papain, STI, and lysozyme on n-butyl silica gel surfaces used in RP-HPLC. In all cases, native and denatured peaks widely separated from one another are observed. The results reveal that a slow denaturation step occurs with a half-life of approximately 15 min. In addition, studies of denaturation as a function of the amount of 1-propanol in the initial mobile phase suggest an additional unfolding step when the protein comes in contact with the bonded-phase surface. The authors emphasized that an important factor in the unfolding is the contact time of the protein with the bonded-phase surface. Limiting the time a protein molecule spends in a column would seem to be one means by which to minimize denaturation. The composition of the mobile phase and the column temperature will influence the effect of contact time on unfolding.

These authors proposed the following kinetic model for protein–enzyme unfolding on the bonded-phase surface. The incubation time is the time from the injection of the sample to the time when the gradient is started. By assuming first-order deactivation kinetics and zero incubation time, the model yields

$$P = P_0 \exp(\bar{k}t_e) \tag{5.4}$$

Here P is the amount of native peak eluted from the column, P_0 is the amount of protein injected into the column (native under the injection conditions), t_e is the time from injection until elution of the native peak, and k is the average rate constant for the various conditions involved in the gradient. In the incubation experiments $\bar{k}t_e$ is assumed constant. If the incubation time is varied, then

$$P_0 = I \exp(-kt) \tag{5.5}$$

Here I is the amount of injected material, k is the rate constant, and t is the incubation time. At the end of the incubation time when the gradient is started, a quantity of native protein, P_0 is present. Substitution of Eq. (5.5) in Eq. (5.4) yields:

$$P = I \exp(-kt) \exp(-\bar{k}t_e) \quad (5.6)$$

or

$$\ln P = (\ln I - \bar{k}t_e) - kt \quad (5.7)$$

A plot of ln P versus t allows the calculation of the rate constant for unfolding on the bonded-phase surface, k.

Table 5.3 shows the values of the rate constant for unfolding, k, obtained using Eq. (5.7) for three mobile phase compositions (0%, 1.8%, and 2.7% (v/v) 1-propanol in 10 mM H_3PO_4, pH 2.2) at 15, 20, and 35 °C. First-order kinetics was obtained for the first 20 min. No data are available for later times.

Cohen *et al.* (1983) further suggested that the disappearance of native protein is a consequence of irreversible conformational unfolding on the bonded-phase surface to yield a denatured peak that grows with incubation time. The authors checked for autoproteolysis and noted it to be very slow. Care was taken to minimize this by preparing fresh enzyme solutions. The authors also emphasized that though it is well known from adsorption studies that globular proteins can slowly unfold at the gas-liquid (Graham and Phillips, 1979) and solid–liquid (Sonderquist and Walton, 1980) interfaces, the driving force for unfolding is undoubtedly hydrophobic. The authors stressed that by decreasing the time a protein molecule spends in a column would appear to be one method by which to minimize denaturation.

Cohen *et al.* (1985) also examined the RP-HPLC behavior of ribonuclease A using an *n*-butyl chemically bonded phase and a gradient of 10 mM H_3PO_4 and 1-propanol. At a column temperature of 25°C, a broad band followed by an overlapped late-eluting sharp peak is observed. The authors noted that as the temperature is raised, the sharp peak grows at the expense of the broad band until at 37°C, only a single narrow-eluting band is found. The broad band is the folded or native state of RNase A and the late-eluting band is the denatured state. They stated that based on absorbance ratio changes in the denatured state as a function of time and the known behavior of the protein, reversible folding or renaturation is proposed to take place in the solution. Furthermore, these authors indicated that RNase is denatured on absorbing to

TABLE 5.3 First-Order Rate Constants for Papain Deactivation on the *n*-Butyl-Bonded Phase[a]

	$k \times 10^4$ (s^{-1}) temperature		
Incubation mobile phase	**15°C**	**20°C**	**35°C**
10 mM H_3PO_4 (pH 2.2) + 1.8% 1-propanol in mobile phase A	4.6	8.8	29
10 mM H_3PO_4 (pH 2.2) + 2.7% 1-propanol in mobile phase A	6.8	11.9	—

[a] From Benedek, K. *et al.* (1984). *J. Chromatogr.*, *317*, 227.

the bonded phase, and on migration down the column, reversible folding takes place in the mobile phase. The authors proposed the following mechanism to explain their results

$$\frac{\text{(Mobile)}}{\text{(Stationary)}} \begin{array}{c} E_{\mathrm{d}} \\ \upharpoonleft\downharpoonright \\ E_{\mathrm{d}} \end{array} \frac{\rightleftarrows E \rightarrow \text{column outlet}}{} . \tag{5.8}$$

Here E is the native (folded) state, and E_{d} is the inactive (unfolded) state. When the mobile-phase composition is sufficient to cause denatured protein to begin to migrate, RNase is transferred from the stationary to the mobile phase. Some denatured protein in solution refolds to the native phase. Also, because the solvent strength at this point is greater than that required to elute the native state, all protein molecules that refold at 25°C should elute without readsorption on the stationary phase.

The denaturation mechanisms presented in the preceding studies have all included the two-state mechanism that classically yields first-order kinetics of denaturation. Proteins–enzymes do denature by other mechanisms, for example parallel (Sadana, 1991), series involving intermediate or intermediates (Henley and Sadana, 1985, 1986, 1987), and other mechanisms. Besides, heterogeneity in either the surface or the protein molecule itself (Malhotra and Sadana, 1987) would significantly affect the denaturation characteristics of the different proteins adsorbed on different chromatographic surfaces utilized in their separation. Heterogeneities may also arise due to the adsorption process, which may yield a distribution of protein states adsorbed on the surface. The influence of a distribution in adsorbed protein states would significantly affect the denaturation–inactivation characteristics of proteins–enzymes (Malhotra and Sadana, 1990). This, in turn, would directly affect the quality or consistency of the protein product separated. These factors should also be considered in future studies when analyzing the denaturation behavior of different columns utilized to separate a wide variety of proteins–enzymes, and other biological macromolecules of interest.

Flurer *et al.* (1988) characterized slurry-packed fused-silica microcolumns of 250 μM i.d. for use in HPLC studies of proteins. Their work utilized the reversed-phase and size-exclusion chromatographic modes for the separation of standard protein mixtures. Their studies indicated that microcolumn liquid chromatography (LC) techniques have considerable potential in the investigations of very small protein samples. Some of the proteins studied include thyroglobulin, albumin, ovalbumin, α-lactalbumin, cytochrome c, and insulin. The authors indicated that these microcolumns have the following advantages: (1) significantly less adsorption behavior due to the column materials inherently associated with this column technology, as well as the drastically reduced quantities of the sorption materials contacting the protein molecules (which should be of considerable significance in minimizing protein denaturation, especially for the more labile proteins); (2) enhanced mass sensitivity of the concentration-sensitive detectors (based on UV absorption or native fluorescence) that are typically employed in protein detection; and (3) compatibility with small-scale manipulations. A serious drawback of the system is the limitation on the utilization of a step gradient for the analysis of complex protein samples, due to the large number of segments that would be required.

Lau *et al.* (1984) utilized a series of five synthetic peptide polymers of 8, 15, 22, 29, and 36 residues with the sequence Ac-(Lys-Leu-Glu-Ala-Leu-Glu-Gly)$_n$-Lys-amide where n = 1 to 5 to examine protein denaturation during RP-HPLC. In all cases a linear relationship between the natural logarithm of the monomeric molecular weight and retention volume was obtained for these peptides, indicating that the 29- and 36-residue dimers had been dissociated on binding to the reversed-phase columns. The authors indicate that the vast majority of proteins are denatured on binding to the hydrophobic matrix. The authors further emphasized that even the ultrashort (C_3), 300Å pore matrix with relatively low carbon loading (2.9%) does not prevent denaturation of these extremely stable synthetic two-stranded α-helical coiled coils. They indicated that although they have shown the organic mobile phase used in RP-HPLC can cause denaturation and disruption of the dimer, the hydrophobicity of the matrix is the important factor. If it is desirable to separate proteins in their native conformation, the hydrophobicity of the matrix must be significantly reduced. Also, the organic mobile phases would have to be replaced with nondenaturing solvent systems. The authors also indicated a cautionary statement in that the fact that the enzyme can be found during hydrophobic HPLC and can be recovered with full biological activity, it does not imply that denaturation or partial denaturation has not occurred. Proteins can rapidly renature on release from the support.

RPC proteins are bound to matrices that are generally more hydrophobic than in HIC. This increased hydrophobicity is achieved through a higher density of alkyl ligands on the matrix surface. Entropy drives the interaction of the protein with the hydrophobic surface. In RPC the interaction is so strong that proteins may not be eluted with buffers. Desorption is accomplished through the introduction of an organic additive in the mobile phase. Protein denaturation becomes thermodynamically favorable as organic solvents assist in desorbing the proteins in an unfolded form, and could cause the proteins to fold into inactive or less active structures. The denaturation is generally irreversible resulting in loss of enzyme activity.

E. Hydrophobic Interaction Chromatography

HIC is based on a mild adsorption process, yielding protein fractions in a biologically active state. Note, however, that protein structural changes may occur on any surface depending on the protein, the mobile-phase composition and pH, and the column temperature. Thus, proteins may be structurally changed either on HIC or on RP-HPLC depending on the conditions. Regnier (1987) indicated that displacement phenomena also occur in HIC. Because HIC is carried out on weakly hydrophobic columns in high concentrations of salt such as ammonium or sodium sulfate, there is substantial hydration of the weak hydrophobic coating both on the column and on the surface of the protein (Arakawa and Timasheff, 1984). Melander *et al.* (1984) indicated that the very high surface tension of these solutions provides sufficient driving force so that hydrophobic groups on the external surface of a protein interact with the weakly hydrophobic surface of the column. When this occurs, both the water and the salt are displaced from the protein-sorbent interface (Wu *et al.*, 1986). Regnier (1987) pointed out that this displacement process has been coupled to variations in the three-dimensional structures of lysozyme, β-lactoglobulin A, and cytochrome c.

Column packings in use utilize short hydrophobic ligands or polyethylene glycol grafted to the surface. A number of materials have been developed for the high-performance hydrophobic interaction chromatography of proteins. TSK Phenyl-5PW is a polymeric material with pendant phenyl groups (Kato *et al.*, 1984). Fasnaugh *et al.* (1984) made a polyamine-coated silica react with a series of acid anhydrides or chlorides to produce materials with different ligands in the coating. The protein mixture usually is applied in a buffer containing a high concentration of a salt high in the Hofmeister series (e.g., citrate or sulfate) (Melander *et al.*, 1984). Proteins are "salted out" under these conditions on the surface of the packing. The column is eluted with a gradient of decreasing ionic strength or increasing gradients of glycols, sucrose, or detergents. These conditions cause proteins to be eluted in the order of increasing hydrophobicity.

Alpert (1986) prepared a series of bonded poly (alkyl aspartate) coatings on silica. The products showed a wide range of hydrophobicity. Recovery of protein applied to columns packed with the material was generally in excess of 90%. Most enzymes were eluted with no loss in activity. These enzymes are cytochrome c, ribonuclease A, myoglobin, conalbumin, α-chymotrypsin, and α-chymotrypsinogen A. If a high ionic strength of the medium was used, then the enzyme will be more stable because a high ionic strength will mean a weak hydrophobic interaction. Low recoveries have been observed for labile proteins such as interleukin-1 from human leukemic monocytes (Lachman *et al.*, 1985) and a cytolic factor from *Bacillus* Calmette-Guerlin (BCG)-stimulated murine peritoneal macrophages (Alpert, 1986).

Regnier (1987) indicated that generally the protein surface area involved in the adsorption process during chromatography is no more than a few hundred square angstroms. The alteration of a single amino acid in the spatial orientation of amino acids within this region can have a major impact on adsorption. Bott and Sarma (1976) stated that there are seven variants of avian lysozyme, all of which have nearly identical structure by X-ray crystallography. When these isoenzymes are analyzed by HIC, Fasnaugh and Regnier (1986) indicated that one particular region of the molecular surface opposite the catalytic cleft dominates chromatographic retention. This area is from residue 41 to 102 and from residue 75 to the α-helical region that starts with residue 89. Regnier (1987) emphasized that amino acid substitutions within this region alter chromatographic behavior, whereas a substitution at the other external faces of the protein had no influence on chromatographic retention.

Wu *et al.* (1986) studied the thermal behavior of a series of standard proteins in HIC. Lysozyme maintains a stable conformation over the temperature range 10 to 40°C, and β-lactoglobulin A has a conformational transition between 25 and 40°C. Calcium-dependent α-lactalbumin, a rather labile species, maintains a stable conformation up to about 20°C and then undergoes structural changes. Also, cytochrome c appears to be relatively stable up to about

65°C. These authors indicated that because cytochrome c undergoes conformational changes at about 35°C on more hydrophobic surfaces, the extent of hydrophobicity of the stationary phase is important for the maintenance of the native state. Their results indicated that HIC appears promising at subambient temperatures. These authors also emphasized that the structural changes observed do not necessarily represent complete unfolding or denaturation. Localized conformational changes can occur (Ingraham *et al.*, 1985) and presumably many of these are reversible.

Ingraham *et al.* (1985) analyzed the denaturation and the effects of temperature on the hydrophobic interaction and RP-HPLC of proteins. A comparison of the temperatures at which various polypeptides underwent denaturation on the column with their normal melting temperatures (where half the molecules are unfolded) demonstrated that the hydrophobic column itself, rather than the temperature, was primarily responsible for denaturation. Hence, even relatively gentle hydrophobic columns can promote denaturation of protein structure. These authors emphasized that because the tertiary and quaternary structures of most proteins are stabilized by hydrophobic interactions, the possibility of denaturation must always be taken into account when a hydrophobic column is used.

Goheen and Engelhorn (1984) described protein separations by means of a new hydrophobic interaction HPLC column. Proteins such as cytochrome c, myoglobin, and lysozyme were eluted as sharp peaks. Proteins such as albumin and β-lactoglobulin were not eluted as sharp peaks. The authors indicated that the broadening of some peaks may depend on the influence of the eluent on the structure of those proteins. The authors emphasized that small adjustments in chromatographic conditions can greatly improve resolution when band spreading exists. They also indicated that lowering the temperature is in many cases the most favorable means of improving resolution for the isolation of purified enzymes or other natural products for which retaining the native structure is crucial.

Fasnaugh *et al.* (1984) compared HIC and RPC of proteins. Selectivity of the HIC column was easily manipulated by changing mobile phase conditions. Protein retention was increased by decreasing the pH from neutrality or by using a salt with a greater "salting-out" ability. Also, chemical modification of the matrix surface alters selectivity. The authors noted several differences between HIC and RPLC columns. Hydrophilic proteins such as cytochrome c and myoglobin were weakly retained on the reversed RPLC column. The hydrophobic protein, β-galactosidase, was strongly retained on the HIC column and only weakly retained on the reversed-phase liquid chromatogrphic (RPLC) column. Other proteins were retained equally by RPLC and HIC columns. These authors emphasized that surface hydrophobicity is apparently the most important factor in the extent of protein retention, although size does contribute.

From the HIC column the recovery of lactate dehydrogenase exceeded 90%, while the recovery of α-chymotrypsin exceeded 86%. From the RPLC column only 54% of lactate dehydrogenase, 91% of α-chymotrypsin, and none of the activity of β-galactosidase were recovered. The authors indicated that this is due to the fact that protein separation on an HIC column is based on native hydrophobicity, while that on a RPLC column is based on the hydrophobicity of the denatured protein. The authors emphasized that in HIC the native structure is not altered through interaction with the column or the mobile phase. In the RPLC column, the proteins are at least partially denatured through interaction with the mobile phase and the column material.

The preceding few studies have attempted to address a major problem in the effective utilization of chromatographic techniques, that is, the extent of protein denaturation encountered on different columns. Dunnill (1983) felt that chromatographic techniques do exhibit the potential of being successfully applied on an industrial scale. This was going to happen with the advent of some appropriate information. This is now illustrated as follows for the scale-up of HIC purification of the antitumor antibiotic SN-07.

Example 5.8

Provide an analysis of the scale-up of HIC purification of the antitumor antibiotic SN-07 (Ishida *et al.,* 1989).

Solution

Ishida *et al.* (1989) analyzed the scale-up of HIC purification of SN-07. The height equivalent to a theoretical plate of the adsorbent gel (Sepabeads FP-PH12) was kept constant for various superficial liquid velocities and column diameters. The scale-up of HIC purification of SN-07 was possible by increasing column diameters from 16-mm to 113-mm i.d. without a decrease in the efficiency. After scale-up the purity and yield (or recovery) of SN-07 were almost unchanged. Besides, the process time was constant.

SN-07, a macromolecular antibiotic, was isolated from the cultural supernatant of a soil bacterium *Actinomadura roseoviolacea* ar. miuraensis nov. var. (Kikuchi *et al.,* 1985). SN-07 consists of a high-molecular-weight nucleic acid and a low-molecular-weight chromophore (Yajima *et al.,* 1987). Ishida *et al.* (1989) indicated that an aqueous solution of SN-07 exhibits the maximum absorption at 260 and 505 nm. Also, SN-07 exhibits antibacterial activity against Gram-positive bacteria, and is effective against lymphocyte leukemia P388 and melanoma B16 *in vivo* (Yajima *et al.,* 1987).

Figure 5.3 shows the successive purification stages involved in the separation of SN-07. Ishida *et al.* (1989) analyzed the effect of the HIC column size on the purity and yield of the SN-07 separation. Sepabeads FP-PH12 were packed in columns (16 and 113 mm i. d., bed height 400 mm). The columns were kept at 5 ± 1°C.

The height equivalent to a theoretical plate (HETP) of the columns was determined by the impulse method using NaCl solution (Hamilton *et al.,* 1960). From the impulse-response curve

$$\text{HETP} = LW^2/4t^2, \tag{5.9}$$

where t is the retention time in minutes, L is the height of the gel bed in cm, and W is the bandwidth of the impulse-response curve in minutes at a height of the peak height multiplied by exp (-1/2).

The authors noted that there was no appreciable change in the HETP values, which implied that the velocity distribution in the column was uniform. Also, the HETP value at a low superficial liquid velocity was nearly twice the average diameter of the gel bead (Nakanishi *et al.,* 1977).

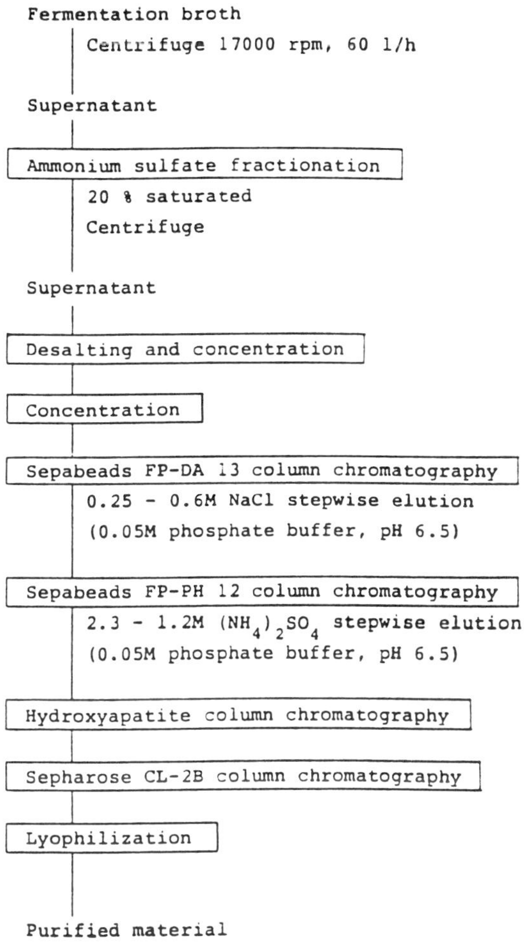

FIGURE 5.3 Purification stages involved in the separation of SN-07. [From Ishida, S. *et al.* (1989). *Bioprocess. Eng.*, *4*, 163.]

The authors emphasized that the scale-up of HIC was readily attained due to:

1. The yields in the absorbance unit at 505 nm/dm^3 of broth were 3.5 for the 16-mm i.d. column and 3.3 for the 113-mm i.d. column, respectively.
2. There was no significant change in the elution profile and absorption spectra of the effluent from either of the two columns. The primary reason for the similarities in product from the two columns is the uniform flow distribution. Because there was no significant change in the absorption spectra, it is reasonable to assume that the activity of the product obtained from both the columns exhibited corresponding antibacterial activities.

The preceding study presented by Ishida *et al.* (1989) is of significant interest because it provides information for a 50-fold scale-up of the recovery of an antibiotic SN-07. More such studies are required for the recovery and isolation of other antibiotics. Though the yields were high, the authors did not speculate on some possible denaturation mechanisms that may cause a decrease in either the yield or the purity (activity) of the recovered antibiotic; perhaps there was no need to do so. Nevertheless, it is of interest to collect and obtain information (mechanistic and otherwise) for the loss of antibiotic yield and purity so that one may attempt to reasonably explain such losses when they occur during the recovery of other antibiotics.

F. Mechanistic Considerations

Wetlaufer and Koenigbauer (1986) subjected three proteins to HIC in the presence of submicellar concentration of the surfactant (3-[3-cholamidopropyl]dimethylammonio-1-propane sulfonate (CHAPS). At several concentrations of CHAPS below the critical micelle concentration, CHAPS increased the retention of lysozyme and pancreatic trypsin inhibitor, but decreased that of ribonuclease A. The authors presented a scheme of multiple equilibria and their kinetic components as a basis for formulating the dependence of retention on surfactant concentration.

Figure 5.4 shows the principal processes involved in surfactant-mediated chromatography. The situation is analogous to proposals for ion-pairing mechanisms in the chromatography of small molecules (Bidlingmeyer *et al.*, 1979). Even a small protein has the potential for associating with the surfactant molecules. The formation of protein-surfactant molecules is described by a series of equilibria involving the successive binding of 1,2, . . . m surfactant molecules. Surfactant ligands will often form micelles. The protein can undergo a significant structural disorganization, concomitant with the binding of additional surfactant molecules. This significant structural disorganization represents denaturation.

Later, Wetlaufer (1988) analyzed the reversible and irreversible denaturation of proteins in chromatographic systems from kinetic and equilibrium view-

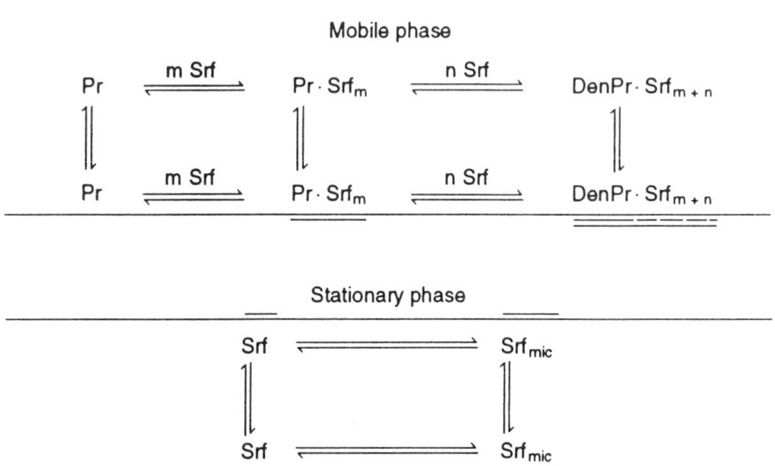

FIGURE 5.4 Scheme for displaying equilibria in surfactant-mediated chromatography. Species bound to the stationary phase marked with double-line sub- or superscripts, for example, Pr · Srf. Pr = native or active protein, Den · Pr = denatured protein, Srf = surfactant monomer, and Srf_m = micellar aggregaten surfactant. [From Wetlaufer, D. B. and Koenigsbauer, M. R. (1986). *J. Chromatogr.*, *359*, 55.]

points, preliminary to consideration of the binding of proteins to stationary phases. This author indicated that protein binding to a stationary phase may result in concomitant stabilization or destabilization. Experiments to clarify mechanisms or to lead to stabilization of proteins against denaturation are discussed.

More studies like those of Wetlaufer and coworkers (1986, 1988) are urgently required that clearly delineate the mechanisms of protein dentauration during not only HIC but also other modes of chromatographic separation-purification of proteins. Understandably this information is difficult to get. The whole area of protein activity loss (recovery) and mechanisms of such loss are hardly available, because, in general, researchers are loathe to present the negative aspects of their separation procedure that will detract from the efficiency of their process. Studies on purity (structural changes, if any) of the protein recovered are even rarer. This includes researchers in academia as well as in industry for different reasons. Nevertheless, this information is critical. Thus, it should be presented, be readily available, and be disseminated widely in the literature so that researchers working with a wide variety of proteins may greatly benefit besides being provided with a complete picture of the protein separation process. For example, significant changes in either the upstream or the downstream processes may become essential if the protein activity-structure upon completion of the purification stages does not match standards, or if the product specifications become more stringent. Also, as more competitors enter the market manufacturing the same protein antibiotic, or other biological macromolecule, more emphasis will need to be shifted to improve recoveries and purity of these biological products during their separation. This would significantly influence upstream or downstream processes. What will really be excruciating is to satisfy the demand to improve both recovery and purity of the biological macromolecules separated. It is thus essential to develop a knowledge base for this.

G. Subunit Exchange Chromatography

Because the isolation of biomolecules such as enzymes, hormones, and other active proteins and biological macromolecules has been continuously growing, several new techniques have been developed. Affinity chromatography, which exploits the specific interaction between an immobilized ligand and a protein in solution, is an important and effective purification tool for small-scale separations. Affinity chromatography on immunosorbents is widely used in downstream processing. Dissociation of the antigen–antibody complex, however, presents a serious denaturation problem. Porath (1985) reviewed the high-performance immobilized metal-ion affinity chromatography of peptides and proteins. He indicated that all possible interactive contributions from the metal-ion microenvironment should be taken into account. These include the matrix, spacer arm, and chelating ligand; and of course, the properties of the surrounding liquid medium and its dissolved solute species. This author demonstrated that high resolution of peptides and protein mixtures could be obtained. Under specified conditions, more than 1000 runs could be made repeatedly on a single column without a major change in pattern or recovery. However, no information was provided on the extent of protein denaturation encountered in the column.

Kang and Ryu (1991) analyzed the scale-up parameters of an immunoglobulin separation system using protein A affinity chromatography. These authors emphasized that because both the ligand and the product are biomolecules of high-molecular weight, they are both subject to significant conformational changes during either the immobilization process (ligand) or during actual process operations. Thus, it is difficult to generalize the scale-up criteria from one system to another. Nevertheless, the authors indicated that the scale-up should consider (1) the characteristics and stability of the product related to the separation conditions and process time, and (2) the tolerance of the gel matrix to the pressure. The authors noted that the dimensional scale-up of the column in the radial direction increased the total purification capacity linearly as the cross-sectional area increased without increasing pressure drop; there was, however a dilution in the product concentration. Similarly, a scale-up of the axial column dimension increased the concentration of the product in the eluate, although the retention time increased linearly with the increase in gel height.

Heparin, an anticoagulant, has significant biomedical uses. It is a sulfuric ester of a complex carbohydrate containing glucosamine glucuronic acid (Fareed, 1985). Heparin acts mainly by selective binding to antithrombin III, a plasma proteinase inhibitor, thereby increasing its inhibitory activity. Dyr and Suttnar (1991) indicated that heparin also helps in inactivating several other vitamin K-dependent coagulation factors. Andersson et al. (1974, 1975) suggested that due to its polyanionic nature, heparin interacts with many cationic biological compounds and the interactions have been used as a basis for the purification by affinity chromatography on soft gels.

Example 5.9

Provide an example for heparin HPLC separation of proteins (Dyr and Suttnar, 1991).

Solution

Dyr and Suttnar (1991) examined the separation of human coagulation factor IX concentrate (which contains at least 50-fold more factor IX than fresh plasma, 0.85 IU of factor IX in 1 mg of protein) on heparin-Glc-Spheron. The concentrate contained approximately 2% of factor IX; and other factors II, VII, and X were also present. Factors II, VII, and X were eluted with the starting buffer. Factor IX was retained on the column and was eluted with a sodium chloride gradient. The purification factor calculated as compared with factor IX concentrate was approximately 60%, and the recovery was 80%. The preparation was homogeneous according to anion-exchange chromatography and immunoblotting (Dyr *et al.*, 1987) after SDS-PAGE.

Kato *et al.* (1986) described the immobilized metal-ion affinity chromatography (IMAC) of proteins on TSK-gel chelate—5PWE. They did not present any information on the mechanisms and the extent of denaturation encountered. Yip *et al.* (1988), too, utilized the same IDA (iminodiacetate) TSK-gel to undertake a methodological study of peptide on immobilized Cu^{II}, Ni^{II} and Zn^{II}. These authors claimed satisfactory resolution. Besides, the column lifetime was acceptable. Human serum was fractionated on zinc and cadmium IDA and tris(carboxymethyl)ethylenediamine chelates. Zinc IDA was found to be useful in the isolation of hemopexin and α_2-macroglobulin. However, like in the previous study no information was provided about: (1) the mechanisms or extent of denaturation of proteins–enzymes, and (2) the conformational changes encountered by the protein during the subunit exchange chromatographic separation process. Either the authors, in general, do not study this aspect, or they minimize the importance of it while reporting their studies. This is not quite appropriate—for surely conformational changes–protein denaturation do occur during the different separation processes, chromatographic or otherwise. These conformational changes or losses in activity should be reported. Once this type of information is collectively available, then remedial measures may be undertaken that help minimize these conformational changes or losses in protein activity. It is understandable that authors, in general, are reluctant to report negative aspects of their studies, but some of these may be important enough to provide feedback that makes it mandatory to change significantly or otherwise the variable conditions during either upstream or downstream processing.

Andersson (1984) noted that the chromatographic behavior of serum proteins on zinc IDA would be modulated by the nature of the eluent; a sulfate-containing solvent increased the binding of serum proteins to immobilized metal ions. Once again, no information was provided as in the previous two IDA studies concerning the mechanisms or extent of protein denaturation.

El-Rassi and Horvath (1986) made an extensive study of the importance of many parameters involved in the interaction of proteins with Cu^{II}, Zn^{II}, Fe^{II}, and Fe^{III}. They confirmed earlier findings and also presented new and interesting results with methanol in the mobile phase. IMAC is a major step in the purification of tissue plasminogen activator (t-PA). According to Porath (1988) the future of IMAC is bright. It may be used for the immobilization and isolation of polysaccharides, nucleotides, and nucleic acids; and also for structures of higher order, such as virus particles. We recognize that affinity chromatography including IMAC and IDA are widespread and have developed their own potential. Many more examples could be provided, however, none, at least to the best of the author's knowledge discussed or even mentioned protein denaturation during chromatographic separation.

The major intent of this chapter is to focus on, and hopefully convince others to analyze and to report, the conformational changes and the state of the protein as it is separated. Only after the establishment of a reasonable pool of such information and the necessary framework, will one be able to reasonably address this problem of minimizing protein conformational changes–losses in activity encountered during the bioseparation of proteins, antibiotics, and other biological macromolecules. Furthermore, as far as this chapter is concerned, we have not assessed the techniques in a disadvantageous fashion but merely wish to point out the current lack of information on protein and other biological macromolecular denaturation during separation. We believe that such information is definitely required to further evaluate and place this technique in proper perspective.

Antonni *et al.* (1979) proposed a method that in some aspects resembles affinity chromatography. In their method, subunit exchange chromatography has been employed for the study of subunit interactions and for the purification of several self-associating proteins (Antonni *et al.*, 1975; Bodge and Wagner, 1975; Takahashi and Gross, 1978; Carrea *et al.*, 1979). The principle is to immobilize the monomers of a polymeric self-associating protein on a solid matrix and to use it for extracting the soluble subunits of the same protein from a multicomponent system. Due to the great specificity of subunit interactions, extraneous proteins and compounds should not be extracted.

Carrea *et al.* (1983) investigated the purification of chymotrypsin by subunit exchange chromatography on the denatured protein. Chymotrypsin subunits immobilized under denaturing conditions and packed in a column proved to be suitable for the purification of chymotrypsin from both bovine and porcine extracts. The subunit exchange chromatography of such extracts, carried out between pH 2.5 and 4, gave an eight fold purification with a 93% recovery of chymotryptic activity. Note that column capacity after seven to eight runs was about 70% of the original one. A partial recovery was obtained by washing the matrix with 6 M urea or 2.5% Tween 80. Column stability under operating conditions was good because no leakage was observed over several days storage at pH 2.5 or 4.2 at room temperature. Also, the interaction between immobilized and free enzyme was shown to be specific because extraneous proteins and compounds were not bound by the derivatives.

Carrea *et al.* (1985) purified glucagon by subunit exchange chromatography. The authors immobilized glucagon onto sepharose matrices activated with CNBr or tresylchloride, as a function of several parameters, including pH of coupling, concentration of added polypeptide, and presence or absence of urea. The subunit exchange chromatography of the extract gave a 90% pure product. The overall recovery of the process was approximately 60%.

Subunit exchange chromatography seems to give reasonable values of protein recovery and purity. More effort needs to be spent in further evaluating the potential of this technique for the separation of proteins and other biological macromolecules of interest.

Monoclonal antiobodies that recognize conformation-dependent epitopes may be used for the purification of proteins from *Escherichia coli* fermentations using immunochromatography. In this procedure the purification of enzymes and proteins is possible where: (1) these biological macromolecules are involved in receptor-ligand interactions, and (2) correct folding of the protein is essential for activity.

Example 5.10

Provide an example of a large-scale immunoaffinity purification of recombinant soluble human antigen (sCDS) from *E. coli* cells (Wells *et al.*, 1993).

Solution

Wells *et al.* (1993) developed a large-scale immunoaffinity (IA) purification process for the bioseparation of recombinant soluble antigen CDS (sCDS) from *E. coli* fermentations. Their process utilized a monoclonal antibody that recognized a conformation-dependent epitope on the surface of domain 1 of CD4. Their procedure was utilized to purify sCD-183 (that contains the N-terminal 183 amino acids of human CD4), and sCD4-PE40 (a fusion protein that consists of the N-terminal 178 amino acids of CD4 and amino acids 1 to 3 and 253 to 613 of *Pseudomonas* exotoxin A (PE40). These authors obtained high recoveries for sCD4-183 (71%) and for sCD4-PE40 (79%). Their results emphasized clearly that directing immobilized antibodies against conformational epitopes is a useful procedure, and can be used for the quick purification of correctly folded protein from a mixture of (oxidized) *E. coli* proteins.

Figure 5.5 shows the flow diagram utilized for the purification of sCD-PE40 from *E. coli* fermentations. Wells *et al.* (1993) indicated that the IA-chromatographic step was placed downstream of a protein oxidation step and preliminary chromatography steps. These preliminary chromatography steps consisted of immobilized metal-affinity ion-exchange chromatography steps.

Wells *et al.* (1994) emphasized that monoclonal antibodies directed against three-dimensional conformation-dependent epitopes select for *correctly* folded proteins in the prior oxidation step. If the proteins are not correctly folded, then the protein lacks the epitopes that are recognized by the conformation-dependent monoclonal antibodies. In the folding of domain 1 of CD4, the formation of a disulfide bond between cysteine residues 16 and 84 is critical.

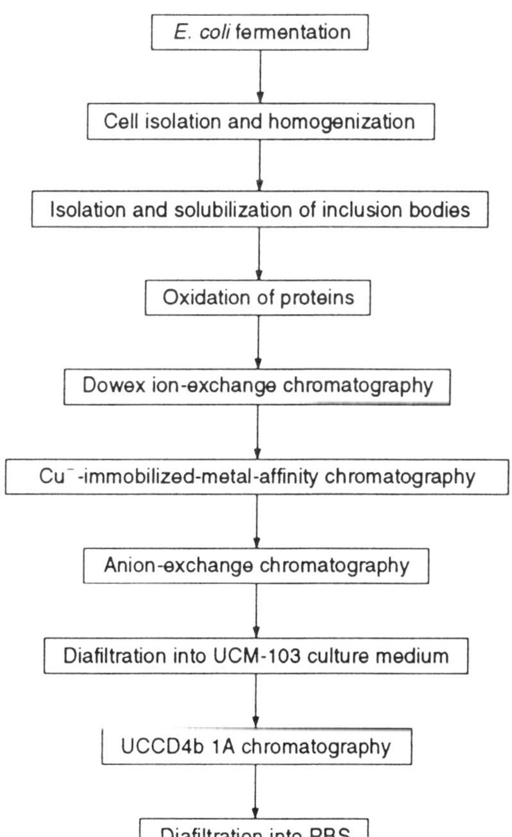

FIGURE 5.5 Flow diagram for the purification of sCD-PE40 from *E. coli* fermentations. [From Wells, P. A. *et al.* (1993). *Biotechnol. Appl. Biochem.*, *18*, 341.]

They indicated that the correct folding of domain 1 guarantees the correct folding of the two-domain sCD4.

Also, the IA columns were effectively utilized for hundreds of purification runs. Thus, this procedure may be economically feasible for some projects. The strong positive points for this procedure are the rapid process and the high recovery of the protein. A major (economic) disadvantage of this process, of course, is the high cost of the large-scale IA columns. However, the authors emphasized that prior to selecting alternative processes, cost comparisons should be made between IA purification and other processes, along with product losses that may occur in these processes.

Wajant *et al.* (1993) described a one-step chromatographic method to purify (S)-hydroxynitrile lyase (hydroxymandelonitrile lyase; EC 4.1.2.11) to electrophoretic homogeneity from *Sorghum bicolor* L. (sorghum). Monoclonal antibodies (IgG1) that recognize one or both of the M_r 33,000 and 22,000$_r$ subunits of HNL were coupled to an Affi-Gel 10 column.

The HNL is used for the stereoselective synthesis of α-hydroxynitriles (Effenberger *et al.*, 1987; Jansen *et al.*, 1992). Wajant *et al.* (1993) indicated that because of the practical importance of (S)-HNL from *S. bicolor* L. in the synthesis of cyanohydrins, quite a bit of effort has been spent in purifying the S-HNL (Smitskamp-Wilms *et al.*, 1991; Jansen *et al.*, 1992). Therefore, Wajant *et al.* (1993) utilized the purification of (S)-HNL from *S. bicolor* L as a model compound to help separate related enzymes from other species.

Sorghum seedlings were homogenized in liquid N_2 and extracted with two volumes (v/w) of 150 mM Tris (pH 7.7)/500 mM NaCl for 1 h at 4°C. The extract was filtered for 40 minutes in a MSE centrifuge. Thereafter, the supernatant was sent through the Affi-Gel 10 column.

The crude extract (100 ml) from the centrifuge contained 217.5 U of HNL activity, and the specific activity was 0.914 U/mg. Three rounds of purification were performed. After the first round of purification, the total activity recovered in the eluate was 80.5 U. The specifiic activity of the HNL was 222.8 U/mg giving a purification factor of 244. The recovery percentage was 80.5/217.5 = 43%. The extract from the column on the first passage contained 93.6 U of total activity and was applied again to the column as the second round of purification. The total activity recovered in the eluate was 56.6 U, yielding a recovery percentage of 56.6/217.5 = 26%. The specific activity in this case was 266.4 U/mg yielding a purification factor of 247. The extract from the column on the second passage contained 8.7 U of total activity, and was applied again to the column as the third round of purification. The total activity recovered in the eluate was 8.7 U, yielding a recovery percentage of 8.7/217.5 = 4%. The total activity recovered in the pooled eluate was 43 + 26 + 4 = 73%.

Wajant *et al.* (1993) emphasized that their procedure leads to a loss of activity that is only a half of what has been reported with other FPLC techniques for purifying HNL from sorghum. Smitskamp-Wilms *et al.* (1991) indicated a 50% recovery, whereas Jansen *et al.* (1992) recovered 40% HNL activity.

Example 5.11

Briefly describe a method to purify α-amylase by immunoaffinity chromatography with a cross-reactive antibody (Katoh and Terashima, 1994).

Solution

Katoh and Terashima (1994) purified two isoenzymes of rice α-amylases expressed and secreted by recombinant yeast by immunoaffinity chromatography using cross-reactive antibody. These authors stated that often concentrations of secreted products in fermentation broths genetically engineered microorganisms are in the range of micrograms to milligrams per liter. Immunoaffinity chromatography represents a convenient means of purification because of the high affinity and specificity between antigens and antibodies. These authors proposed using a cross-reactive antibody as the ligand in immunoaffinity chromatography. This overcomes the selection of a monoclonal antibody showing suitable binding characteristics with a target protein. Katoh and Terashima (1994) indicated proteins that exhibit similar functions from various species often have high homology in their amino acid sequences. Thus, antibodies against these proteins would exhibit cross-reactivity with each other. Thus, an antibody that cross-reacts with the target protein could be obtained by immunizing easily available proteins, which exhibit homology with the target protein.

Katoh and Terashima (1994) obtained antibody ligands by immunization of commercially available barley α-amylases for the purification of rice α-amylases. Rice α-amylases were concentrated and purified by a single immunoaffinity chromatograpic step. SDS-PAGE indicated a single band, and hence the high purity of the α-amylase purified. The single-step immunoaffinity chromatographic procedure yielded a protein with a specific activity of 220 U/mg with a recovery of 75%. This specific activity represented a 2000-fold purification, because the fermentation broth concentration of α-amylase was 0.11 U/mg solid. Furthermore, these authors indicated that the purification of α-amylase by the antibarley α-amylase column showed several bands by SDS-PAGE indicating impurities. Also, the specific activity and recovery were 120 U/mg and 20%, respectively. Katoh and Terashima (1994) emphasized that using a selective and high affinity immunoaffinity column during the early stages of the purification process greatly simplifies the process, besies yielding a high purity product along with high product recovery.

H. Other Chromatographic Techniques

Kralova *et al.* (1986) employed column chromatography for the removal of contaminating proteinases from crude preparations of microbial enzymes. These authors recommended the removal of contaminating proteinases in the early stages of purification. These authors employed column chromatographic methods based primarily on bioaffinity of proteinases, the purified enzyme remaining unbound. In the case of thermally stable enzymes, the proteinases were denatured by heating. The crude preparation of glucose isomerase, obtained after disintegration and extraction of *S. nigrificans* contained proteins with proteinase activity that caused considerable losses of glucose isomerase activity. By considering the thermal stability of glucose isomerase, the authors tried denaturation as a method of proteinase inactivation, and recommended heating at 60 °C for 30 min in the presence of stabilizing ions, Mg^{2+} and Co^{2+}. After about 30 min, the glucose isomerase activity is unchanged, whereas the proteinase activity drops to about 15% of its original value. This, however, cannot be used in the case of uricase from *Candida utilis* having low thermal stability. Instead, the authors recommend removing the majority of proteinase activity (92%) chromatographically on a column of uric acid, on which practically no uricase is bound.

Safarik (1984) isolated trypsin by column chromatography on tea particles. Because of autoproteolysis, proteolytic enzymes are usually isolated with low yields when a multistep isolation procedure is used. The yield can be increased when one-step techniques are used for their isolation and purification. Safarik (1984) used inexpensive materials for the rapid isolation of various proteases. He showed that tea particles are suitable for the isolation of trypsin, which was used as a model protease.

Frey (1991) indicated that preparative chromatography is an attractive means of purifying biomolecules (Garg *et al.*, 1990; Sofer and Nystrom, 1989). Frey (1991) analyzed methods for regulating process performance and product quality in preparative chromatography. Input–output models using discrete variables were provided utilizing control theory aspects to analyze relationships between output variables (yield, purity, and production rate) and input variables (cut point locations and feed slug size). The author clearly demonstrated that the two important parameters of product separation, the yield and purity, can be effectively regulated in elution chromatography by using the feed slug size and two cut points (as determined by a nonspecific detector or timer).

We have examined different HPLC techniques for the purification of some proteins–enzymes from dilute solutions such as fermentation broths and bioreactor fluids. The techniques demand high selectivities and a mild operating environment. The HPLC purification technique does have two major limitations:

1. The solid support used as the stationary phase often has a negative effect on delicate substances such as proteins and antibiotics because they may irreversibly adsorb, denature, and destroy biological activity or even otherwise alter molecular integrity.
2. Operating costs for HPLC are prohibitively high due to the regulatory requirements for frequent replacement of support materials to prevent cross-contamination of product batches.

Cazes (1988) indicated that centrifugal partition chromatography (CPC) eliminates these problems by eliminating the solid stationary phase. In CPC, separation is achieved by liquid–liquid partition and countercurrent distribution of product streams between two immiscible liquid phases. The stationary phase in CPC is a solvent that is retained in the column by centrifugal force. The liquid mobile phase flows through the stationary phase, as in conventional chromatography.

Example 5.12

Describe the separation of enzymes and long-chain fatty acids by CPC (Cazes, 1989).

Solution

Cazes (1988) indicated that CPC can isolate and purify a wide range of biochemicals such as natural products (plant extracts, coenzymes, and vitamins), biopolymers (enzymes, proteins, DNA, and RNA), fermentation products (antibiotics, amino acids, and peptides), pharmaceuticals, and fine chem icals.

Long-chain fatty acids show promise as the therapeutic agents for the prevention of arteriosclerosis and stroke. The CPC process may be used to extract oleic, linoleic, and linolenic acid from cereal oils, and eicosapentanoic, docosapentanoic, and arachidic acid from fish oils. Initially, a preliminary reversed-phase elution of unwanted fatty acids and impurities is followed by a second-stage, normal-phase elution of the desired fatty acids. The choice of the solvents and the specific operating conditions are adjusted to suit the composition of the feedstock. The authors emphasized that because irreversible adsorption of highly retained components cannot occur, the total recovery of the feedstock components is the rule. This author emphasized that to preserve the biological activity of enzymes, separation systems should provide an extremely mild environment, free of catalytic surfaces and organic solvents. The CPC process is biocompatible because the mild operating conditions prevent decomposition and denaturation of valuable components and the absence of a solid support rules out catalytic support effects.

Figure 5.6 shows the isolation of L-leucine dehydrogenase from *Bacillus sphericus* in a two-stage process, with two CPC systems connected in series. Cazes (1988) utilized the first unit as a continuous extractor to isolate and concentrate the enzyme from cell debris and gross impurities. The second unit accepts crude enzyme from the extraction stage and converts it to a pure, biologically active product.

The author emphasized that by using the CPC process:

1. High recoveries are possible because irreversible adsorption cannot occur. Total recovery of feedstock components is the rule.
2. Scalability from laboratory to production scale is simple.

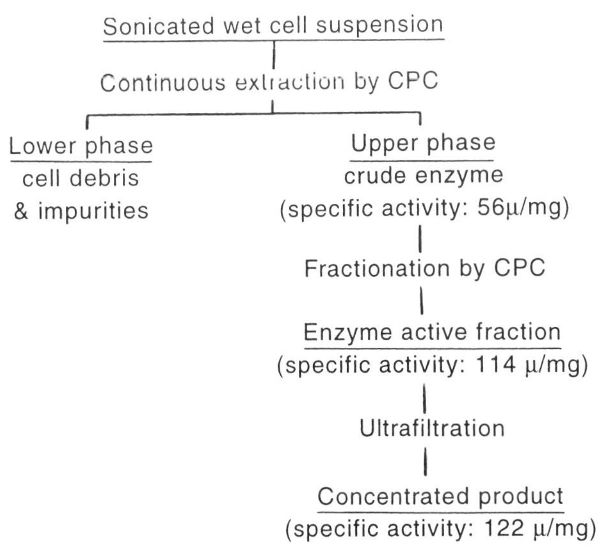

FIGURE 5.6 Isolation and purification of L-leucine dehydrogenase. [From Cazes, J. (1988). *Biotechnology,* December 1988.]

3. Versatility is provided because it can be performed with any two-phase solvent system.
4. Biocompatibilty exists because the mild operating conditions prevent decomposition and denaturation of valuable components.
5. Rapidity is possible because the processing times are relatively short due to the high mobile flow rates.

III. CONCLUSIONS

The analysis of the separation of proteins and other biological molecules of interest by different chromatographic techniques clearly indicates the scarcity of detailed information on the quantity (mass recovery) and the quality (structure) of the product recovered. In spite of an overall general agreement of the need for such knowledge, the availability of this type of information is difficult to get in the literature. The different HPLC techniques for the separation of proteins and other biological macromolecules and the denaturation encountered: (1) provide some numbers on the quality and quantity of the product recovered, and (2) help focus on the dearth of data required to assist in the control of these two aspects of product recovery. Such availability of data for protein product recovery is rare; data for other biological macromolecules such as antibiotics, peptides, and fatty acids are apparently (for all practical purposes) nonexistent. Even though, in general, chromatographic techniques have evolved to be rapid and effective resolution techniques, these type of data need to be made available so that one may correctly examine the true effectiveness of such techniques. In the extreme, it does not make much sense to separate a biological product from a dilute solution if it does not adhere to some strict product specifications, especially with regard to structural variations. More studies are urgently required that delineate and report the effect of processing conditions on both the quality and the quantity of the biological product separated.

The few mechanistic studies for protein denaturation and conformational transitions experienced by proteins during chromatographic separation presented here provide significant physical insights into the protein structural changes occurring on chromatographic surfaces. More such studies are urgently required because we need to probe the phenomena that cause denaturation of not only of proteins but also of other biological macromolecules such as antibiotics, peptides, and fatty acids.

Because chromatographic separation exhibits considerable potential for the industrial-scale separation of proteins and other biological macromolecules, significant effort needs to be spent on scale-up parameters. This is a particularly vexing problem because the biomolecules to be separated, in general, have high molecular weight and are delicate in structure, thus, generalizations will be difficult and prone to error. Nevertheless, a systematic approach is required to elucidate the effect of scale-up parameters on the quality and quantity of biological macromolecular product separated. With an increasing competitiveness in the market and an ever increasing demand for purer products, there will be considerable pressure on improving not only the quantity but also the quality of product separated. If process conditions exist that decrease either of these, then the quality-quantity of the product separated may even dictate changes in either the downstream or the upstream processing conditions.

REFERENCES

Ackland, K. E., Berndt, W. G., Frezza, J. E., Landgraf, B. E., Pritchard, K. W., and Ciardelli, T. L. (1991). *J. Chromatogr., 540*, 187.

Agarwal, V. P. (1989). *J. Liq. Chromatogr., 12*(4), 613.

Albert, E., Drysdale, J. W., Isselbacher, K. J., and Schur, P. H. (1972). *J. Biol. Chem., 247*, 3792.

Alpert, A. J. (1983). *J. Chromatogr., 266*, 23.

Alpert, A. J. (1986). *J. Chromatogr., 359*, 85.

Andersson, L. O., Miller-Andersson, M., and Borg, H. (1974). *Thromb. Res., 5*, 439.

Andersson, L. O., Borg, H., and Miller-Andersson, M. (1975). *Thromb. Res., 7*, 451.

Andersson, L. O. (1984). *J. Chromatogr., 315*, 167.

Antonni, E., Carrea, G., Cremonisi, P., Pasta, P., Rossi Fanelli, M. R., and Chiacone, E. (1979). *Anal. Biochem., 95*, 89.

Arakawa, T. and Timasheff, S. N. (1984). *Biochemistry, 23*, 5912.

Arora, G. and Lee, B. H. (1994). *Biotechnol. Appl. Biochem., 19*, 179.

Badar, S. and Edward, H. A. (1985). *J. Assoc. Of Anal. Chem., 68*, 5.

Baird, A. and Klagsbrun, M. (1991). *Cancer Cells, 3*, 239.

Beckman, L. and Beckman, G. (1967). *Biochem. Genet., 1*, 145.

Benedek, K., Dong, S., and Karger, B. L. (1984). *J. Chromatogr., 317*, 227 .

Benzonna, G. and Desnuelle, P. (1965). *Biochim. Biophys. Acta, 105*, 123.

Bidlingmeyer, B. A., Deming, S. N., Price, Jr., W. P., Sachok, B., and Petrusek, M. (1979). *J. Chromatogr., 186*, 419.

Bodge, J. and Wagner, K. G. (1975). *Biochem. Biophys. Res. Commun., 62*, 868.

Bott, R. and Sarma, R. (1976). *J. Mol. Biol., 106*, 1037.

Brokerhoff, H. and Jensen, R. G. (1974).*Lipolytic Enzymes*, Academic: New York.

Burgess, W. H. and Maciag, T. (1989). *Annu. Rev. Biochem., 58*, 575.

Carrea, G., Lugaro, G., Ninda, R., Vecchini, P., and Antonni, E. (1979). *FEBS Lett., 104*, 393.

Carrea, G., Pasta, P., and Antonni, E. (1983). *Biotechnol .Bioeng., 25*, 1331.

Carrea, G., Pasta, P., and Antonni, E. (1985). *Biotechnol. Bioeng., 27*, 704.

Cazes, J. (1988). *Biotechnology*, December 1988.

Clonis, Y.D. (1987). *Biotechnology, 5*, 1290.

Cohen, S. A., Dong, S., Benedek, K., and Karger, B. L. (1983). In *Symposium Proceedings, 5th International Symposium on Chromatography and Biological Recognition*, Chaiken, I. M., Wilchek, M., and Parikh, I., Eds., Academic: New York, p 479.

Cohen, S. A., Benedek, K. P., Dong, S., Tapuhi, Y., and Karger, B. L. (1984). *Anal. Chem., 56*, 217.

Cohen, S. A., Benedek, K., Tapuhi, Y., Ford, J. C., and Karger, B. L. (1985). *Anal. Biochem., 144*, 275.

Cohen, S. A., Benedek, K., Tapuhi, Y., Ford, J. C., and Karger, B. L. (1986). *J. Chromatogr., 359*, 19.

Colvin, J. R., Smith, D. B., and Cook, W. H. (1954). *Chem. Rev., 54*, 687.

Costa, M. H. B., Ho, P. L., da Silva, A. M., Baptista, G. R., Leite, L. C. G., Cabero-Crespo, J., Venturini, K. M., Katz, M., Liberman, C., da Silva, A. R., and Raw, I. (1993). *Biotechnol. Appl. Biochem., 17*, 155.

Cramer, S. M., El-Rassi, Z., and Horvath, Cs. (1987). *J. Chromatogr., 394*, 305.

Davis, H. G., Green, R. H., Kelly, D. R., and Roberts, S. M. (1990). Biotransformations in Preparative Organic Synthesis, Academic: New York.

Dijke, P. T. and Iwata, K. K. (1989). *Biotechnology, 7*, 793.

Dunnill, P. (1983). *Process Biochem., 18*(10), 9.

Dyr, J. E., Fortova, H., Suttnar, J., and Vorlova, Z. (1987). *Thromb. Haemst., 58*, 565.

Dyr, J. E. and Suttnar, J. (1991) *J. Chromatogr., 563*, 124.

Effenberger, F., Horsch, B., and Forster, S. (1987). *Angew. Chem.*, *99*, 491.
El-Rassi, Z. and Horvath, Cs. (1986). *J. Chromatogr.*, *359*, 241.
Fareed, J. (1985). *Semin. Thromb. Hemost.* *11*, 1.
Fasnaugh, J. L., Kennedy, L. A., and Regnier, F. E. (1984). *J. Chromatogr.*, *317*, 141.
Fasnaugh, J. L., Pfannkoch, E., Gupta, S., and Regnier, F. E. (1984). *Anal. Biochem.*, *137*, 464.
Fasnaugh, J. L. and Regnier, F. E. (1986). *J. Chromatogr.*, *359*, 131.
Flurer, C. L., Borra, C., Andreolini, F., and Novotny, M. (1988). *J. Chromatogr.*, *448*, 73.
Folkman, J., Szabo, S., Stovroff, M., McNeil, P., Li, W., and Shing, Y. (1991). *Ann. Surg.*, *214*, 414.
Frey, D. D. (1991). *Biotechnol. Prog.*, *7*, 213.
Garg, V. K., Costello, M. A. K., Czuba, B. A. (1990). Protein Purification: Chromatographic Considerations and Strategies for Attaining Therapeutic Grade Purity. In *Purification and Analysis of Recombinant Proteins*, Seetharam, R., and Sharma, S. K., Eds., Marcel Dekker: New York.
Geoffery, P., Legrand, M., Hermann, C., and Fritig, B. (1984). *J. Chromatogr.*, *315*, 333.
Goheen, S. C. and Engelhorn, S. C. (1984). *J. Chromatogr.*, *317*, 55.
Gospodarowicz, D., Cheng, J., Lui, G. M., Fujii, D. K., Baird, A., and Bohlen, P. (1985). *Biochem. Biophys. Res. Commun.*, *128*, 554.
Graham, D. E. and Phillips, M. C. (1979). *J. Colloid Interface Sci.*, *70*, 403.
Gueguen, Y., Chemardin, P., Arnaud, A., and Galzy, P. (1994). *Biotechnol. Appl. Biochem.*, *20*, 185.
Habibi-Najafi, M. B., and Lee, B. H. (1993). Proline Specific Peptidases of *Lactobacillus casei* LLG, In *Book of Abstracts*, Institute of Food Technologists, Eds., Hamilton, P. B., Bogue, D. C., and Anderson, R. A., IFT: Chicago, IL, p 153.
Habibi-Najafi, M. B. and Lee, B. H. (1994). *Appl. Microbiol. and Biotechnol.* *42*, 548.
Hamilton, P. B., Bogue, D. C., and Anderson, R. A. (1960). *Anal. Chem.*, *32*, 1782.
Hearn, M. T. W., Hodder, A. N., and Aguilar, M. I. (1985). *J. Chromatogr.*, *327*, 47.
Henley, J. P. and Sadana, A. (1984). *Biotech. Bioeng.*, *26*, 959.
Henley, J. P. and Sadana, A. (1986). *Biotech. Bioeng.*, *28*, 977.
Henley, J. P. and Sadana, A. (1987). *Biotech. Bioeng.*, *30*, 1041.
Hodges, R. S., Burke, T. W. L., and Mant, C. T. (1988). *J. Chromatogr.*, *444*, 349.
Hoffman, N. E. and Liao, J. C. (1977). *Anal. Chem.*, *49*, 2231.
Horvath, Cs., Frenz, J., and El-Rassi, Z. (1983). *J. Chromatogr.*, *255*, 273.
Horvath, Cs., Nahum, A., and Frenz, J. H. (1981). *J. Chromatogr.*, *218*, 365.
Huang, J. X. and Horvath, Cs. (1987). *J. Chromatogr.*, *406*, 285.
Huang, J. X., Guiochon, G., and Schudel, J. (1990). *J. Chromatogr.*, *504*, 335.
Ingraham, R. H., Lau, S. Y. M., Taneja, A. K., and Hodges, R. S. (1985). *J. Chromatogr.*, *327*, 77.
Ishida, S., Saeki, J., Kumazawa, E., Kawanishi, G., Sada, E., and Jones, J. B. (1986). *Tetrahedron*, *42*, 3351.
Ishida, S., Saeki, J., Kumazawa, G., Sada, E., and Katoh, S. (1989). *Bioprocess Eng.*, *4*, 163.
Jackson, B. A. (1980). *J. Am. Vet. Med. Assoc.*, *176*, 1141.
James, B. N. and Do, D. D. (1991). *J. Chromatogr.*, *542*, 19.
Jansen, J., Woker, J., and Kula, M. R. (1992). *Biotechnol. Appl. Biochem.*, *12*, 90.
Kalasz, H. and Horvath, Cs. (1981). *J. Chromatogr.*, *215*, 295.
Kang, K. A. and Ryu, D. D. Y. (1991). *Biotechnol. Progr.*, *7*, 205.
Kato, Y., Kiramura, T., and Hashimoto, T. (1984). *J. Chromatogr.*, *298*, 407.
Kato, Y., Nakamura, K., and Hashimoto, T. (1986). *J. Chromatogr.*, *354*, 511.
Katoh, S. (1989). *Bioprocess. Eng.*, *4*, 163.
Katoh, S. and Terashima, M. (1994). *Appl. Microbiol. Biotechnol.*, *42*, 36.
Kikuchi, Y., Niwano, M., Yajima, N., Nakamura, G, and Miyata, N. (1985). *J. Antibiot.*, *38*, 1670.
Klibanov, A. M. (1990). *Acc. Chem. Res.*, *23*, 114.
Kopaciewicz, K. and Regnier, F. E. (1983). *Anal. Biochem.*, *133*, 251.
Kopaciewicz, K., Rounds, M. A., Fasnaugh, J., and Regnier, F. E. (1983) *J. Chromatogr.*, *266*, 3.
Kralova, B., Demnerova, K., Dobransky, T., and Lahejckova, R. (1986). *J. Chromatogr.*, *366*, 427.
Kunitani, M., Johnson, D., and Snyder, L. R. (1986). *J. Chromatogr.*, *359*, 147.
Lachman, L. B., Shih, L. C. N., and Brown, D. C. (1985). *Methods Enzymol.*, *116*, 467.

Lau, S. Y. M., Taneja, A. K., and Hodges, R. S. (1984) *J. Chromatogr., 35*, 3.
Lesins, V. and Ruckenstein, E. (1989). *J. Colloid Interf. Sci., 132*, 566.
Liautaud, J. (1986). *Asiatic Symposium on Blood and Its Derivatives*, Institut Merieux, Eds., Macau, December 1–4, 1986.
Lin, B., Ma, Z., and Guichon, G. (1991). *J. Chromatogr., 542*, 1.
Lobb, R. R., Harper, J. W., and Fett, J. W. (1986). *Proc. Natl. Acad. Sci. USA, 83*, 2091.
Lu, X. M., Benedek, K., and Karger, B. L. (1986). *J. Chromatogr., 359*, 19.
Makino, K., Ozaki, H., Matsumoto, T., Imaishi, H., Takeuchi, T., and Fukui, T. (1987). *J. Chromatogr., 400*, 271.
Malhotra, A. and Sadana, A. (1987). *Biotechnol. Bioeng., 30*, 1041.
McCormick, D. (1988). *Biotechnology, 6*, 158.
Melander, W. R., Lin, H. J., Jacobsen, J., and Horvath, Cs. (1984). *J. Phys. Chem, 88*, 4527.
Milne, R. W., Nation, R. L., Reynolds, G. D., Somogyi, A. A., and van Crugten, J. T. (1991). *J. Chromatogr., 565*, 457.
Moscatelli, D., Presta, M., and Rifkin, D. (1986). *Proc. Natl. Acad. Sci. USA, 83*, 2091.
Nakanishi, K., Yamamoto, S., Matsuno, R., and Kumikubo, T. (1977). *Agric. Biol. Chem., 41*, 1465.
Nguyen, P. N., Bradley, J. L., and McQuire, P. M. (1982). *J. Chromatogr., 236*, 508.
Nouvs, J. F. M. (1981). *Arch. Lebensmittelhyg., 32*, 103.
Osborne, R. J., Joel, S. P., and Slevin, M. L. (1986). *Br. Med. J., 292*, 548.
Omar, I. C., Nishio, N., and Nagai, S. (1987). *Agric. Biol. Chem., 51*, 2145.
Parente, S. S. and Wetlaufer, D. B. (1984). *J. Chromatogr., 314*, 337.
Parikh, I. and Cuatrecasas, P. (1985). *Chem. Eng. News*, August 26, 17.
Perrone, P. A. and Brown, P. R. (1984). *J. Chromatogr., 317*, 301.
Porath, E. (1985). *J. Chromatogr., 443*, 3.
Regnier, F. E. (1987). *Science, 238*, 319.
Robinson, D. and Stirling, J. L. (1968). *Biochem. J., 107*, 321.
Rounds, M. A. and Regnier, F. E. (1984). *J. Chromatogr., 283*, 37.
Sadana, A. (1991). *Biocatalysis. Fundamentals of Enzyme Deactivation Kinetics*, Prentice Hall: Englewood Cliffs, N.J.
Sadek, P. C., Carr, P. W., Bowers, L. D., and Haddad, L. C. (1985). *Anal. Biochem., 144*, 128.
Sadler, A. J., Micanovic, R., Gatzenstein, G. E., Lewis, R. V., and Middaugh, C. R. (1984). *J. Chromatogr., 317*, 93.
Safarik, I. (1984). *J. Chromatogr., 315*, 478.
Sawe, J. and Odar-Cederlof, I. (1987). *Eur. J. Clin. Pharmacol., 32*, 377.
Schott, H., Semmler, R., and Eckstein, H. (1987). *J. Chromatogr., 389*, 165.
Smith, C. J. P. and Kelleher, P. C. (1980). *Biochim. Biophys. Acta, 605*, 1.
Smitskamp-Wilms, E., Brusee, J., and van der Gen, A. (1991). *Recl. Trav. Chim. Pays-Bas, 110*, 209.
Snyder, J. R., Dolan, J. W., and Cox, G. B. (1991). *J. Chromatogr., 540*, 21.
Sofer, G. K. and Nystrom, L. E. (1989). *Process Chromatography. A Practical Guide*, Academic: New York.
Sonderquist, M. E. and Walton, A. G. (1980). *J. Colloid Interface Sci., 75*, 386.
Steindl, F., Jungbauer, A., Wenisch, E., Himmler, G., and Katinger, H. (1987). *Enzyme Microb. Technol., 9*, 361.
Subramanian, G., Phillips, M. W., and Cramer, S. M. (1988). *J. Chromatogr., 439*, 341.
Sugihara, A., Senoo, T., Enoki, A., Shimada, Y., Nagao, T., and Tominaga, Y. (1995). *Appl. Microbiol. Biotechnol., 43*, 277.
Sugihara, A., Tani, T., and Tominaga, Y. (1991). *J. Biochem. Tokyo, 109*, 211.
Suzuki, T., Mushiga, Y., Yamane, T., and Shimuzu, S. (1988). *Appl. Microbiol. Biotechnol., 27*, 417.
Takahashi, M. and Gross, E. L. (1978). *Biochemistry, 17*, 806.
Torres, A. R., Dunn, B. E., Eldberg, S. C., and Peterson, E .A. (1984). *J. Chromatogr., 316*, 125.
Ueda, T. and Ishida, V. (1987). *J. Chromatogr., 386*, 273.
Valko, K., Slegel, P., and Bati, J. (1987). *J. Chromatogr. 386*, 345.
Vanacek, G. and Regnier, F. E. (1980). *J. Chromatogr., 109*, 38.
van Enckevart, H. J., Dass, D. V., and Langdon, A. G. (1984). *J. Colloid Interface Sci., 98*, 138.

van der Mast, C. A., Hekstra, D., and Voorma, H. O. (1991). *J. Chromatogr., 564*, 115.
Veeragavan, S., Bernier, A., and Braendli, E. (1991). *J. Chromatogr., 541*, 207.
Vigh, G., Varga-Puchany, Z., Szepesi, G., and Gazdog, M. (1987). *J. Chromatogr., 386*, 353.
Vulfson, E. N. (1994). Industrial Applications of Lipases, In *Lipases*, Wooley, P. and Petersen, S. B.,Eds., Cambridge University Press: London, p 271.
Wajant, H., Botinger, H., and Mundry, K. W. (1993). *Biotechnol. Appl. Biochem., 18*, 75.
Wells, P. A., Biedermann, B., Garlick, R. L., Lyle, S. B., Martin, J. P., Herberg, J. T., Meyer, H. F., Henderson, S. L., and Eckenrode, F. M. (1993). *Biotechnol. Appl. Biochem., 18*, 341.
Wetlaufer, D .B. and Koenigsbauer, M. R. (1986). *J. Chromatogr., 359*, 55.
Wetlaufer, D. B. (1988). *Proc. Int. Symp. Affinity Chromatogr. Interfacial Macromol. Interact. Macromol. Chem., 17*, 17.
Wong, L. T., Hsia, J. C., and Tinker, D. O. (1985). 5th International Symposium of HPLC of Proteins, Peptides, and Polynucleotides, Toronto, Canada, November 4–6.
Wong, L. T., Lu, C. Y., Tinker, D. O., and Hsia, J. C. (1988). *J. Biochem. Biophys. Methods, 15*, 267.
Wong, L. T., Persaud, K., Tinker, D. O., and Hsia, J. C. (1988). In Biological Activities of α-Fetoprotein, Mizejewski, G. J. and Jacobson, H. I., Eds., Vol 2, CRC Press, Boca Raton, FL.
Wu, S. L., Benedek, K., and Karger, B. L. (1986). *J. Chromatogr., 35*, 3.
Yajima, N., Miyata, N., Kawanishi, G., Katayama, S., and Tsukagoshi, S. (1987). *Jpn. J. Cancer Res., 78*, 1121.
Yip, T. T., Nakagawa, Y., Below, M. M., and Porath, J. (1988). *Anal. Biochem., 168*, 75.
Zimmermann, E. F. and Madappally, M. M. (1973). *Biochem. J., 134*, 807.
Zimmermann, E. F., Vooting-Hawkins, M., and Michael, J. G. (1977) *Nature (London), 265*, 354.

6 PROTEIN INACTIVATIONS DURING NOVEL BIOSEPARATION TECHNIQUES

I. INTRODUCTION

Advances in genetic engineering, rDNA technology, cell fusion techniques, and biotechnology, in general, have made it possible to produce proteins, pharmaceuticals, drugs, vaccines, and other bioproducts in sufficient quantities. The apparent bottlenecks now appear to be in the separation of these required bioproducts in the desired conformational form where they are, in general, not only ultrapure but also stable. The separation of required amounts to fulfill the current demand of bioproducts, especially life-saving drugs such as tissue plasminogen activator (t-PA), and streptokinase, requires continual improvement in processing conditions. Note that the cost of separation of bioproducts may be as high as 90% of the total cost of the process (Kadam, 1986). The purification percentage will tend to be a significant fraction of the total cost as long as very high quality bioproducts are required. For example, for the separation of recombinant proteins this high purity is essential to minimize side reactions, cross-reactivity, and undesirable responses in pharmaceutical applications and in diagnostics (e.g., biosensors) (Kelley and Hatton, 1991).

Because the purification steps are so cost intensive, there is now considerable emphasis to either improve previous purification strategies, or analyze and develop new and more promising strategies for the separation of valuable bioproducts. The problem is compounded by the need for the separation of these bioproducts (e.g., proteins and drugs from a dilute mixture of similar components), which are also, in general, of a delicate nature. Besides, the upstream nature of the production of these bioproducts requires extremely careful and precautionary steps. For example, most proteins must be folded into a specific three-dimensional conformation to express their biological activities and specificity, which further complicates the process of separating and purifying them. Also, each separation process is specific for a particular process due to the rather unique characteristics of each bioproduct separated. This often tests the knowledge of engineers, chemists, and others to the very limit.

Initially, upstream processes have been developed, analyzed, and optimized without any special regard for the downstream processing that necessarily had to follow. Gradually, researchers and industrial workers realized that because upstream conditions significantly affect the downstream processing conditions, the optimization of the entire process would require the optimization of the whole process itself and not just the upstream or the downstream processes taken alone (Wheelwright, 1991; Kelley and Hatton, 1991). Furthermore, the high cost of separation and purification coupled with the difficulty of getting highly purified products prevents some biotechnological processes with applications in medicine, agriculture, and industry from becoming viable, cost-effective, and successful.

People working in biotechnology realized this, and subsequently many of them got involved in proteins, pharmaceuticals, and other bioproduct separation and purification. As a result of their research, novel and imaginative techniques sprang up. Some researchers modified existing procedures such as chromatography, electrophoresis, and precipitation. However, not all the techniques developed have the potential to be applied extensively. Some promising techniques developed in the past few years may have been overlooked because so many new techniques appear in the literature. In this chapter, we will examine two techniques that have demonstrated significant potential for the successful separation of a wide variety of bioproducts.

At the outset, some broad guidelines are helpful to place the problem in proper perspective. The purification techniques for these proteins, pharmaceuticals, drugs, and other bioproducts should be simple, easily scalable, continuous, and low cost; and (of course) should facilitate obtaining the bioproducts in the required state or conformation. For example, the process utilized should not inactivate the protein. Even though continuous processes may appear, in general, to be more economical, they are not always essential; for example, high-value therapeutic proteins are produced in a batch mode for various reasons, including quality assurance(QA)/ quality control (QC), cost, and risk (Sadana and Raju, 1990). One needs to keep in mind that although a variety of bioseparation procedures exist, they can be classified into four distinct steps: removal of insolubles, isolation of product, purification, and polishing (Cussler, 1987). Furthermore, because there is a need for optimizing the upstream and downstream processes taken together care must be taken to select only those processes in both of these steps that are compatible with each other. It is hoped that as the art and science of downstream processing develops gradually and reaches the state of development of upstream processing, the selection of compatible downstream and upstream processes will become a lot less difficult. This is bound to be a major task faced by most groups that plan to bring a reasonable number of valuable drugs to the marketplace. This is also a formidable problem considering the different types and the exquisite specificities of each bioproduct produced, even among its own class (e.g., proteins). This is where the knowledge of a wide variety of individuals from varying disciplines can effectively come into play.

We will present a couple of techniques that either have demonstrated significant potential or seem to exhibit potential and require further development and attention so that they can contribute to the effectiveness of downstream processing. The two techniques to be examined are liquid–liquid extraction using reverse micelles (Woll *et al.*, 1987; Goklen and Hatton, 1985; Dekker, 1986), and partitioning of proteins in two-phase aqueous polymer systems (Kula, 1987). Further research is urgently required to remove the inherent drawbacks, refine the procedures, and enhance the feasibility for the bioseparation of the various proteins present in different environments. Besides, a lot more attention needs to be paid to not only the quantity of the product separated but also the quality (e.g., purity, conformational state, and heterogeneity). In many instances, especially in the pharmaceutical industry this may prove to be a critical factor.

Other techniques (that may appear to be unconventional or a combination of the more conventional techniques) include the purification of the compounds, for example, proteins in the unfolded form (Knuth and Burgess, 1987); the solution-controlled gel sorption similar to aqueous extraction (Gehrke *et al.*, 1991); and the affinity purification of insoluble recombinant fusion proteins containing glutathione-S-transferase (Hartman *et al.*, 1992). Some of these techniques are beginning to appear in the literature and they deserve to be mentioned so that their potential may become more fully known. This will then permit individuals to screen, and either develop and analyze these procedures more fully or reject them. Because the serious analysis and development of downstream processing of bioproducts is rather new, it seems appropriate to carefully examine and pay attention to quite a few of the new techniques that appear, at least in the open literature. This does not, of course, mean that one should ignore or minimize the importance of the more established and developed techniques that have already proved their worth in downstream processing.

II. LIQUID–LIQUID EXTRACTION

Liquid–liquid extraction is a well-developed process for chemical engineering applications. Treybal (1980) indicated that liquid–liquid extraction operates continuously in large-scale processes with high throughputs. Besides, not only is liquid–liquid extraction easy to operate but also it is flexible in its operation. Also, Ichikawa *et al.* (1992) indicated that because liquid–liquid extraction is not accompanied by a phase change, it is recognized as an energy-saving process that provides milder conditions for the separation of bioproducts. This is of particular importance when one wants to maintain the integrity of the product. Liquid–liquid extraction has been investigated thoroughly by way of reverse micelles, and by way of partitioning between two aqueous phases for bioseparation purposes. By basing their techniques on the knowledge available, Sadana and Raju (1990) initially recommended using the reverse micelles for isolating a product from fermentation broths; and liquid–liquid partitioning for the purification and polishing steps, which may require stage-by-stage procedures. More detailed analysis is required of the different bioseparation techniques to help propose more general guidelines of wider applicability. Feedback of the information concerning the application of different techniques for the separation of a wide variety of bioproducts is essential in the buildup of a suitable framework. This framework would assist different workers to not only help select the different stages involved in the separation but also assist in placing the process in the proper sequence of the entire downstream processing train. This is where the knowledge gained by experience and by heuristics can be of immeasurable value.

We will initially examine liquid–liquid extraction by reverse micelles. This will be followed by a discussion on liquid–liquid partitioning in two-phase aqueous systems. As we analyze each technique particular emphasis will be paid to not only the quantity but also the quality of the bioproduct separated.

A. Liquid–Liquid Extraction Using Reverse Micelles

Reverse micelles are aggregates of surfactant molecules in organic solvents. These surfactant aggregates consist of an inner core of water, an inner layer made up of the surfactant molecule's polar ionic groups, and an external layer composed of a hydrocarbon chain. Figure 6.1(a) shows protein solubilization in reverse micelles. Because proteins are hydrophilic molecules, they cannot generally be solubilized directly in apolar solvents. In a less polar medium Jolivalt *et al.* (1990) indicated that the delicate equilibrium between hydrogen and hydrophobic bonds (Ghelis and Yon, 1982; Kauzmann, 1959) is destroyed. The molecular conformation changes and the hydrophobic residues initially buried inside the molecule come to the surface. The utilization of reverse micelles in apolar solvents permits one to solubilize proteins while maintaining the aqueous environment.

The formation of micelles in organic solvents minimizes the interactions between the polar heads of the surfactant and the apolar medium. Jolivalt *et al.* (1990) emphasized that the ability of these aggregates to solubilize water

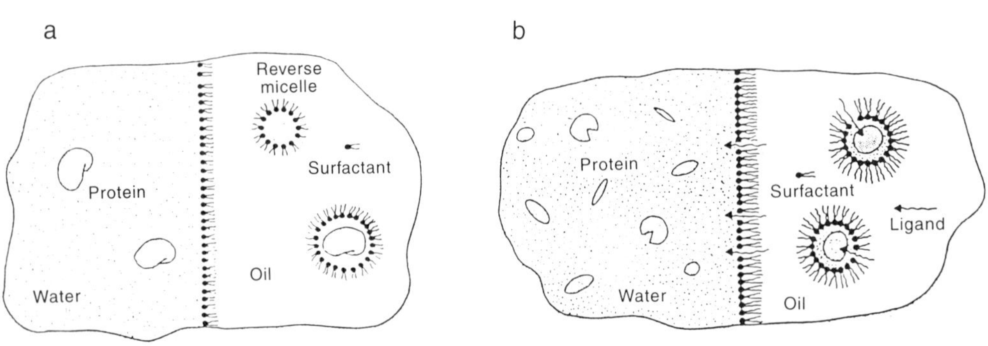

FIGURE 6.1 (a) Protein solubilization in reverse micelles; (b) principle of affinity partitioning in reverse micelles. [From Woll, J. M. *et al.* (1987).]

varies widely with the molecular structure of the surfactant and the nature of the solvent. Battistel *et al.* (1988) indicated that the properties of the water present in the micellar core differ from those of the bulk water. The mole ratio of the water to the surfactant, W_0, significantly affects the properties of the water in the inner core of the reverse micelles (Eicke and Kvita, 1984; Wong and Luisi, 1979). Battistel *et al.* (1988) emphasized that in the inner core the water molecules surrounding the protein molecules protect it from denaturation.

Liquid–liquid extraction using reverse micelles can be made selective, is easily scalable, and can be run continuously. Enzymes have been extracted from fermentation broth (Rahaman *et al.*, 1988), and from vegetable meal (Leser *et al.*, 1989). There is a continuous need to enhance the selectivity of a bioseparation process. If it is possible to include in the micelles a molecular recognition capability of the kind used in conventional affinity separations, for example, it is then possible to attain enhanced selectivity. Figure 6.1(b) shows the principle of affinity partitioning in reverse micelles. Woll *et al.* (1987) added a cautionary statement about utilizing reverse micelles for protein extraction from a fermentation broth compared with extraction from a "clean solution." Additional difficulties should be anticipated in real-life fermentation broth situations.

Proteins can be transferred from an aqueous phase into micelle-containing organic solvents and vice versa by manipulating various parameters. Thus, the transferred proteins are protected from the inactivating effects of the organic solvents. Protein partitioning between the two phases depends on relative volumes of the two phases, polar solvent, ionic strength, pH, and presence of biologically active surfactants (Woll *et al.*, 1987; Goklen and Hatton, 1985).

Woll *et al.* (1987) and Dekker (1986) indicated that proton transfer is dominated by electrostatic interaction between the protein and the surfactant head layer. This is because the solubilization of proteins in reverse micelles is observed only when the pH is below the protein isoelectric point. Solubilizaton is expected only when the surfactant head layer and the protein are of opposite polarity. The isoelectric point of the protein is that point at which there is a change from total transfer of protein from the aqueous phase to the organic phase to no transfer. At the pH at which the transfer is essentially complete, the concentration factor equals the ratio of the aqueous phase to the organic phase. Researchers have noted that protein concentration does not affect the protein transfer. However, the protein concentration increases with increasing surfactant concentration.

It is of interest to know the state of the enzyme molecule inside the micellar system. Is it more active? How about its stability? What are the conformational changes, if any? Researchers have indicated that inside the reverse micelles the biomolecules, such as enzymes (Wolf and Luisi, 1979; Martinek *et al.*, 1981; Hilhorst *et al.*, 1982) and nucleic acids (Imre and Luisi, 1981) may be subjected to a minimum of denaturation. The activity and the stability of these biomolecules may be controlled by changing the W_0 in the micelle. It would be of interest to examine the behavior of these biomolecules, for example, enzymes in the interior of the micelles. The following example analyzes briefly the thermodynamic behavior of enzymes in the interior of the micelles. Basically, the influence of W_0 on the activity and on the stability is analyzed.

Example 6.1

Describe a thermodynamic analysis of the activity and stability of globular proteins in the interior of reverse micelles (Battistel *et al.,* 1988).

Solution

A thermodynamic analysis is available of the properties of proteins in the interior of micelles with particular emphasis on their activity and stability (Battistel *et al.,* 1988). The reverse micelles were formed in the organic solvent (isooctane) by the negatively charged surfactant bis(2-ethylhexyl)sodium sulfosuccinate (AOT). These authors analyzed the conformational changes (inactivation) undergone by lysozyme, ribonuclease A, and cytochrome c by differential scanning calorimetry (DSC).

They noted that the water content in the micellar core influences the thermal stability of the proteins enclosed in the micelles. An increase in W_0 decreases the thermal stability. For example, as the water content and the size of the micelles increase, the temperature of the midpoint of the melting process T_m (that is, the stability) decreases. Fletcher *et al.* (1984) indicated that most globular proteins exhibit a maximum in enzyme activity as W_0 ranges from 8 to 12. Thus, an increase in W_0 greater than this range would tend to decrease the stability as well as the activity. For example, for ribonuclease A T_m is 48.1°C in aqueous solution. For W_0 values of 15.5 and 25, T_m decreases from 48.1 to 37.5 and 27.7°C, respectively. If the amount of water is more than the optimum W_0, then destabilization and inactivation of the enzyme occurs. A monolayer of water around the protein molecules apparently is sufficient to preserve the activity and stability of the protein.

Furthermore, the authors suggested that qualitative calculations indicate the free energy change for the conformational changes at 25°C at $W_0 = 11.1$ for ribonuclease A is lower than the change in aqueous solution (Table 6.1). This may be due to a subtle conformational change-rearrangement experienced by the protein in the micellar interior after solubilization. Basically, they emphasize that the difference in free-energy change (for conformational changes-unfolding) is mainly due to an unfavorable increase in the $-T\Delta S$ term for aqueous solution compared with that of the micellar core.

Besides, Battistel *et al.* (1988) indicated that the conformational flexibility

TABLE 6.1 A Comparison of Estimated Thermodynamic Parameter Values for Ribonuclease A Transition in Aqueous Solution and in AOT–Reverse Micelles of AOT–Isooctane[a,b]

	ΔH (kJ/mol)	TΔS (kJ/mol)	ΔG (kJ/mol)
Water	178	160	18
$W_o = 25.5$	198	196	2.0

[a] From Battistel, E. *et al.* (1988). *J. Phys. Chem.*, 92, 6680.
[b] $T = 25°C$, pH = 3.3.

of globular proteins decreases in reverse micelles compared with that in aqueous solution. More studies like this analysis are required that shed novel physical insights into the structural changes undergone by the protein during the transfer in and out of the reverse micelles, and also during their presence in the micellar interior. These conformational changes would then directly provide the information on the quality of the protein product separated. Mild or significant deformations may be encountered by the protein molecules as they either enter or exit the reverse micelles. These conformational changes may be occurring at the interface or in the interior of the micelles. Enzymes are known to inactivate at interfaces. Thus, it is often recommended to minimize the contact time an enzyme molecule spends at an interface. It would be desirable, of course, to avoid contact with the interface, but that cannot be avoided due to entry and exit of the protein molecule in the reverse micelle. If estimates of the deformation are available along with the location where they either first originate or grow (propagate), then corrective measures may be contemplated. These corrective measures may either attempt to stop these conformational changes or better still help to guide them in appropriate directions.

It is of interest to note the methods or mechanisms by which proteins reach the interior core of reverse micelles. Also, how do the different methods utilized affect the solubilization of the protein, if at all? What are the structural changes, if any, and how do these affect the overall inactivation of the enzymes-proteins?

Matzke *et al.* (1992) indicated that there are three different methods by which proteins may be solubilized in reverse micelles (Figure 6.2). However, the mechanisms of solubilization by these three techniques are not well understood. These authors attempted to provide some insights into the processes and the parameters involved. They studied the solubilization of α-chymotrypsin and alcohol dehydrogenase in AOT.

The authors noted that for solubilizaton to occur when a dry protein powder is added to a solution containing reverse micelles the size of the reverse micelle should roughly be the size of the protein molecule to be solubilized. Otherwise, the energetics is unfavorable. When the reverse micelle is about the size of the protein, minimal rearrangement in the reverse micelle is required to accommodate the protein. This minimal rearrangement apparently leads to a minimal energy barrier that permits the solubilization. Similar results were obtained by Leser *et al.* (1987). No information was provided by Matzke *et al.* (1992) concerning the protein structural changes, if any, as the protein enters the reverse micelle. It is presumably reasonable to assume a minimal energy barrier that permits solubilization would lead to a minimum of protein structural changes as the protein enters the reverse micelle.

In contrast to this result, Matzke *et al.* (1992) indicated that the micelle size does not appreciably influence protein solubilization when an aqueous protein solution is injected into an organic solution. In this case these authors noted that there is no sudden change in protein solubilizaton as the micelle diameter increases. They emphasized that in this case the reverse micelles are forced to form with the protein already inside. One would thus anticipate a lesser degree of protein structural changes in this case compared with those of the previous case (dry protein powder addition).

The phase-transfer method is most often used for the extraction of proteins

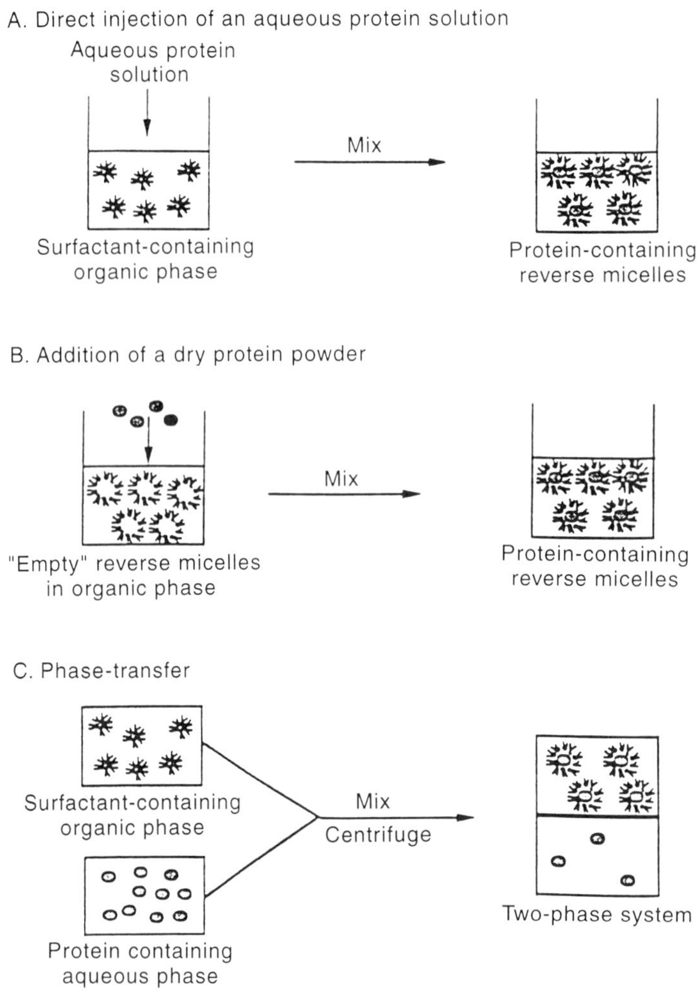

FIGURE 6.2 Protein solubilization methods in reverse micelles. [From Matzke, S. F. *et al.* (1992). *Biotechnol. Bioeng.*, *40*, 91.]

from a fermentation solution (Matzke *et al.*, 1992). In this case, the organic and the aqueous phase are in equilibrium, and the proteins are transferred from the aqueous to the organic phase (with surfactants) under appropriate conditions. These authors emphasized that when a positively charged surfactant is used, protein transfer occurs above the isoelectric point (negative charge on the protein), as expected. When a negatively charged surfactant is used, protein transfer below the isoelectric point (positive charge on the protein) is hindered due to protein–surfactant interactions at the interface. This hinders the formation of reverse micelles too. Apparently, no information was presented by these authors on the protein structural changes involved at the interface during the protein–surfactant interactions.

It would be of interest to note and to analyze the conformational changes in both of the above cases when protein transfer is possible and when it is hindered, especially, the analysis of the structural changes encountered by the protein during its hindered transfer and as it interacts with the negatively charged surfactant at the interface. This should provide novel physical insights into the protein structural changes that are required as the protein enters and exits from the reverse micelles. Fletcher *et al.* (1987) indicated that there are collisions between the reverse micelles and also with the protein molecules. These are bound to lead to conformational changes at the interface for the protein molecules as they either enter or exit the reverse micelles. The analysis of surfactant–protein interactions as they hinder the protein transfer at the interface provides a novel technique by which to provide some insights into this process.

Extraction of enzymes by reverse micelles seems to exhibit the potential to be used successfully (Luisi and Majid, 1986; Luisi *et al.*, 1986; Hatton, 1987). Nevertheless, some aspects of this technique need a little more attention. For example, more information needs to be made available concerning the amount of protein extracted. How is the rate of extraction affected by the different parameters? Mechanisms about the extraction and the desolubilization (or back extraction) need to be clearly delineated. Selectivity of extraction is also an important aspect that needs to be examined. It is intended to use the technique in a fermentation broth where quite a few different types of biomolecules will be present that have similar and delicate structures. The technique needs to be able to selectively solubilize–desolubilize only the required protein or bioproduct. A predominant consideration in the process will be the quality of the product extracted. What is the structure of the final protein state? What is the extent of denaturation, if any? Is the protein homogeneous? Is it heterogeneous? Some applications may require one protein conformational form or the other. How does one modify the process to attain the desired characteristics of the final protein product? These and other similar questions should apparently be answered before the reverse micelle technique can really begin to be effectively applied on a small-scale and an industrial-scale level. The next example attempts to answer at least a couple of these questions.

Example 6.2

Provide information pertaining to (1) the amount of enzyme/protein recovered (Jolivalt *et al.*, 1990); (2) the loss of activity (Sarcar *et al.*, 1992); and (3) the structural changes, if any, (Samana *et al.*, 1984) exhibited by enzymes when subjected to the reverse micelle technique.

Solution

Case One. The amount of enzyme recovered was studied. Jolivalt *et al.* (1990) extracted α-chymotrypsin from an aqueous phase into an organic phase using a quaternary ammonium salt (Aliquat 336) dissolved in isooctane by means of isotridecanol. Aliquat 336 is a mixture of trialkyl methyl ammonium chloride, $R_3CH_3N^+Cl^-$. The protein is significantly extracted above its isoelectric point where it is negatively charged and at low ionic strength. These authors noted the maximum solubilization of α-chymotrypsin of 90% in 1%

v/v solution of isotridecanol in isooctane at an Aliquat concentration of 0.04 M. No information was provided by these authors on the structural changes, if any, experienced by α-chymotrypsin on extraction by the preceding system.

Case Two. The time-dependent enzyme stability was studied by Sarcar *et al.* (1992), who analyzed the actvity and stability of yeast alcohol dehydrogenase (YADH) enclosed in AOT reverse micelles in isooctane. These authors noted that up to about 22°C the enzyme was more active in aqueous solution than in the reverse micellar solution. Then between 22 and about 50°C the enzyme was more active in the micellar solution compared with the activity in the aqueous solution. They emphasized that above 25°C the increase in temperature increases not only the rate of reaction but also the interdroplet interaction in the reverse micellar case. This leads to a higher rate of reaction in the reverse micelle case compared with that in aqueous solution.

Yeast alcohol dehydrogenase lost about 20% of its activity in 30 min in aqueous solution at 17.5°C. At 35°C in 30 min the enzyme lost about 45% of its activity in aqueous solution. These authors noted that the loss of activity and the rate of loss of activity were more in reverse micelles as compared with those in aqueous solution. For example, at 14.5°C the enzyme lost 85% of its activity in 30 min. At 26°C, this enzyme apparently had no activity in reverse micelles after 30 min. Furthermore, the loss in activity in reverse micelles may be due to the interactions of the surfactant head group with the enzyme at the interface in accord with the suggestion of Steinmann *et al.* (1986).

Case Three. The effect of nonionic–ionic surfactant and water content on enzyme stability was studied by Samana *et al.* (1984), who analyzed the behavior of horse liver alcohol dehydrogenase (HLADH) in reverse micelles using the ionic detergents sodium dodecyl sulfate (SDS) and cetyltrimethylammonium bromide. Later, this same group (Lee and Biellmann, 1987) described the stability of HLADH in the nonionic surfactant, Triton X-100. Lee and Biellmann (1987) noted that since the critical micelle concentration is lower for non-ionic surfactants compared with that of ionic surfactants (Makino *et al.*, 1973), the stability of HLADH was better in Triton X-100 than in the ionic surfactant, as expected.

Lee and Biellmann (1987) also studied the effect of water content on the activity-time curves for HLADH in Triton X-100(surfactant)-1-butanol (cosurfactant)-cyclohexane (organic phase) systems. These authors noted that as the water content of the reverse micelles increased the stability of this enzyme increased. For example, for a 0 w/w water content, HLADH loses about 87% of its activity in 2 days in a 50 mM Tes buffer, pH 7.5. For a 10 (w/w) water content for the same system (keeping the surfactant-cosurfactant concentrations, the same), the enzyme lost only 35% of its activity in 2 days. There was a gradual decrease in: (1) the loss of activity, and (2) the rate of loss of activity as the water content of the reverse micelles was increased. These authors suggested that at the low water content in the reverse micelles the binding and the dissociation of the coenzyme [positively charged nicotinamide adenine dinucleotide (NAD^+)] may be perturbed which then presumably leads to a lower activity.

Case Four. The structural changes of enzymes in reverse micelles have been examined. The activity and conformational changes undergone by lysozyme in different reverse micellar systems is available (Steinmann *et al.*, 1986). The systems investigated were AOT-isooctane, cetyltrioctylammoniumbromide (CTAB)-$CHCl_3$, and tetraethyleneglycoldodecylether (EO_4C_{12})/isooctane/water. These authors noted no significant conformational changes in lysozyme in the EO_4C_{12} and CTAB systems when compared with those of water. Circular dichroism (CD) spectrum studies indicated significant conformational changes in lysozyme in the AOT system when compared with those of water. Furthermore, these authors noted that in the presence of NAG, NAG_3, or the substrate, lysozyme retained its conformational structure and its activity. In the absence of the substrate the enzyme loses its activity rather rapidly. Steinmann *et al.* (1986) suggested that in the reverse micelle the enzyme has more helix content due to the micellar environment that favors hydrogen bonding. More studies like this analysis are required that shed physical insights into the denaturation of enzymes in micellar environments, the causes and the extent of this denaturation, and how this denaturation may be either prevented or minimized. Do recognize that conformational changes in themselves are not necessarily always deleterious to activity. If the enzyme denatures or undergoes conformational changes by a series-type mechanism, the intermediate state may be more active or stable than the initial enzyme state (Sadana, 1991).

It is of interest to note that the denaturation of proteins in reverse micelles depends on both the protein and the micelle. For example, Giovenco and Verheggen (1987) indicated that β-hydroxybutyrate dehydrogenase or isocitrate dehydrogenase do not undergo denaturation on extraction in reverse micelles in the hexanol-isooctane-CTAB system. However, glucose-6-phosphate dehydrogenase is denatured in this same system. It would be of significant interest to examine and to analyze these reverse micellar systems where protein stability problems are a minimum or are alleviated.

Ayala *et al.* (1992) indicated that even though nonionic surfactants alleviate protein stability problems in reverse micellar systems (Ayala and Mendoza-Hernandez, 1990; Lee and Biellmann, 1987; Sanchez-Ferrer *et al.*, 1988), they have not been studied in any significant detail. It would be of interest to analyze the nonionic surfactant reverse micellar systems. α-Chymotrypsin and hemoglobin were reextracted in reverse micelles formed by the nonionic surfactant Tween-85 in hexane (Ayala *et al.*, 1992). These authors noted that the α-chymotrypsin extracted into the reverse micelles was not denatured. This was supported by absorption spectra. At 14% Tween-85 and 0.01 μ ionic strength, they noted that 54% of the hemoglobin and 100% of the α-chymotrypsin were extracted. Because the optimum conditions for the extraction of the two proteins were different, this could be used to separate these proteins from a solution. The authors emphasized that the amount of water and protein that could be solubilized in the reverse micelles depends on the concentration of the detergent-cosurfactant, the ionic strength of the aqueous phase, and the protein properties. In general, as the surfactant concentration increased the amount of protein extracted in the reverse micelles increased. Above 14% Tween 85, however, the amount of protein extracted decreased due to the protein-surfactant interactions at the interface that lead to denaturation. The analysis of Ayala *et al.* (1992) is instructive because, like others, it begins to lay down the guiding principles for "solvent engineering" that would assist in not only obtaining optimal quantities of protein extracted, but also attaining the required quality or desirable standards. More such studies should be vigorously pursued in the future.

Enough emphasis has been placed on the parameters in the aqueous phase (such as pH and ionic strength) that affect the reverse micellar extraction. It is also important to analyze the influence of organic phase parameters, for example, the surfactant concentration. Enough concentration of the surfactant (amphiphilic molecule) is required to form the reverse micelles. An inordinate excess of the surfactant molecules will hinder phase separation (Ichikawa *et al.*, 1992). These authors emphasized the need for utilizing the appropriate or optimum amount of surfactant for the reverse micellar extraction. Other parameters that could be affected by not utilizing the optimum amount of surfactant could be increased protein denaturation and lower rates of extraction (back extraction or desolubilization).

Example 6.3

Provide information concerning the amount of surfactant and solubilizing water required to extract a given amount of protein using reverse micellar systems (Ichikawa *et al.*, 1992).

Solution

Ichikawa *et al.* (1992) attempted to determine the amount of surfactant and the solubilizing water required to completely solubilize α-chymotrypsin in an AOT-isooctane reverse micellar system under suitable pH and salt concentration. These authors noted that as the AOT concentration increased the amount of α-chymotrypsin extracted increased. For 100% α-chymotrypsin extraction, the authors required at least 5.4, 15, and 45 gmol/m^3 of AOT for α-chymotrypsin concentration in aqueous solution of 0.009, 0.078, and 0.42 gmol/m^3, respectively. They observed no protein denaturation for these respective protein concentrations if their minimum respective AOT concentrations were used in the organic phase. If lower AOT concentrations than the preceding were used, then denaturation would be observed for α-chymotrypsin (Ichikawa *et al.*, 1992), as well as for lysozyme (Fletcher and Parrot, 1988). In this case, surfactant–protein interactions lead to a gel-like complex that precipitates and contains the denatured protein. If the AOT concentration were larger than the preceding minimum, then the concentration of the protein in the reverse micelles and in the aqueous solution would be the same. These authors also noted that 3500 water molecules of solubilizing water are required to solubilize one protein molecule under suitable pH and salt concentration. Apparently the solubilizing water provides the hydrophilic surroundings necessary for the required protein extraction into the reverse micelles.

The Ichikawa *et al.* (1992) analysis is of interest because it provides some numbers and quantitative estimates for: (1) the amount of surfactant required to minimize protein denaturation in reverse micelles, and (2) the minimum amount of solubilizing water required for the efficient extraction of α-chymotrypsin in an AOT-isooctane reverse micellar system. Similar analysis should be conducted for other protein–surfactant–organic phase systems to provide numbers for the minimum amounts of surfactant and solubilizing water required for the efficient extraction of proteins in reverse micelles, accompanied by a minimum of protein conformational change. If a significant amount of protein conformational change (protein quality) is observed, then every attempt should be made to minimize the protein conformational change without compromising the quantity of the protein extracted.

Protein denaturation in reverse micellar systems is an important aspect that needs to be carefully considered. Now we have examined the influence of system parameters such as pH, ionic strength, and surfactant concentration on the rate of protein extraction and on protein denaturation. It is also of interest to examine the influence of external parameters such as temperature and agitation on protein denaturation in reverse micellar systems.

Hayes and Gulari (1991) analyzed the glycerol fatty acid esterification with lipase in AOT-water-isooctane reverse micellar systems. 1-Monoglyceride (a food emulsifier) was the major product. These authors noted that at 38°C lipase inactivates rapidly in reverse micelles of water-glycerol-AOT-lauric acid/isooctane, whereas there is only minimum lipase inactivation at room temperature. Agitation also increases the long-term lipase inactivation in reverse micellar systems (Hayes and Gulari, 1991; Han and Rhee, 1985).

Example 6.4

Describe the influence of temperature on protein desolubilization from reverse micelles (Dekker *et al.,* 1990).

Solution

The back transfer of proteins from the reverse micelles to the aqueous phase is a slow step (Dekker *et al.*, 1990). The slow mass transfer was due to the interfacial resistance between the organic and aqueous phases. Dekker *et al.* (1991) proposed a novel increased temperature method to enhance the yield and activity of α-amylase from a (TOMAC)-isooctane reverse micellar system. The higher temperature resulted in a change in the phase behavior where the aqueous phase was expelled from the reverse micelles containing the enzyme. During the forward extraction process at pH 10.1 and at 10°C 95% of α-amylase was transferred. The back transfer was performed at 35°C. This resulted in an enzyme recovery of 83%. The overall yield of enzyme activity was 73%. This is a novel technique for promoting the back transfer or desolubilization of the protein from the reverse micelle. The only disadvantage is the higher temperature required, which leads to enhanced inactivation. Apparently there is a compromise in that the back transfer is presumably facilitated at the expense of some deactivation. Recovery of other enzymes by this increased temperature method should be attempted to see if this technique enhances the back transfer of the enzyme, in general. It would be useful if the enzyme denaturation could be minimized.

We have been considering the reverse micellar extraction of enzymes in a batch mode. It is of interest to be able to perform these reverse micellar extractions in a continuous mode and to note the enzyme-protein inactivations encountered, if any. The difference in the flow patterns in batch and continuous flow processes is bound to significantly affect not only the quantity but also the quality of the protein extracted. Among other things, a continuous extraction would involve a significant amount of mixing and agitation. These are known to denature proteins, especially at air–water interfaces.

Example 6.5

An analysis of the continuous extraction of an enzyme by reverse micelles (Dekker *et al.*, 1986).

Solution

The continuous extraction of α-amylase has been analyzed (Dekker *et al.*, 1986). TOMAC was used as the surfactant and 0.1% v/v octanol in isooctane was the organic phase. The continuous extraction was performed by two mixer-settler units with the reverse micelles going back and forth from the two units. α-Amylase was extracted into the organic phase containing the reverse micelles. Then these reverse micelles were extracted into a second aqueous phase. This procedure resulted in an eightfold extraction of the enzyme from the original solution accompanied by a 30% loss in enzyme activity.

These authors indicated that during the first extraction into the reverse micellar phase, 90 ± 2% of the α-amylase is extracted at pH 9.9. Furthermore, there is about 2 to 8% inactivation of α-amylase during the first extraction. To permit a continuous extraction, a continuous mixing-agitation was required. This further inactivated the enzyme. The authors obtained a first-order inactivation rate constant of $2.4 \times 10^{-4}\ s^{-1}$.

At steady state during the first extraction 25% of the enzyme is recovered. During the second extraction, at steady state, 45% of the enzyme is recovered. Because most of the enzyme is extracted into the reverse micellar phase, this indicates that the balance of 30% of the enzyme undergoes inactivation either during the first extraction in the reverse micelles, or during the second extraction of the enzyme in the aqueous phase. Dekker *et al.* (1986) emphasized that the modeling and the experimental results are in close agreement if the enzyme inactivation is taken into account. Also, enzyme-surfactant interactions at the interface lead to inactivation and precipitation. Finally, the process suggested by these authors is useful because they have demonstrated the feasibility of a continuous process. An aspect that needs to be examined further is the loss of enzyme–surfactant at the interface. Methods or modifications are required that minimize the protein–surfactant interactions at the interface. This will not only enhance enzyme activity recovery and quality, but also minimize surfactant loss.

The reverse micellar extraction of proteins from a typical fermentation broth involves both an extraction step (into the reverse micelles), and a reextraction step (out of the micelles). The extraction of a protein (e.g., α-amylase) is controlled by the diffusion in the aqueous phase (Dekker *et al.*, 1989, 1990). The back extraction is a particularly slow process due to the reactions involved at the interface (Dekker *et al.*, 1990; Dungan *et al.*, 1991; Bausch, 1989). Bausch *et al.* (1992) have recently analyzed the reextraction of hydrophilic solutes (amino-acids-chymotrypsin) out of AOT reverse micelles, and noted that the reextraction was controlled by an interfacial process and was independent of the solute. It would be of interest to examine other techniques for the reextraction of proteins–solutes from reverse micelles.

A novel technique has been proposed to desolubilize proteins from reverse micelles without resorting to changing the pH or ionic strength (Phillips *et al.*, 1992). These authors showed that gas (C_2H_4) solubilization in the reverse micelles led to a decrease in the liquid density, and a precipitation of the enclosed proteins. Clathrate hydrates, which have crystalline inclusions of water and gas (Berecz and Balla-Achs, 1983) are formed under appropriate conditions. This leads to a precipitation of the protein and the water in the solid phase. As expected, this facilitates the recovery of the protein. They admitted that though they have demonstrated the feasibility of the process, they have not analyzed in any detail the quality or conformational state of the protein recovered. These authors emphasized that their process does not involve changing significantly the local environment of the protein during its recovery. This should apparently minimize the inactivation. Nevertheless, the phase changes involved due to gas pressurization and recovery of the protein in the solid state could lead to significant conformational changes of the protein. This contributes to qualitative as well as quantitative losses. This aspect needs to be carefully examined to help determine the overall feasibility of the process.

Example 6.6

Describe the effect of water content and reverse micellar extraction on protein extraction from an aqueous phase into a reverse micellar phase (Hilhorst *et al.*, 1995).

Solution

The suitability of micellar extraction as a separation process is strongly dependent on the different factors governing the transfer of the protein in and out of the reverse micelles. Electrostatic interactions are involved in the transfer process. Hilhorst *et al.* (1995) indicated that the size and properties of the protein also play an important part in the transfer process. Wolbert *et al.* (1989) previously showed that the amount of protein transferred depends on the distribution of charged groups on the surface. Surprisingly, the more symmetrical the charge distribution is, the less is the amount of protein transferred.

Hilhorst *et al.* (1995) indicated that larger micelles include or enclose proteins in their core more easily. Furthermore, because the enlargement of the micelles is dependent on electrostatic interaction between the surfactant head groups and the charged groups on the proteins, one may reasonably anticipate that higher density inside the micelles may facilitate the protein transfer. These authors studied the effect of micellar size and charge density on the transfer of α-amylase. The system utilized was made up of cationic surfactant TOMAC (10 mM), the nonioinc (co)surfactant Rewopal HV5 (2 mM), and octanol (0.1% v/v) in isooctane that was in contact with an aqueous phase containing 50 mM ethylene diamine.

These authors varied the micellar size (based on the water content of the organic phase) by changes in the percentage of octanol and by the type of the counterion in the aqueous phase. They noted that an increase in the micellar size resulted in the transfer of α-amylase at lower pH values. Furthermore, an increase in the charge density of the micelles facilitated the transfer by yielding a broader transfer profile.

The amount of transfer is directly connected to the dynamics of phase separation of the reactants (Hilhorst *et al.*, 1995). A poor phase separation leads to increasing enzyme inactivation. The electrostatic interactions are between the charged surfactant head groups and the charged groups on the enzyme. These authors emphasized that electrostatic interactions are involved in the beginning of the transfer of the protein into the reversed micelle. If this transfer is slow, then enzyme inactivation results. This analysis is of significant interest because (1) it helps delineate the factors involved in the protein transfer process; and (2) it provides logical reasoning to elucidate enzyme inactivation during the process, and how it may be minimized.

Vulfson *et al.* (1991) indicated that in spite of activity and stability characteristics exhibited by enzymes in reverse micelles (Luisi and Laani, 1988; Luisi *et al.*, 1988), this technique, as of yet, has not been applied on an industrial scale. This is primarily due to the poor recovery of bioproducts from these reverse micelles containing a high concentration of surfactants. These reverse micelles or microemulsions may be stabilized by some lower alcohols added to hexane or toluene (Smith *et al.*, 1977; Lund and Holt, 1980). Vulfson *et al.* (1991) analyzed polyphenoloxidase activity and stability in detergentless microemulsions and in ternary solvent mixtures.

The rapid apparently irreversible inactivation was reduced substantially in detergentless (or surfactant-free) microemulsions when compared with that of aqueous (buffered) media (Vulfson *et al.*, 1991). Also, although the activity of polyphenoloxidase was lower in the toluene-isopropanol-water microemulsion than in water at pH 7 (buffer), the stability characteristics of this enzyme were better in the microemulsion. A microemulsion formed with hexane-isopropanol-water exhibited activity and stability characteristics in between the previous microemulsion and water. These authors particularly noted that the activity and stability characteristics were dependent on the water content of the microemulsions. For the hexane-isopropanol-water microemulsion the enzyme exhibited both a higher initial rate and a faster rate of inactivation for a 6% water content compared with those of a 2% water content.

These authors concluded that solvent engineering principles can be effectively utilized to obtain useful activity and stability characteristics of enzymes, at least for polyphenoloxidase. This technique is worth attempting for the extraction of other enzymes and bioproducts, in general, by appropriate detergentless microemulsions. Although often the activity and stability characteristics of enzymes are of opposing nature (Henley and Sadana, 1984, 1986), it would be desirable to see if these characteristics could be improved by a suitable modification of these detergentless microemulsions.

Finally, Kelley and Hatton (1991) emphasized combining the purification stages into a single step, if possible. This minimizes the contact time for the processing of a given amount of fluid. Besides, not only the equipment size is reduced, but also the chances for denaturation are decreased. Of course, care has to be taken to combine the different and appropriate purification stages. For example, during the AOT reverse micelle removal of two dehydrogenases from *Azotobacter vinelandii* the steps of disruption, clarification, and purification were effectively combined. There was a six fold increase in the specific activity of the dehydrogenases. However, no data were presented by these authors concerning the denaturation of the enzymes. It would be useful to obtain or gather data about the quality and quantity of the bioproduct separated when the different stages either are combined or used as is. Is the enzyme more or less denatured when the stages are combined? What about the heterogeneity of the bioproduct? In some cases, it may be critical. Sometimes a heterogeneous product may be required. Perhaps it is the combination of stages that yields the desired quality of product. Another aspect that should not be ignored is the stability of the product. The bioseparation process may yield the required quantitative amounts. Even the quality may be acceptable. However, the product conformational state should also be stable. That is, the product should exhibit a reasonable activity for an appropriate duration of time. These and similar questions need to be answered, and a framework should be developed that would assist the different workers in selecting a bioseparation process that matches their ever changing requirements.

Selectivity can be increased considerably if affinity ligands are attached to the long alkyl chain (Woll *et al.*, 1987). Then the ligand binds selectively to the desired protein and pulls the protein into the reverse micellar core. The long alkyl chain will be positioned in the surfactant layer, and the protein ligand will be in a micellar core. These authors used reverse micelles to purify an extracellular alkaline protease from *Bacillus* sp. ATCC 21536. They noted a trade-off in the recovery of protein mass and the recovery of protein activity as the pH of the solution was varied. Lower pHs increased the driving force for protein transfer because of increasingly favorable electrostatic interactions between the protein and the surfactants, but below a pH of 5 there was a dramatic falloff in protein activity. For example, at pH 5.5 there is 23% protein recovery.

Paradkar and Dordick (1991) emphasized that not only should the bioseparation process be efficient by way of quantity and quality of the bioproduct separated, but also it should be highly selective. The demands on a particular bioseparation process are generally not simple. These authors indicated that in the separation of carbohydrates and carbohydrate-containing compounds, this is critical. Affinity-type separations possess enhanced selectivity, and may be used to purify sugar-containing compounds (Mohr and Pommerening, 1985; Cuatrecasas and Wilchek, 1968). Paradkar and Dordick (1991) emphasized that concanavalin A (con A) has been used in affinity partitioning (Mattiasson and Senstad, 1989) and in affinity escort systems (Herak and Merrill, 1989) for glycoprotein purification. Selectivity may be enhanced by using the affinity ligand as a cosurfactant (Coughlin and Baclaski, 1990). Paradkar and Dordick (1991) utilized con A to analyze affinity partitioning of an enzyme in reverse micelles.

Example 6.7

Describe an analysis for the affinity partitioning of glycoproteins in reverse micelles (Paradkar and Dordick, 1991).

Solution

Con A is utilized to selectively separate horseradish peroxidase (a glycoprotein) in AOT reverse micelles (Paradkar and Dordick, 1991). The con A was extracted into the reverse micelles. Then this con A was selectively coupled to horseradish peroxidase present in aqueous solution. These authors noted an 85% extraction of horseradish peroxidase in the AOT-isooctane reverse micelles in 15 min at pH 5.08. Without the con A there was less than 1% extraction of the enzyme. Surprisingly enough, during the con A extraction of the enzyme, neither did shaking increase the extraction nor did it affect the stability of the product.

These authors emphasized that their con A-AOT system does provide some useful numbers. For example, they used extraction times of 15 min each for: (1) extraction of horseradish peroxidase into the organic phase, and (2) extraction of horse radish peroxidase into a second aqueous phase. Eighty-three percent of the horseradish peroxidase was extracted. Because 14% of the horseradish peroxidase remained in the initial aqueous phase, there was only an insignificant amount of inactivation when considering error in the analysis, etc. The selectivity separation factor obtained was approximately 40. Thus, the process not only is efficient and selective but also exhibits a minimum denaturation of horseradish peroxidase. More studies like the present analysis are required that determine the feasibility of affinity-type partitioning in reverse-micellar solutions for the effective separation of other enzymes. Particular attention should be paid to the selectivity, quantity, and quality of the bioproduct separated.

It is useful to examine a wide variety of examples to analyze how effective the reverse micelle procedure is as a bioseparation process. One of the major parameters that helps determine the effectiveness of a bioseparation process is the extent of denaturation undergone or exhibited by the bioproduct separated. Fundamental information on this aspect is scarcely available in the open literature. Even if the information is available, the authors provide only a bare minimum of denaturation information.

A new method has been utilized to monitor the interfacial inactivation of enzymes (Ghatore *et al.*, 1994). Enzyme inactivation by solvent molecules can be both reversible and irreversible by nature (Khmelnitsy *et al.*, 1988; Singer, 1962; Butler, 1979). Enzyme inactivation can occur either at the water-organic solvent interface or in the dissolved solvent itself. It is of interest to be able to separate the contributions to inactivation made by these two effects. The technique utilized by Ghatore *et al.* (1994) permitted them to separate these two effects. These authors noted that although greater dissolved solvent inactivation occurred with the more polar solvents, the more generally accepted simple relationship with solvent polarity (e.g., log P) usually was not adequate, especially if interfacial inactivation were determined specifically. Thus, based on their results on the inactivation of urease, lipase, ribonuclease, and chymotrypsin by different solvents, these authors were unable to correlate the degree of inactivation to a single parameter.

They noted that the best general trend was observed with the Hildebrand solubility parameter. These authors emphasized that the enzymes selected for analysis differ in many aspects. For example, urease is the largest enzyme, besides being a metallic enzyme and a hexamer, lipase is interfacially active; ribonuclease is very small; and chymotrypsin is a protease. In spite of these differences in the enzymes studied, it is to be commended that the authors were able to find some general correlation of the inactivation exhibited by urease, lipase, ribonuclease, and chymotrypsin with the solubility parameter. More studies like this analysis are required for other enzymes and other organic solvent–aqueous systems. These would help provide the basis and the predictive relationships for the extent of denaturation exhibited in different aqueous-solvent systems. Such information would be of considerable value in the design of such bioseparation systems.

Surfactant-based bioseparation has developed into an attractive technique for the separation of biomolecules utilizing (1) micellar separation with ionic surfactants (Wolbert *et al.*, 1989; Hatton, 1989; Goklen and Hatton, 1987; Dekker *et al.*, 1989), and (2) nonionic microemulsion separation systems (Wienchek *et al.*, 1989; Qutubuddin *et al.*, 1994; Ayala *et al.*, 1992). It would be of interest to analyze the influence of surfactants on protein separations using nonionic microemulsions.

Vasudevan *et al.* (1995) analyzed the influence of surfactant structure effects on protein separations using nonionic microemulsions. These authors indicated that a mixture of oil, water, and surfactant yields thermodynamically stable microemulsions (Rosen, 1989). Furthermore, they stated that Winsor II microemulsions are oil-continuous phases. These contain a dispersion of water droplets stabilized by water molecules at the droplet interface. Reverse micelles are a subset of the broader class of Winsor II microemulsions.

These authors indicated that in the nonionic microemulsion system utilized for protein separation two difficulties emerged (Qutubuddin *et al.*, 1994; Wienchek, 1989). The first was slow phase disengagement, and the second was excess aqueous phases. These difficulties constrain the economic feasibility of these processes. Surfactants, if used, in microemulsions for biseparation should: (1) display a significant affinity or specificity for the different proteins vis-a-vis surfactant–protein interactions; (2) should assist in the phase disengagement process, besides exhibiting good microemulsion characteristics; and (3) assist or simplify easier the protein recovery from the microemulsion phase (Vasudevan *et al.*, 1995).

These authors utilized different surfactants, (such as linear alcohol ethoxylate (Neodol 91-2.5), two alkyl phenol ethoxylates (Igepal CO-520, Trycol 6985), and a series of alkyl sorbitan esters that are either ethoxylated (Tweens) or unethoxylated (Spans). These were utilized to extract cytochrome c. They noted that in this case the partition coefficients were an order of magnitude higher than those obtained for comparable sorbitan systems. According to these authors, the sorbitan group plays an important role in the extraction via a weak electrostatic protein-surfactant interaction.

The authors emphasized that surfactant structure plays a crucial role in the extraction of proteins in nonionic microemulsion bioseparation systems. They indicated that the utilization of Neodol 91-2.5 provides the following two distinct advantages: (1) phase disengagement is trivial, and (2) increased length scale allows the extraction of much larger species including protein-protein and protein-cell complexes. Very little, if any, information was provided by the authors with respect to the inactivation of cytochrome c on using the different surfactant-based nonionic microemulsion bioseparation systems. This type of information is required to provide the complete picture that is necessary to help evaluate the different microemulsion systems utilized for the bioseparation of proteins.

Reverse micelles offer the potential for large-scale continuous purification of proteins (Dekker *et al.*, 1986, 1987, 1989, 1991). A critical problem that still remains is the recovery of the proteins from the reverse micelles. Different methods have been considered to help remove the proteins from the reverse micelles. Some of these methods employ the interactions of proteins with surfactants. These interactions may be manipulated by changing the pH and the ionic strength (Gupta *et al.*, 1994).

These authors utilized a novel technique to help recover the proteins from the reverse micelles. They utilized the principle of Leser *et al.* (1993), who pressurized the reverse micelle with ethylene to convert the micellar water pool to a clathrate hydrate. Similarly, the loss of water in the reverse micelle caused the protein to precipitate as a solid phase.

Gupta *et al.* (1994) preferred to use an ambient pressure technique. They utilized 4-Å aluminosilicate molecular sieves to remove the water selectively from the reverse micelles. The small-pore molecular sieves prevented the adsorption of the surfactant on the protein. These authors utilized this method to recover with high yields α-chymotrypsin, cytochrome c, and tryptophan from (AOT)-isooctane-water reverse micelle solutions. They emphasized that their method does offer some unique advantages compared with those of the traditional techniques. The hydrophiles or proteins are recovered directly from the reverse micelle as a fine powder; this precludes the necessity of a crystallization step. It is important to note that this technique is useful for the recovery of proteins because pH, temperature, pressure, and ionic strength are not any of the variables involved. In other words, denaturation of proteins during the recovery of the proteins from the reverse micelle is minimized. Gupta *et al.* (1994) recommended studying and analyzing the amount of surfactant adsorbed to the different precipitated proteins. This would help estimate the protein activity after recovery.

B. Two-Phase Aqueous Polymer Systems

Liquid–liquid extraction can be especially useful for recovering biochemicals and for many years was an important technique in antibiotic purification. High-molecular-weight polymers can be used to form immiscible phases under appropriate conditions. They should be sufficiently different and soluble in the same solvent. Kula (1987) indicated that above a critical minimum concentration of each polymer in water, two immiscible phases form. Kula (1990) further stated that the cost of these phase-forming chemicals may be high; however, this cost can be decreased by recirculating these chemicals. Veide *et al.* (1983, 1984) emphasized that this technique can effectively combine the purification and concentration steps if the partition coefficient, K, is high. This is defined as the ratio of the product in the top phase to that in the lower phase. Besides, this technique effectively removes cell particles and nucleic acids. Strandberg *et al.* (1991) indicated that this is not always the case, and many times DNA techniques have to be used to facilitate the protein purification. These authors emphasized the compromise that must be made between recovery and maximizing capacity in the design of two-phase aqueous partition systes. For example, to obtain a threefold increase in protein concentration and 95% yield, a K value of > 50 is required. Thus, the manipulation of the partition coefficient must be done carefully. Note that for a given K different yields can be obtained by adjusting the volume ratio of the two phases. This, of course, has not taken into account the quality of the protein separated. Often this is an important factor and should be carefully examined. It is expected that protein conformational changes may take place at the liquid–liquid interface. It would be of interest to know the effect of the partition coefficient on the separation of bioproducts between the top and the bottom phases in these types of systems.

Huddleston *et al.* (1991) presented the yields of different bioproducts during the fermentation and purification of β-galactosidase from *Escherichia coli.* Figure 6.3 shows the process scheme for this fermentation and separation. The yields of β-galactosidase, DNA, RNA, protein, and proteases in the top phase are 85 to 100, <0.1, 2, 13, and 3%, respectively. The yields in the bottom phase in the same order are <15, 100, 98, 87, and 50%, respectively. Note that most of the β-galactosidase is extracted in the top phase. Also, most of the DNA, RNA, protein, and proteases are extracted in the lower phase. Most of the cell debris also remains in the lower phase. The authors did not provide any information on the quality of the bioproducts separated either in the top or in the lower phase. Also, in the case of proteases, the numbers do not add up to 100%.

Example 6.8

Provide an example of a large-scale fermentation and separation of a recombinant protein from *E. coli* (Strandberg *et al.*, 1991).

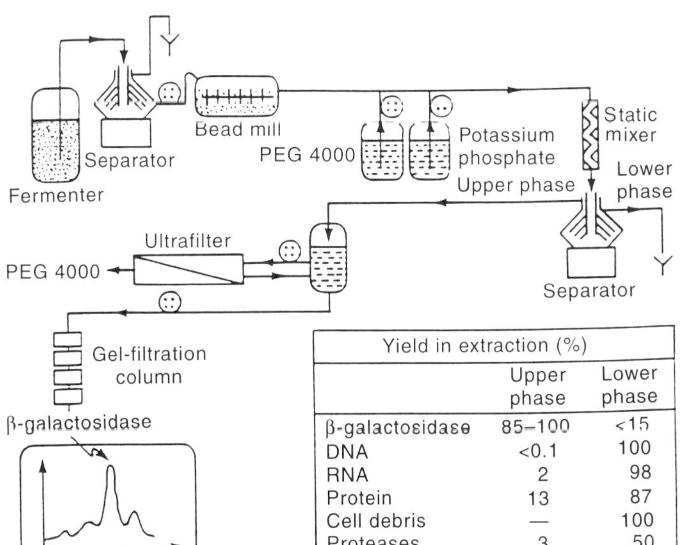

Yield in extraction (%)		
	Upper phase	Lower phase
β-galactosidase	85–100	<15
DNA	<0.1	100
RNA	2	98
Protein	13	87
Cell debris	—	100
Proteases	3	50

FIGURE 6.3 Continuous isolation and purification of *E. coli* β-galactosidase by aqueous two-phase extraction. [From Huddleston, J. *et al.* (1991). *TIBTECH, 9,* 381.]

Solution

Strandberg *et al.* (1991) analyzed a complete process for the separation of a recombinant protein from *E. coli.* In their pilot-scale process these authors included the following steps: fermentation, cell harvesting, disintegration, two-phase aqueous extraction, diafiltration, and freeze-drying. These authors extracted the fusion protein AGβgal, which consists of five immunoglobulin G (IgG)-binding regions of staphylococcal protein A, two IgG-binding regions from streptococcal protein G, and β-galactosidase from *E. coli.* The overall yield (recovery) was 37% when using a two-phase aqueous system with polyethylene glycol (PEG) 4000-potassium phosphate.

Figure 6.4 shows the process flow diagram for the recovery of AGβgal from *E. coli.* The cells were disintegrated using a bead mill. The authors indicated a recovery of 81 $\pm$ 4% of Agβgal. During the aqueous-phase extraction the authors obtained a recovery of 66% in the top phase. The partition coefficient was 3.9. There was a 20–25% variation in the value of the partition coefficient. This, the authors suggested is due to the perturbations in the flow.

The top-phase volume is the parameter that can be adjusted to meet requirements (Strandberg *et al.,* 1991). A higher value of the partition coefficient decreases the top-phase volume required to achieve a particular separation. This is of advantage for the subsequent filtration process. Higher values of the partition coefficient increase the concentration of the protein extracted in the

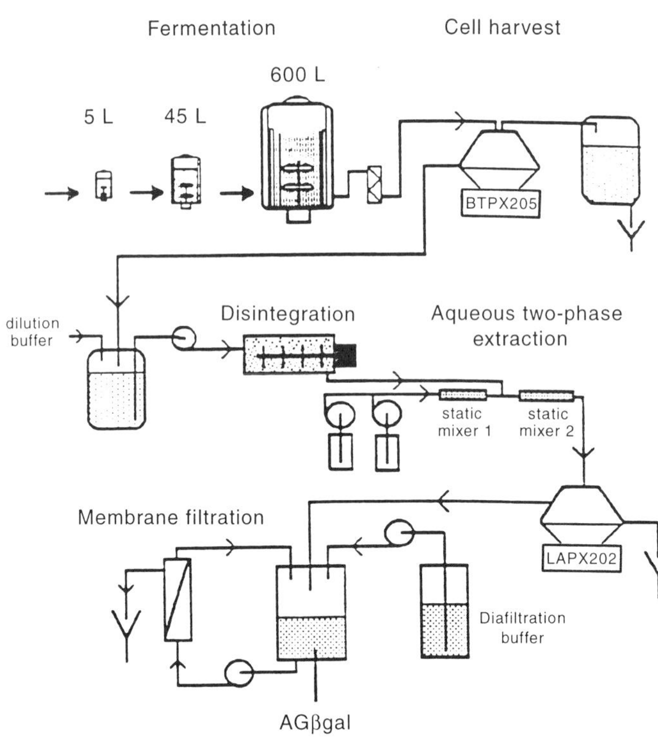

FIGURE 6.4 Fermentation and purification of the recombinant protein *E. coli* AGβgal. [From Strandberg, L. *et al.* (1991). *Process Biochem.,* *26,* 225.]

top phase. Kohler *et al.* (1989, 1991), however, indicated that these high values of the partition coefficient and the subsequent high values of the protein concentration lead to protein precipitation (at the interface). This then lowers the recovery value. Thus, one has to carefully optimize the partition coefficient value. It would be of significant advantage to understand what causes the preciptiation of this protein so that this precipitation could be minimized. This would then considerably assist in increasing the recovery of AGβgal from *E. coli.*

The authors indicated that they used the diafiltration step to remove the PEG 4000 from the top phase. These authors noted that β-galactosidase either adsorbs on the filter membrane or inactivates on the membrane surface. In any case, this recovery of the enzyme from the diafiltration step was only 69%. Thus, the total recovery of the protein was 0.66 (extraction in the top phase) $\times$ 0.69 (diafiltration step) $\times$ 0.81(disintegration step) = 0.37. This breakdown of the recoveries in the different steps is of interest, because it provides an overall picture of the recoveries, and where possible improvements may be attempted. It would be of considerable help if similar numbers for conformational changes of the protein occurring were made available at the different stages involved in the bioseparation process.

Albertsson (1986) indicated that the two phases exhibit different physicochemical properties. This permits the partition of biological macromolecules such as proteins and nucleic acids. The purification achieved by partition of biochemicals between the two phases depends on if it is possible to steer the partition in a selective way. There are quite a few advantages with this technique. Huddleston *et al.* (1991) indicated that two-phase aqueous systems possess the ability to combine a few of the initial stages (the primary steps) involved during the bioseparation process. Furthermore, because the water content of the two immiscible phases is high (80 to 99% w/w), the phases are excellent environments for the different and delicate bioproducts that often need to be separated. Besides, these systems are relatively easier to scale-up. Huddleston *et al.* (1991) indicated that often there is a linearity in the scale up of such systems from the laboratory (bench) scale. These authors provided a cautionary statement in that the mechanisms of partition are still not fully understood. This still places this technique on an empirical basis, and there is considerably more emphasis on expertise than on science, per se. They attempted to provide insights into the molecular processes that drive these partition types of systems, so that they may become more viable for application at the commercial level.

The most frequently used two-phase systems are composed of PEG, dextran, and water. Partitioning of proteins between the lower dextran-rich phase and the upper PEG-rich phase is influenced by pH, temperature, salts, molecular weight, charged groups, hydrophobicity, and affinity ligands bound to the polymers. The effects of salts on the partition coefficient can be summarized in the equation

$$\log k = \log k_0 + \gamma z \tag{6.1}$$

where k_0 is the partition coefficient at the isoelectric pH, γ is a factor that depends on the salts, and z is the charge (Kula, 1987). Kula (1987) indicated that most favorable conditions occur when the charge difference and the γ value are as large as possible. Kula and coworkers have been pioneers in this area of two-phase extraction. The next example mentions some of the enzymes extracted by them using the two-phase aqueous system.

Example 6.9

Describe the two-phase aqueous extraction of enzymes (Kula, 1987).

Solution

A PEG-salt two-phase aqueous system has been utilized to extract glucose-6-phosphate dehydrogenase from *Saccharomyces cerevisiae* (Kula, 1987). This author obtained a yield of 91%. The purification factor was 1.8. No numbers on the denaturation of the enzyme were provided. Nevertheless, this author emphasized that the method exploits the behavior of hydrophobic polymers in solution and provides a gentle and fast separation technique well suited for large-scale operation. Alcohol dehydrogenase was extracted from *S. cerevisiae* using polyethylene glycol (PEG)-salt. This author obtained a 96% yield, and a purification factor of 2.5. For both of these enzymes extracted the yield is very reasonable. The picture would have been more complete if qualitative data (on the conformational states) had been available.

An attempt was also made to extract human serum albumin (HSA) with a two-phase aqueous polymer system using novel acrylic copolymers. A 95% recovery of HSA and a purification factor of 2.2 were obtained. For the protein recovered the author obtained a value of 0.61 for the ratio of initial specific activity (HSA)-total protein. Because most of the protein (95%) was recovered, this indicates that there was about a 40% activity unaccounted for. Some of the factors that contribute to this number (40%) would be protein denaturation due to conformational changes either in solution or at the solvent-solvent interface, errors in analysis, protein in the other phase, etc. More detailed information on the accountability and the mechanisms of the loss of protein activity would be desirable. If such information is made available, then measures can be taken to perhaps minimize this loss in activity or conformational changes. Some conformational changes may even prove to be beneficial as far as the specific activity or the quality of the product is concerned (Sadana, 1991).

Kula (1987) also attempted scale-up studies for the aqueous-phase extraction of formate dehydrogenase from *Candida boidinii* using PEG and either salt or crude dextran. A 95% yield was obtained in a 10-ml scale when PEG and a salt were used. Success was obtained in scaling up to a 250-liter process that gave a 94% yield of formate dehydrogenase. The yields obtained by Kula (1987) are reasonable. However, once again no numbers were presented on the quality of the product separated either at the 10 ml or at the 250 liter process. It is of particular interest to see how the quality of the product is affected as the process is scaled up. This can often become a critical factor, perhaps requiring process modifications, even if the required yields (quantitative aspects) are met.

More studies like this one are required that provide some numbers on the quality of the protein extracted. In this case, there is a considerable loss in the initial specific activity of the protein. Care must be taken in this and other liquid–liquid partitioning systems to improve the quality of the protein extracted by minimizing the conformational changes that are presumably predominantly occurring at the interface of the two solvents. If conformational changes and subsequent loss in activity leading to poor protein quality are occurring, then it is appropriate to have some sensing devices or on-line monitoring to have the information available quickly. This will considerably assist in providing the immediate remedial action necessary.

Example 6.10

Briefly describe on-line monitoring of protein activity and concentration during a two-phase aqueous extraction (Papamichael *et al.*, 1991).

Solution

Papamichael *et al.* (1991) emphasized the importance of on-line monitoring systems during the transfer of two-phase extraction systems for the large-scale recovery of proteins and other bioproducts (Hustedt *et al.*, 1985, 1987). Monitoring is required of not only the system parameters but also product parameters such as protein activity and concentration. This leads to an efficient, reliable, and reproducible process (Papamichael *et al.*, 1991). If the conformational state of the protein could be monitored, then that would also considerably assist in product quality control. These authors incorporated the assay procedures for fumarase concentration and activity into their process for the automatic determination of these parameters. Fumarase activity was determined by analyzing the L-malate to fumarate reaction. The biuret reaction was used to assay for the protein content.

Papamichael *et al.* (1991) emphasized that aqueous-phase extraction using disrupted cells hinders the analysis. Reproducibility is also poor in these systems due to phase turbidity and solid entrainment. Nevertheless, the method described by these authors is reasonably accurate and also reproducible. Standard deviations for the enzyme activity and protein concentration were obtained. They were 3.2 and 2%, respectively. The authors emphasized that their technique can be applied for the determination of other enzymes provided that the reaction can be followed spectrophotometrically. More such studies should be carried out in the future that help the on-line determination of protein concentration, activity, and conformational states. This would provide the necessary information on the quantity as well as on the quality of the protein or other bioproduct extracted.

It is of interest to be able to enhance the separation of proteins using the two-phase aqueous extraction technique. The next example analyzes charge-directed partitioning (Luther and Glatz, 1995). In this method the charges on the protein are genetically altered to enhance the separation of a protein using two-phase aqueous systems.

Example 6.11

Briefly describe the genetically altered charge modification utilized to enhance the electrochemical partitioning of a β-galactosidase and T4 lysozyme in aqueous two-phase systems (Luther and Glatz, 1994).

Solution

The influence of genetically charged modifications on the electrochemical partitioning of β-galactosidase and T4 lysozyme in dextran-PEG two-phase aqueous systems containing potassium phosphate has been analyzed (Luther and Glatz, 1994). These authors wanted to enhance the partitioning effects in these types of systems, because: (1) the main component of these systems is water, and is biologically friendly, (2) the phase-forming polymers help stabilize the proteins (Albertsson, 1986), (3) the short processing times are required due to rapid intraphase transport (Fauquex *et al.*, 1989; Kroner *et al.*, 1978), and finally (4) the process is easily scalable and uses readily available conventional equipment (Sikdar *et al.*, 1991).

Luther and Glatz (1994) indicated that in these types of phase systems an uneven distribution of salts or polyelectrolytes generates an interfacial potential difference. These authors utilized the principles of electrochemical partitioning to analyze the influence of genetically engineered charge modifications on overall partitioning effects. The charge modifications were made on two proteins: β-galactosidase by fusing on charged tails of polyaspartic acid, and T4 lysozyme by charge point mutations by replacing positive lysine residues by negative glutamic acids.

They indicated that for β-galactosidase, though the model is in qualitative agreement, the increase in partitioning was less than predicted, even though it exhibited changes of more than two orders of magnitude. Similarly, for T4 lysozyme, though there was qualitative agreement, there was a smaller than expected dependence of partitioning on charge difference. Some oversimplified assumptions may have contributed to the lack of quantitative fit. Nevertheless, this analysis is of interest. Future modified models, with better assumptions would presumably lead to a better quantitative effect. No information was provided by these authors on the degree of inactivation exhibited by either β-galactosidase or T4 lysozyme as the charges were changed to enhance electrochemical partitioning. This type of information is required, or should be presented, to help evaluate whether one should enhance electrochemical partitioning by genetically engineered modifications, or otherwise.

Charge-directed partitioning has been utilized to enhance the separation of β-galactosidase and T4 lysozyme in PEG-dextran systems (Luther and Glatz, 1995). These authors emphasized that specific affinity binding groups can be incorporated into the phase system to modify the partition behavior (Godbole *et al.*, 1991; Wuenschell *et al.*, 1990; Shanbhag and Axelsson, 1975).

Luther and Glatz (1995) indicated that fusion tails (that may be small peptides or even large proteins) may be genetically fused to a protein. This assists in the separation of the protein by any of the following mechanisms that include substrate and immunoaffinity binding, hydrophobic and charge interactions, and metal chelation (Ford *et al.*, 1991; Hammond *et al.*, 1991; Kohler *et al.*, 1991; Zhao *et al.*, 1990). Luther and Glatz (1995) noted that the high charge density of the tails for β-galactosidase had provided sites for enhanced binding of the protein during polyelectrolyte precipitation (Niederauer *et al.*, 1993; Parker *et al.*, 1990), and ion-exchange membrane separations (Heng and Glatz, 1993). Therefore, Luther and Glatz (1995) wanted to investigate the charge-based mechanism for partitioning. This is the direct interaction of a charged solute with a polyelectrolyte that has been confined to one phase.

These authors noted that attractive interactions between polymers and proteins with opposite charges had a significant effect on the partitioning behavior.

As a result, polymer aggregates between β-galactosidase and (diethylamino)ethyl (DEAE)-dextran were formed. They also noted that polymers having the same charge as that of the protein had little effect on the partitioning.

No information was provided by these authors about the relative amount of denaturation exhibited by β-galactosidase during charge-directed partitioning, and when these fusion tails were not utilized. It would be of interest to evaluate and to analyze whether the fusion tails alleviate or exacerbate β-galactosidase denaturation exhibited on using two-phase aqueous systems for bioseparation.

Let us now attempt to expand the scope of the two-phase aqueous extraction technique. It would be useful to obtain a predictive technique or a model that would help estimate the partitioning in other systems. One would like to be able to extract not only proteins or enzymes, but also other biomolecules of interest, such as peptides, etc.

Example 6.12

Describe a theory that helps predict the partitioning of biomolecules in two-phase systems (Diamond and Hsu, 1989).

Solution

There is a need for the prediction of partitioning of biomolecules in two-phase aqueous systems where data are unavailable for certain parameter ranges of interest. Diamond and Hsu (1989) indicated that some attempts have been made to predict protein distribution between the two phases (Albertsson, 1986; Baskir *et al.*, 1987). However, applications are lacking. Diamond and Hsu (1989) were interested in the two-phase extraction of dipeptides and low-molecular-weight proteins. They have attempted to use the Flory-Huggins theory (Flory, 1941; Huggins, 1941) to help estimate the partitioning of biomolecules of interest.

Diamond and Hsu (1989) obtained a simple linear relationship that helps estimate the partitioning of low-molecular-weight proteins and dipeptides in PEG dextran water systems. These authors emphasized that by knowing the partitioning at one tie-line composition, one can use the Flory-Higgins theory to predict the partitioning at another tie-line composition. Kang and Sandler (1987) and Gustafsson *et al.* (1986) applied the Flory-Higgins theory to two-phase systems, and they indicated that this technique provides both a quantitative and a qualitative description. Diamond and Hsu (1989) emphasized that two assumptions must be made before one can apply the Flory-Higgins theory to two-phase aqueous systems. The biomolecules and the polymers are to be considered as a homogeneous, polymeric species with a random coil structure. Also, the partitioning biomolecule concentration must be small compared with that of the phase-forming polymers.

These authors were able to obtain the simple linear relation

$$\ln K_{\rm b} = k_{\rm DH}\,(w_{\rm PEG,t} - w_{\rm PEG,b}) \tag{6.2}$$

to predict the partitioning of low-molecular-weight proteins such as horse heart cytochrome c, ribonuclease A (from bovine pancreas), chicken egg lysozyme, and bovine serum albumin (BSA). The dipeptides used included glycylglycine, glycylalanine, α-aminobutyric acid, and glycylnorvaline. Here K_b is the partition coefficient of the biomolecule. w_{PEG} is the weight fraction of the polyethylene glycol, either in the top (t) or the (b) bottom phases. k_{DH} is the Diamond-Hsu coefficient (Diamond and Hsu, 1989) that depends on the biomolecule of interest, the polymers used, and the interactions between the biomolecule-water-polymer system.

Equation (6.2) could be used to predict the partitioning of the low-molecular-weight proteins as well as the dipeptides in different water-dextran-PEG systems by knowing the partitioning of the biomolecule of interest in just one system (Diamond and Hsu, 1989). Though quantitative estimates of partitioning are possible using the Diamond-Hsu technique (1989), there are no means that permit the estimation of the qualitative nature of the biomolecule extracted. This is not too difficult to circumvent, however. Qualitative protein extraction will have to be determined by some other technique. Once this information is made available, then the Diamond-Hsu technique (1989) can be used to estimate the quantitative aspects of biomolecule partitioning. It would, of course, be useful if a single technique or modeling procedure is made available that helps estimate both the qualitative and the quantitative aspects of biomolecule extraction. Nevertheless, this technique is useful because it begins to provide the predictive framework required to help assess the total effectiveness of the two-phase aqueous extraction technique.

The purification of biological macromolecules from solution may be carried out by two-phase aqueous systems using polyoxyethylene detergents (Bordier, 1981). By using the polyethylene detergent, Triton X-1114, this author was the first to show that integral membrane proteins partitioned preferentially into the coacervate (detergent-rich) phase, whereas water-soluble proteins (e.g., BSA, hemoglobin) were excluded from the detergent-rich phase. The coacervate phase is formed due to the inverse temperature-solubility behavior exhibited by nonionic detergents carrying polyoxyethylene groups as the hydrophobic moeity. An increase in temperature yields a two-phase separation. One of the liquid phases (detergent-rich) is the coacervate phase where large detergent aggregates may form. Terstappen *et al.* (1992) indicated that some water-soluble extracellular proteins (such as cholesterol oxidases, lipases) may also partition predominantly in the detergent-rich phase (Ramelmeier *et al.*, 1991). It would be of interest to analyze the partition behavior of extracellular lipases in detergent-based two-phase aqueous systems.

Example 6.13

Describe the partition behavior of the extracellular protein, lipase from *Pseudomonas cepacia* using detergent-based two-phase aqueous systems (Terstappen *et al.*, 1992).

Solution

The polyethylene detergent C14EO6 (hexaethyleneglycol mono n-tetradecyl ether) has been used to purify lipase from *P. cepacia* in a two-phase aqueous system (Terstappen *et al.*, 1992). These authors noted that prokaryotic lipases preferentially partitioned into the detergent-rich coacervate phase. In contrast, eukaryotic lipases were excluded from this phase. They were able to extract 76% of the lipase in just one purification step. This leads to a fourfold concentration of lipase. The purification factor obtained is 24. The authors mentioned that the recovery efficiency could easily be increased by the addition of an extra extraction step. Some further advantages of the system utilized are a low phase volume ratio of 0.18 and rather low volumes of detergent (< 1% w/v) are required to facilitate the phase separation at the appropriate temperature.

The authors indicated that the prokaryotic lipases are preferentially partitioned into the coacervate phase due to their surface hydrophobicity. This surface hydrophobicity is "masked" by glycosylation in eukaryotic lipases. In this case they indicated that the carbohydrate chains on the protein surface influence the protein–detergent interaction. This prevents the significant partitioning of the eukaryotic lipases into the coacervate phase. This analysis is of interest, especially their explanation of the difference in partition behavior for prokaryotic and eukaryotic lipases in detergent-based two-phase aqueous systems.

It is also of interest to analyze the bioseparation of biological molecules other than proteins using the two-phase aqueous system technique. Paquet *et al.* (1994) indicated that initially the two-phase aqueous systems were utilized (Albertsson, 1986) for the separation and purification of biomolecules (such as enzymes and other biologically active proteins, nucleic acids, and cells). This technique has found application in analytic measurements (Walter and Johannson, 1986), extractive bioconversion (Kaul and Mattiasson, 1991), and extractive fermentation (Stredansky *et al.*, 1993; Persson *et al.*, 1991). Yang and Chu (1990) indicated that few data are available for the partitioning of low molar mass metabolites.

Modeling of the partitioning of the biomolecules in two-phase aqueous extraction systems would provide insights into the partitioning process and help facilitate the effective separation of the bioproducts. The next example provides a model for the affinity partitioning of proteins using metals.

Example 6.14

Describe briefly a mathematical model for the metal affinity partitioning of proteins (Suh and Arnold, 1990).

Solution

A model for the partitioning of proteins using Cu(II)PEG-dextran systems is available (Suh and Arnold, 1990). The proteins extracted were horse myoglobin, sperm whale myoglobin, *C. krusei* cytochrome c, and tuna heart cytochrome c. Wuenschell *et al.* (1990) indicated that Cu(II)PEG-iminodiacetic acid (IDA) used in PEG-dextran systems increases the partitioning of proteins, if these proteins contain accessible histidine residues. These authors also noted that the partition coefficients of proteins were increased by the addition of IDA. Suh and Arnold (1990) emphasized that the affinity between metal chelates and metal-coordinating residues on protein surfaces has been utilized to advantage in chromatographic separations (Sulkowski, 1985; Van Dam *et al.*, 1989), and also in metal affinity precipitation (Martell and Smith, 1974).

Flanagan and Barondes (1975) initially proposed a mathematical model for the affinity partitioning of S-23 myeloma protein in a PEG-dextran two-phase system. These authors noted that the partition coefficient was increased on the addition of a polymer affinity ligand (dinitrophenyl-PEG). One of the limitations of the Flanagan-Barondes model (1975) was that analysis was limited to the case where the polymer-bound ligand is in sufficient excess to saturate all the binding sites. Cordes *et al.* (1987) later removed this restriction, and their expression was valid for all ligand concentrations. Suh and Arnold (1990) extended the Cordes *et al.* (1987) analysis to account for the inhibition of the protein-ligand interaction. The inhibition of ligand binding was due to the hydrogen ion. The Suh-Arnold analysis (1990) also takes into account the case where more than one type of ligand binding site contributes to the affinity partitioning.

Suh and Arnold (1990) emphasized that their model has general applicability, and can and should be applied for the partitioning of other biomolecules in other systems. These authors indicated that their model is applicable to the systems where proteins have accessible histidine and cysteine sites. No information was provided by Suh and Arnold (1990) about the quality of the proteins separated. Both quantitative and qualitative information about protein recovery is required for a comparison of the cases when IDA is used and when it is absent. Quantitative results are definitely better when IDA is used compared with those when it is not used. However, what about the quality of the product? How is the quality of the bioproducts affected when IDA is used compared with when it is not used? Emphasis needs to be placed on both the quantitative and the qualitative aspects of protein partitioning in these types of systems to fully evaluate the effectiveness of these processes. Future model development should also, if possible, include some qualitative aspects of protein recovery. Together the qualitative and quantitative aspects of protein recovery should provide a reasonable and appropriate framework for the bioseparation of useful and valuable bioproducts by two-phase aqueous extraction.

Because the two-phase extraction technique provides a gentle environment and ease of scale-up (Kula, 1979; Kroner *et al.*, 1985), it is instructive to develop a means for the continuous extraction of proteins. These authors emphasized that demixing of the fine droplets in these types of systems can be slow and time consuming. An agitated vessel-centrifuge that assists the mixing–demixing process may be considered as a single stage. They indicated that this is, however, expensive. These authors believed that more than a single stage can be obtained in spray columns for the protein extraction in two-phase systems.

The spray column extraction of BSA in a laboratory-scale spray column using a PEG-maltodextrin two-phase aqueous polymer system has been analyzed (Raghav Rao *et al.*, 1991). For this system, these authors indicated that data are needed for the mass transfer of the proteins, holdup of the dispersed phase, dynamics of the droplet, and extent of mixing in both phases. In their experiments these authors dispersed the PEG (light phase) into the maltodextran phase (heavy phase). They noted that the mass transfer coefficients for BSA were higher in the PEG-maltodextran systems compared with those in the PEG-dextran systems. Spray columns of 22- and 33-mm diameter were used. Empirical correlations were provided for the mass transfer and holdup of BSA in this system. They emphasized that their correlations are valid for a wide range of column diameters, and are independent of the sparger design.

This analysis is of interest because the authors emphasized the determination of the mass transfer coefficients and the dispersed phase holdup. However, no numbers were provided by these authors on the quantitative and qualitative nature of the protein extracted. Perhaps this will follow in subsequent studies. The dispersed-phase holdup, mass transfer of proteins, droplet dynamics, and extent of mixing in both the phases are bound to affect both the quantitative and the qualitative nature of the protein extracted. A thorough understanding of the influence of the preceding parameters on the quality of protein extracted would provide significant physical insights into the continuous extraction of proteins in two-phase aqueous polymer systems.

Huddleston *et al.* (1994) analyzed the similarities in the partition of proteins between a polymer-bonded solid phase and the corresponding mobile phase, and the partition in PEG-salt two-phase aqueous systems. These authors indicated that because the molecular mechanisms underlying the partitioning behavior in a two-phase aqueous system are not well understood, the development of this method is to a large extent empirical (Huddleston and Lyddiatt, 1990). This prevents the rapid commercialization of the process, and the intense secrecy involved in revealing the information. Huddleston *et al.* (1994) wanted to explore the possibility of characterizing protein surface properties using hydrophobic interaction chromatography (HIC) as a means to design partitioning systems. These protein surface properties would, of course, be of relevance to two-phase aqueous systems. At the outset, these authors cautioned that any relationship that develops on comparing the partitioning between a solid phase and a liquid phase, and between two liquid phases would expected to be only qualitative and not quantitative.

They utilized the principles of solvophobic theory that were extended to include hydrophobic interaction chromatography (Melander *et al.*, 1984) to compare the behavior of proteins separated in hydrophobic interaction chromatography using mildly hydrophobic ligands with partitioning in two-phase aqueous systems. Huddleston *et al.* (1994) indicated that the information available about model proteins from the relatively rapid HIC method may be utilized to make practical judgments in two-phase aqueous systems. Some of the qualitative information includes (1) the degree of resolution available, (2) the effective range of PEG molecular weights chosen, and (3) the phase preference under the condition selected. Furthermore, the comparision and analysis made by these authors suggest that only those proteins exhibitimg unusual surface properties would be extracted to high degrees of purity by two-phase aqueous systems, employing a limited number of equilibrium stages. This is consistent with the conceptual model of partitioning proposed.

Finally, Huddleston *et al.* (1994) cautioned that even though their analysis is helpful, additional investigation of an empirical nature will still be required to achieve optimum results. In any case, despite its shortcomings, this analysis is a step in the right direction. This is because it indicates a way to make two-phase aqueous partitioning more scientific and less empirical. These authors emphasized that two-phase aqueous partitioning is uniquely sensitive to protein surface properties and subtle differences. This contributes to the empirical nature of the partitioning process.

III. CONCLUSIONS

Novel bioseparation techniques need to be continuously researched and developed for the effective removal of proteins and other bioproducts of interest from often dilute solutions that contain delicate and similar structures. Quite a few techniques and analysis for their application as useful bioseparation techniques are under investigation. In this chapter we have analyzed the two new and novel techniques that we believe have exhibited tremendous potential, and have been emphasized more in the literature as being good candidates for commercial application. The two techniques discussed are the reverse micelle technique and the two-phase aqueous extraction technique.

Though only a few of the studies available in the literature are presented in the preceding analysis, one fact is clear. Hardly any emphasis has been placed on the quality of the protein or other bioproducts separated. There are quite a few studies that emphasize the quantity of the protein or other bioproduct separated. More studies are definitely required that further delineate the influence of different parameters on the quantity of the protein or other bioproduct separated. Such studies will continue to appear that will shed further insights into either the reverse micelle or the two-phase aqueous extraction technique. As the knowledge of mass transfer in these systems grows and the interactions at the interfaces are further delineated, it is expected that both of these techniques will become closer to commercial application.

However, enough emphasis needs to be placed on the quality of the product as it is separated by these and other techniques that are presently under various stages of development. Until this is done it appears that these techniques cannot be fully and completely evaluated as to their effectiveness as a bioseparation technique. It is possible that some of this information is available and has been missed in this analysis. Also, some of this type of information may be available in industrial sources. It is essential to have data on the quantity, quality, and stability of the product separated to completely and fully evaluate a bioseparation technique. Such data are required for a wide variety of proteins and other bioproducts before a suitable framework can be developed to assist in the choice of an appropriate bioseparation technique. This framework and modeling when developed can also assist in the selection of parameter values (at least ranges of them) for the optimum removal of the bioproducts that satisfy the quantity, quality, and stability requirements. This framework and modeling should also be of considerable assistance in selecting parameter values to match not only the continuously changing demands of the market but also the increasingly stringent requirements placed by controlling agencies on the bioproducts separated.

REFERENCES

Albertsson, P. A. (1986). *Partitioning of the Cells and Macromolecules*, 3rd ed., Academic Press: New York.

Ayala, G. A. and Mendoza-Hernandez, G. (1990). *Biochem. Int.*, *22*, 717.

Ayala, G. A., Kamat, S., Beckman, E. J., and Russell, A. J. (1992). *Biotechnol. Bioeng.*, *39*, 806.
Baskir, J. N., Hatton, T. A., and Suter, U. W. (1987). *Macromolecules*, *20*, 1300.
Battistel, E., Luisi, P. L., and Rialdi, G. (1988). *J. Phys. Chem.*, *92*, 6680.
Bausch, T. (1989). Diploma Thesis, Technical University, Munich, Germany.
Bausch, T. E., Plucinski, P. K., and Nitsch, W. (1992). *J. Colloid and Interface Sci.*, *150*, 226.
Berecz, E. and Balla-Achs, M. (1983). *Studies in Inorganic Chemistry-4: Gas Hydrates*, Elsevier: Amsterdam.
Bordier, C. (1981). *J. Biol. Chem.*, *256*, 1604.
Butler, L. G. (1979) *Enzyme Microb. Technol.*, *1*, 253.
Cordes, A., Flossdorf, J., and Kula, M. R. (1987). *Biotechnol. Bioeng.*, *30*, 514.
Coughlin, R. W. and Baclaski, J. B. (1990). *Biotechnol. Prog.*, *6*, 307.
Cuatrecasas, P. and Wilchek, M. (1968). *Biochem. Biophys. Res. Commun.*, *33*, 235.
Cussler, E. L. (1987). Supercritical Gels for Protein Concentration, In *Protein Purification; Micro to Macro*, Burgess, R., Ed., Alan R. Liss: New York, p 307.
Dekker, M,. Van't Riet, K., Weijers, S. R., Baltussen, J. W. A., Lanne, C., and Bijsterbosch, B. H. (1986). *Chem. Eng.*, *33*, B27.
Dekker, M., Hilhorst, R., and Laane, C. (1989). *Anal. Biochem.*, *178*, 217.
Dekker, M., Van't Riet, K., Bijterbosch, B. H., Fijneman, P., and Hilhorst, R. (1990). *Chem. Eng. Sci.*, *45*, 2949.
Dekker, M., Van't Riet, K., Bijsterbosch, B. H., Wolbert, G., and Hilhorst, R. (1990). *Chem. Eng. Sci.*, *45*(9), 29.
Dekker, M., Van't Riet, K., Van der Pol, J. J., Baltussen, J. W. A., Hilhorst, R., and Bijsterbosch, B.H. (1991). *Chem. Engg. J.*, *46*, B69.
Dekker, M., Baltussen, J. W. A., Van't Riet, K., Bijsterbosch, B. H., Laane, C., and Hilhorst, R. (1987). Reverse Micellar Extraction of Enzymes: Effect of Nonionic Surfactants on the Diffusion and Extraction of α-amylase. In Laane, C. *et al.*, eds., *Biocatalysis in Organic media*, Elsevier: Amsterdam.
Diamond, A. D. and Hsu, J. T. (1989). *Biotechnol. Bioeng.*, *34*, 1000.
Dungan, S. R., Bausch, T., Hatton, T. A., Plucinski, P., and Nitsch, W. (1991). *J. Colloid Interf. Sci.*, *145*, 33.
Eicke, H. F. and Kvita, P. (1984). In *Reverse Micelles*, Luisi, P. L. and Straub, B. E., Eds., Plenum: New York, p 21.
Fauquex, P. F., Hustedt, H., and Kula, M. R. (1985). *J. Chem. Technol. Biotechnol.*, *35B*, 51–59.
Flanagan, S. D. and Barondes, J. H. (1975). *J. Biol. Chem.*, *250*, 1484.
Fletcher, P. D. I., Freedman, R. B., Mead, J., Oldfield, C., and Robinson, B. H. (1984). *Colloids Surf.*, *10*, 193.
Fletcher, P. D. I., Howe, A. M., and Robinson, B. (1987). *J. Chem. Soc. Faraday Trans.*, *183*, 985.
Fletcher, P. D. I. and Parrot, D. (1988). *J. Chem. Soc.*, *84*, 1131.
Flory, P. J. (1941). *J. Chem. Phys.*, *9*, 660.
Ford, C. F., Suominen, I., and Glatz, C. E. (1991). *Protein Expression Purif.*, *2*, 95.
Gehrke, S. H., Robeson, J., Johnson, J. F., and Vaid, N. (1991). *Biotechnol. Prog.*, *7*, 355.
Ghatore, A. S., Guerra, M. J., Bell, G., and Halling, P. J. (1994). *Biotechnol. Bioeng.*, *44*, 1355.
Ghelis, C. and Yon, J. (1982). In *Protein Folding*, Academic Press: San Diego, CA, p 173.
Giovenco, S. and Verheggen, F. (1987). *Enzyme Microb. Technol.*, *9*, 470.
Godbole, G. P., Tsai, R., and Clark, W. M. (1991). *Biotechnol. Bioeng.*, *38*, 535.
Goklen, K. E. and Hatton, T. A. (1985). *Biotechnol. Progr.*, *1*(1) 69.
Goklen, K. E. and Hatton, T. A. (1987). *Sep. Sci. Technol.*, *22*, 831.
Gupta, R. B., Han, C. J., and Johnston, K. P. (1994). *Biotechnol. Bioeng.*, *44*, 830.
Gustafsson, A., Wennerstrom, H., and Tjernfeld, F. (1986). *Polymer*, *27*, 1786.
Hammond, P. M., Atkinson, T., Sherwood, R. F., and Scawen, M. D. (1991). *Biopharm*, April, 16.
Hartman, J., Daram, P., Frizell, R. A., Rado, T., Benos, D. J., and Sorscher, E. J. (1992). *Biotechnol. Bioeng.*, *39*, 828.
Han, D. and Rhee, J. S. (1985). *Biotechnol. Lett.*, *7*, 651.
Hatton, T. A. (1987). *ACS Symp. Ser.*, *9*, 170.

Hatton, T. A. (1989) Reverse Micelle Extraction of Proteins. In Surfactant-Based Separation Processes, Surfactant Science Series, Vol. 33, Scamehorn, J. F., and Harwell, J. H., Eds., Marcel Dekker: New York, pp 57–90.

Hayes, D. G. and Gulari, E. (1991). *Biotechnol. Bioeng., 38,* 507.

Heng, M. H. and Glatz, C. E. (1993). *Biotechnol. Bioeng, 42,* 333.

Henley, J. P. and Sadana, A. (1984). *Biotechnol. Bioeng., 26,* 959.

Henley, J. P. and Sadana, A. (1986). *Biotechnol. Bioeng., 28,* 256.

Herak, D. C. and Merrill, E. W. (1989). *Biotechnol. Progr., 5,* 9.

Hilhorst, R., Laane, L., and Veerger, C. (1982). *Proc. Natl. Acad. Sci. USA, 79,* 3927.

Hilhorst, R., Sergeeva, M., Heering, D., Rietveld, P., Fijneman, P., Wolbert, R. B. G., Dekker, M., and Bijsterbosch, B. H. (1995). *Biotechnol. Bioeng., 46,* 375.

Huddleston, J. G. and Lyddiatt, A. (1990). *Appl. Biochem. Biotechnol., 26,* 249.

Huddleston, J., Veide, A., Kohler, K., Flanagan, J., Enfors, S. O., and Lyddiatt, A. (1991) *TIBTECH, 9,* 381.

Huddleston, J. G., Wang, R., and Lyddiatt, A. (1994). *Biotechnol. Bioeng., 44,* 626.

Huddleston, J. G. and Lyddiatt, A. (1990) *Appl. Biochem. and Biotechnol., 26,* 249.

Huddleston, J. G., Ottomar, K. W., Ngonyani, D., and Lyddiatt, A. (1991). *Enzyme Microb. Technol., 13,* 24.

Huggins, M. L. (1941). *J. Chem. Phys., 9,* 440.

Hustedt, H., Borner, B., Kroner, K. H., and Papamichael, N. (1987). *Biotechnol. Tech., 1,* 49.

Hustedt, H., Kroner, K. H., and Papamichael, N. (1988). *Process Biochem., 23,* 129.

Ichikawa, S., Imai, M., and Shimuzu, M. (1992). *Biotechnol. Bioeng., 39,* 20.

Imre, E. V. and Luisi, P. L. (1981). *Biochem. Biophys. Res. Commun., 107,* 538.

Jolivalt, C., Minier, M., and Renon, H. (1990). *J. Colloid Interface Sci., 135*(1), 85.

Kadam, K. L., (1986). *Enzyme Microb. Tech., 8,* 266.

Kang, C. H. and Sandler, S. I. (1987). *Fluid Phase Equilibria, 38,* 245.

Kaul, R. and Mattiasson, B. (1991). Extractive Bioconversions in Aqueous Two-Phase Systems. In *Extractive Bioconversions,* Mattiasson, B., and O. Holst, O., Eds., Marcel Dekker: New York, pp 173–178.

Kauzmann, W. (1959). In *Advances in Protein Chemistry,* Anfinsen, Jr., C. B., Anton, M.L., Bailey, K., and Edsall, J. T., Eds., Academic Press: New York, p 1.

Kelley, B. D. and Hatton, T. A. (1991). *Bioseparation, 1,* 333.

Knuth, M. W. and Burgess, R. R. (1987). The Use of Polyethyleneimine in Protein Purification, In *Protein Purification: Micro to Macro,* Burgess, R., Ed., Alan R. Liss: New York, p 75.

Khmelnitsky, Y. L., Levashov, A. V., and Klyachko, N. L. (1988). *Enzyme Microb. Technol., 10,* 710.

Kohler, K., von Bonsdorff-Lindberg, L., and Enfors, S. O. (1989) *Enz. Microb. Technol., 11,* 730.

Kohler, K., Ljungquist, C., Kondo, A., Veide, A., and Nilsson, B. (1991). *Biotechnology, 9,* 642.

Kohler, K., Veide, A., and Enfors, S. O. (1991) *Enz. Microb. Technol., 13,* 204.

Kroner, K. H., Hustedt, H., Granda, S., and Kula, M. R. (1978). *Biotechnol. Bioeng., 20,* 1967.

Kroner, K. H., Hustedt, H., and Kula, M. R. (1985). *Biotechnol. Bioeng., 24,* 1015.

Kula, M. R. (1979). *Applied Biochemistry and Bioengineering,* Vol. 2, Academic Press: New York.

Kula, M.R. (1987). Use of Phase Partitioning to Scale Up Protein Purification, In *Protein Purification: Micro to Macro,* Burgess, R., Ed., Alan R. Liss: New York, p 99.

Kula, M. R. (1990). *Bioseparation, 1,* 181.

Lee, K. M. and Biellmann, J. F. (1987). *FEBS Letts., 223*(1) 33.

Leser, M. E., Luisi, P. L., and Palmieri, S. (1989). *Biotechnol. Bioeng., 34,* 1140.

Leser, M. E., Markoci, K., and Luisi, P. L. (1993). *Biotechnol. Bioeng., 41,* 489.

Luisi, P.L. and Lanne, C. (1986). *Trends Biotechnol., 4,* 153.

Luisi, P. L. and Magid, L. J. (1986). CRC Critical Review In Biochemistry, Vol. 20, No. 4, CRC Press, Boca Raton: Florida, p 409.

Luisi, P. L., Giomini, M., Pileni, M. P., and Robinson, B. H. (1988). *Biochem. Biophys. Acta, 947,* 209.

Lund, G. and Holt, S. L. (1980). *J. Am. Oil Chem. Soc., 57,* 264.

Luther, J. R. and Glatz, C. E. (1994). *Biotechnol. Bioeng., 44,* 147.

Luther, J. R. and Glatz, C. E. (1995). *Biotechnol. Bioeng., 46,* 62.

Makino, S., Reynolds, J. A., and Tanford, C. (1973). *J. Biol. Chem., 248,* 4926.

Martell, A. E. and Smith, R. M. (1974). *Critical Stability Constants*, Plenum: New York.

Martinek, K., Levashov, A. V., Klyatchko, N. L., Pantin, V. I., and Berezin, I. V. (1981). *Biochim. Biophys. Acta, 657,* 277.

Mattiasson, B. and Senstaad, C. (1989). *Biochem. Biotechnol., 11,* 41.

Matzke, S. F., Creagh, A. L., Haynes, C. A., Prausnitz, J. M., and Blanch, H. W. (1992). *Biotechnol. Bioeng., 40,* 91.

Mohr, M. and Pommerening, K. (1985). *Affinity Chromatography,* Marcel Dekker: New York.

Melander, W. R., Corradini, D., and Horvath, C. (1984). *J. Chromatogr., 317,* 67.

Niederauer, M. Q., Suominen, I., Rougvie, M. A., Ford, C. and Glatz, C. E. (1993). *Biotechnol. Prog., 10,* 237.

Papamichael, N., Borner, B., and Hustedt, H. (1991). *J. Chem. Technol. Biotechnol., 50,* 457.

Paquet, V., Myint, M., Roque, C., and Soucaille, P. (1994). *Biotechnol. Bioengg., 44,* 445.

Paradkar, V. M. and Dordick, J. S. (1991). *Biotechnol. Prog., 7,* 330.

Parker, D. E., Glatz, C. E., Ford, C. F., Gendel, S. M., Suominen, I., and Rougvie, M. A. (1990). *Biotechnol. Bioeng., 36,* 467.

Persson, I., Tjernfeld, F., and Hahn-Hagerdal, B. (1989). *Biotechnol. Tech., 3,* 265.

Phillips, J. B., Nguyen, H., and John, V. T. (1992). *Biotechnol. Prog., 7,* 43.

Qutubuddin, S., Wienchek, J. M., Nabi, A., and Boo, J. Y. (1994). *Sep. Sci. Technol., 29,* 923.

Raghav Rao, K. S. M. S., Szlag, D. S., Sikdar, S. K., Joshi, J. B., and Sawant, S. B. (1991). *Chem. Engg. J., 46,* B75.

Rahaman, R. S., Chee, J. Y., Cabral, J. M. S., and Hatton, T. A. (1988). *Biotech. Progr., 4,* 218.

Ramelmeier, R. A., Terstappen, G. C., and Kula, M. R. (1991). *Bioseparation, 2,* 315.

Rosen, M.J. (1989). Surfactants and Interfacial Phenomena. 2nd ed., Wiley Interscience: New York.

Sadana, A. (1991). Biocatalysis: Fundamentals of Enzyme Deactivation Kinetics, Prentice Hall: Englewood Cliffs, NJ.

Sadana, A. and Raju, R. R. (1990). *Biopharm, 3*(5), 53.

Samana, J. P., Lee, K. M., and Biellmann, J. F. (1984). *Eur. J. Biochem., 163,* 609.

Sanchez-Ferrer, A., Bru, R., and Garcia-Carmona, F. (1988). *FEBS Lett., 233,* 363.

Sarcar, S., Jain, T. K., and Mitra, A. (1992). *Biotechnol. Bioeng., 39,* 474.

Shanbhag, V. P. and Axelsson, C. G. (1975). *Eur. J. Biochem., 60,* 17.

Sikdar, S. K., Cole, K. D., Stewart, R. M., Szlag, D. C., Todd, P., and Cazebas, H., Jr., (1991). *Biotechnology, 9,* 253.

Singer, S. J. (1962). *Adv. Protein Chem., 17,* 1.

Smith, G. D., Donelan, C. E., and Barden, R. E. (1977). *J. Colloid Interface Sci., 60,* 488.

Steinmann, B., Jackale, H., and Luisi, P. L. (1986). *Biopolymers, 25,* 1133.

Strandberg, L., Kohler, K., and Enfors, S. O. (1991). *Process Biochem., 26,* 225.

Stredansky, M., Svorc, R., Sturdik, E., and Dercova, K. (1993). *J. Biotechnol., 27,* 181.

Suh, S. S. and Arnold, F. H. (1990). *Biotechnol. Bioeng., 35,* 682.

Sulkowski, E. (1985). *Trends Biotechnol., 3,* 1.

Terstappen, G. C., Geerts, A. J., and Kula, M. R. (1992). *Biotechnol. Appl. Biochem., 16,* 228.

Treybal, R. E. (1980). *Mass Transfer Operations,* 3rd ed., McGraw-Hill, New York.

Van Dam, M., Wuenschell, G., and Arnold, F. H. (1989). *Biotechnol. Appl. Biochem., 11,* 492.

Vasudevan, M., Tahan, K., and Wienchek, J. M. (1995). *Biotechnol. Bioeng., 46,* 99.

Veide, A., Smeds, A. L., and Enfors, S. O. (1983). *Biotechnol. Bioeng., 25,* 1789.

Veide, A., Lindback, T., and Enfors, S. O. (1984) *Enzyme Microb. Technol., 6,* 325.

Vulfson, E. N., Ahmed, G., Gill, I., Kozlov, I. A., Goodenough, P. W., and Law, B. A. (1991). *Biotechnol. Letts., 13*(2), 91.

Walter, H. and Johannson, G. (1986). *Anal. Biochem., 155,* 215.

Wheelwright, S. M. (1991). *Protein Purification : Design and Scale-up of Downstream Processing,* Hanser: Munich, Germany.

Wienchek, J. M. (1989). Liquid Membrane Separations Employing Nonionic Microemulsions, Ph.D. Thesis, Case Western Reserve University, Cleveland, OH.

Wolbert, R. B. G., Hilhorst, R., Voskuilen, G., Nachtesaal, H., Dekker, M., Van't Riet, K., and Bijsterbosch, B. H. (1989). *Eur. J. Biochem., 184,* 627.

Wolf, R. and Luisi, P. L. (1979). *Biochem. Biophys. Res. Commun., 89,* 209.

Woll, J.M., Dillon, A.S., Rahaman, R.S., and Hatton, T. A. (1987). Protein Separations Uisng Reversed Micelles. In: *Protein Purification: Micro to Macro,* Burgess, R., Ed., Alan R. Liss: New York, p 117.

Woll, J. M., Hatton, T. A., and Yarmush, M. L. (1989). *Biotech. Prog.*, *5*, 57.
Wong, R. and Luisi, P. L. (1979). *Biochem. Biophys. Res. Commun.*, *89*, 209.
Wuenschell, G. E., Naranjo, E., and Arnold, F. H. (1990). *Bioprocess Engg.*, *5*, 199.
Yang, W. Y. and Chu, I. M. (1990). *Biotechnol. Tech.*, *4*, 191.
Zhao, J., Ford, C., Glatz, C. E., Rougvie, M. A., and Gendel, S. M. (1990). *J. Biotechnol.*, *14*, 273.

7

ADSORPTION INFLUENCE ON BIOSEPARATION AND INACTIVATION

I. INTRODUCTION

The adsorption of proteins occurs at different types of interfaces and the initial layers of proteins adsorbed significantly affect the processes occurring at these interfaces. Interfacial interactions are important in the bioseparation of proteins and other biological macromolecules of interest. Other areas of interest where these interactions are important are biomedical applications of artificial devices, biosensors (Rechnitz, 1987), immunoassays (Gribnau *et al.*, 1986), and drug delivery systems.

Over the years, significant attention has been paid to the determination of the quantitative aspects of protein adsorption to different surfaces. This effort was justified; however, in addition to the amount of protein adsorbed, the biological consequences of proteins at solid surfaces often depend on the nature and state of the adsorbed layer. In particular, information about the protein conformation and orientation, or more precisely, time-dependent structural changes in the adsorbed layer are urgently required. The lack of information about how proteins are organized has hindered delineation of the role of the interface in protein adsorption studies in spite of three decades of research.

A shortcoming of several studies is that an insufficient number of variables are studied (Elgersma *et al.*, 1990). Some of the parameters that may influence protein-surface interactions include electrostatic interactions (Elgersma *et al.*, 1990; Clark *et al.,* 1988), pH (Bagchi and Birnbaum, 1981; Sonderquist and Walton, 1980), negatively charged surfaces (Norde and Lyklema, 1978a,b; Norde, 1983), surface charge (Hlady and Furedi-Milhofer, 1979), co-adsorption of low-molecular weight ions (Elgersma *et al.*, 1990), and isoelectric point (Abramson, 1942). Other factors that may influence protein adsorption onto surfaces include intermolecular forces between adsorbed molecules, solvent-solvent interactions, strength of functional group bonds, chemistry of the solid surface, topology, and morphology.

Proteins in solution diffuse to the interface. This is thermodynamically favorable because some of the conformational and hydration energy of the protein is lost at the interface (MacRitchie, 1978). Initially, at low protein concentrations there is no barrier to adsorption; and for protein molecules that are readily adsorbed at the interface, the rate of adsorption is diffusion controlled. However, after some time, especially at high surface concentrations, there is an activation energy barrier to adsorption (Graham and Phillips, 1979), which may involve electrostatic, steric, and osmotic effects close to the interfacial or surface layers. Then the ability of protein molecules to interpenetrate and create space in the existing film and to rearrange at the surface is rate determining.

Initially, Joly (1965) suggested that enzymes adsorbed at gas–liquid interfaces are generally present in an unfolded partially active or inactive state as a more or less rigid film. Graham and Phillips (1979) then stated that the capacity of proteins to unfold at an interface depends greatly on the conformational stability of flexible segments of the protein molecule. Interfaces are primarily responsible for protein inactivation as highlighted by the experiments by Virkar *et al.* (1979). By using a partially filled disk reactor these authors noted that shear-associated damage can be severe, but it arises when gas–liquid interfaces are present. Then the replenishment of the interface associated with intense shear causes interfacial denaturation. Dunnill (1983) suggests that this, rather than shear alone, is the explanation for much loss of protein structure and enzyme activity in pumps (Virkar *et al.,* 1979) and in centrifuges and ultrafiltration systems (Narendarnathan and Dunnill, 1982) where air is entrained. Virkar *et al.* (1979) further noted that a decrease in the air–liquid interfacial area by completely filling the reactor vessel minimized the enzyme inactivation. Similar observations may be made with liquid–liquid (e.g., present in liquid–liquid extraction systems) or liquid–solid (for example, high-pressure liquid chromatographic separations) systems.

When proteins are adsorbed at interfaces, they undergo a change from their globular conformation to an extended chain conformation (MacRitchie, 1987). This is surface denaturation of the protein. The analysis of protein adsorption at surfaces including mechanisms, kinetics, time-dependent conformational changes, etc., is a difficult process. The adsorption of proteins–enzymes at the interface is a complex phenomenon that involves the following steps that occur simultaneously (Aptel *et al.,* 1987): (1) transport to the interface by diffusion or diffusion–convection, mixing and shearing action generally enhancing this step; (2) adsorption–desorption at the interface, described by an interfacial chemical reaction and its related kinetic adsorption and desorption mechanisms; (3) structural alterations of molecules in contact with the interface, and at higher occupancy, interactions with other adsorbed molecules; and (4) adsorption competition between molecules of different nature or molecular weight. Lok *et al.* (1983a and b) correctly point out the factors that influence protein adsorption onto surfaces include intrinsic protein adsorption kinetics, chemical equilibrium between surface-adsorbed protein and free solution proteins, and flow of protein past the adsorbing surface. They also hint that the conformation of proteins in the adsorbed layer may be an important factor.

The possible effects of a given surface on a given protein (mixture) would include, among others, permanent or reversible adsorption with or without concomitant denaturation or conformational changes, preferential adsorption of specific proteins, and changes in the microenvironment of enzymes.

The adsorption of proteins on the surfaces of glass is well known in many areas of biochemistry (Silman and Katchalski, 1966; Hummel and Anderson, 1965; Bull, 1956). For example, vibronectin, fibronectin, and laminin bind strongly to glass (Barnes, 1984). Vibronectin is a surface-active protein and readily denatures at solid–liquid interfaces. This is the primary reason for its efficient adsorption at both hydrophobic and hydrophilic surfaces. Fibronectin, however, appears to adsorb more on hydrophobic than on hydrophilic surfaces primarily due to a higher conformational change on hydrophobic surfaces (Grinell and Feld, 1982).

Example 7.1

Briefly describe some of the processes that are influenced both in a favorable and in a deleterious manner by protein adsorption. Also, briefly describe some of the effects that primarily control protein adsorption (Haynes *et al.*, 1994).

Solution

Haynes *et al.* (1994) indicate that protein adsorption occurs both *in vivo* and *in vitro*. Some positive influences or consequences occur in: (1) stabilization of microemulsions, pharmaceutical creams, lotions, formulated foods and foams; (2) protein purification strategy development; (3) drug delivery systems; and (4) biosensors during the *in vivo* monitoring of glucose levels in blood, and *in vitro* immunoassays. Some negative aspects of protein adsorption are observed during: (1) thrombus development on blood vessels and in artificial implant materials, (2) fouling of kidney dialysis membranes and in processing equipment, and (3) plaque formation on teeth and dental restoratives.

These authors emphasize that a better control of these processes requires a better understanding of the protein adsorption process. The following four effects play a significant role during adsorption. These include (1) structural rearrangements in the protein molecule, (2) redistribution of charged groups in the interfacial layer, (3) protein surface polarity, and (4) dehydration of the sorbent surface. More often than not a synergistic combination of some or all the previously mentioned effects come into play during protein adsorption. For example, Haynes *et al.* (1994) indicate that the strong affinity for the two similar-sized globular proteins hen egg-white lysozyme and bovine milk α-lactal bumin in aqueous solution for polystyrene surface arises due to a synergistic combination of dehydration, structural perturbation, and electrostatic subprocesses.

More detailed protein adsorption followed by subsequent denaturation studies is required to delineate the mechanisms involved at the interface. This is of primary importance because the initial protein layer or layers mediate and control further interactions at the interface. For example, the adhesion of blood platelets to glass surfaces is known in the field of clinical chemistry, and blood platelets adhere more on surfaces coated with fibrinogen (Zucker and Broman, 1969). The amounts of protein adsorbed on the surface of glassware have a significant effect on the quantitative analysis of very small amounts of proteins in the case of radioimmunoassay or enzyme immunoassay (Rosselin *et al.,* 1966). Adsorption of proteins on a glass surface is not a specific phenomenon but rather a general phenomenon that is usually neglected because of the low surface area of typical glassware. Though adsorption of proteins to glassware is important, especially in clinical studies, one really needs to know more about the adsorption of proteins in general, and blood proteins in particular, to different polymer surfaces that have significant biomedical usage.

An area of considerable importance where interfacial interactions are important is in biomedical applications of artificial devices. A wide number of clinically important implants and devices exist. Some (e.g., catheters) may only contact the blood once and for a relatively short time; others (e.g., kidney dialyzers) may be exposed to blood for hours, while tissue implants (e.g., heart valves) will hopefully last for the lifetime of the patient (Hoffman, 1982). Leonard *et al.* (1987) have emphasized that even though the basic properties of an ideal blood compatible material cannot be agreed on, it should in principle be effective throughout the lifetime of an individual. This is not surprising because different materials should in general be rather specific for different types of usage. Thus, an understanding of the rapid adsorption of plasma proteins when blood contacts an artificial surface or foreign material is of importance (Baeir and Dutton, 1969; Vroman and Adams, 1969).

Example 7.2

Describe briefly the adsorption of blood proteins to different surfaces.

Solution

A major disadvantage in the use of blood contacting foreign materials is the formation of a thrombus at the blood-polymer interface. Thrombosis involves a series of events beginning with the deposition of a protein layer at the blood-polymer interface. The formation of this protein layer is followed by the adherence of platelets, fibrin, and possibly leukocytes (Young *et al.,* 1982; Ihlenfeld *et al.,* 1979). Further deposition with possible concurrent entrapment of erythrocytes and other blood elements in a fibrin network constitutes thrombus formation. The growth of this thrombus eventually results in partial or total blockage if the thrombus is not sheared off or otherwise released from the surface (Sharma *et al.,* 1982). An understanding of the physics and chemistry of the initial layer of protein adsorption cannot be overemphasized.

The surface-induced coagulation of blood is a critical factor in the design and application of most devices for use with the cardiovascular system. The clotting time of blood is dependent upon the material with which it is in contact. It is important to study the adsorption behavior of proteins that are a major component of plasma such as albumin, γ-globulin, and fibrinogen in relation to the antithrombogenicity of polymer materials. The possible effects of a given surface on a protein (mixture) would include, among others, permanent or reversible adsorption with or without concomitant denaturation or conformational changes, preferential adsorption of specific proteins, and changes in the microenvironment of enzymes. Because the adsorption of proteins on a surface depends on both the protein and the surface, it is important to characterize the nature of both the protein sample and the surface. How homogeneous or heterogeneous is each of these? Does the heterogeneity affect the adsorption and further properties; and if it does, by how much?

Adsorption of proteins at solid–liquid interfaces has been reviewed in the literature (MacRitchie, 1978; Norde, 1986; Hlady and Andrade, 1986; Lundstrom *et al.,* 1987). Not much information is presented in a concise manner concerning the heterogeneity either of the protein adsorbate or of the surface, and the subsequent effects of this heterogeneity on denaturation of the adsorbed protein on the surface. In fact, the previous reviews seem to either neglect or perhaps treat rather lightly the denaturation (after adsorption) of the protein on the surface. Also, although much is known about the qualitative nature of protein adsorption on different surfaces, there are still apparently no rigorous mathematical theories that describe protein adsorption on different surfaces. The studies presented together in the chapter should provide a judicious framework within which one can compare the status of one's work and hopefully stimulate work in appropriate directions in the future. The studies presented should be viewed only as appropriate examples, because surely more studies are available in the literature that would help either further delineate or reinforce the ideas presented later or even perhaps elucidate other ideas or factors of importance. One of the major intents of the chapter is to focus on heterogeneity in adsorption and its subsequent effects on protein denaturation. This is important, and as indicated earlier the initial adsorbed protein layer mediates or controls the adsorption of further layers.

II. ADSORPTION OF PROTEINS AND OTHER BIOLOGICAL MACROMOLECULES

A. Surfaces for Protein Adsorption

Adsorption of blood proteins has been done on different glass and polymeric surfaces. Some examples are polyethylene tubing, silicon rubber tubing, plasticized polyvinyl chloride tubing, and a segmented polyether urethane urea tubing (Young *et al.,* 1988); polyvinyl chloride (PVC), copolymer of methacrylic acid and methacrylate (PMA), and a surface-grafted polyethylene oxide (PEO) films (Golander and Kiss, 1988); silica with two different surface energies (Johnsson *et al.,* 1985); silicon or glass plates (Elwing *et al.,* 1987); and polymeric surfaces with varying surface properties and functionalities that are polydimethylsiloxane (PDMS), polymethyl methacrylate (PMMA), polystyrene sulfonate (PSS), and other polymeric materials (Young, 1984).

Jeon *et al.* (1991) recently indicated that a large number of studies have been done to minimize protein adsorption to different surfaces. This is important in such diverse areas as chromatographic supports, blood-contacting devices, and immunoassays. An excellent protein-resistant surface is PEO. These authors indicate that between two adsorbed PEO surfaces in a good aqueous solvent repulsion forces develop at certain separation distances due to a steric repulsion phenomenon. The protein-resistant character is probably caused by a steric stabilization effect. The protein resistance character of PEO chains terminally attached to a hydrophobic solid substrate was analyzed. They propose that as the protein approaches the PEO surface by diffusion, it is affected by the van der Waals attraction between the PEO surface and protein through water. A further approach of the protein initiates the compression of PEO chains, which induces a steric repulsion effect. Also, an additional van der Waals attraction becomes important between the substrate and protein through the water-solvated PEO layer. Furthermore, Jeon *et al.* (1991) state that the good protein resistance properties of PEO are related to the fact that its refractive index is the lowest among the water soluble synthetic polymers, resulting in a low van der Waals interaction with the protein. Finally, the van der Waals attraction is small compared with the steric repulsion.

In a later study on the effect of protein size on protein-surface interactions in the presence of polyethylene oxide, Jeon and Andrade (1991) determined the PEO surface density conditions for optimal protein resistance. These authors noted that for small proteins (R ~20 Å), D should be small (~10 Å); and for large proteins (R ~60 − 80 Å), D should be larger (~15 Å). Here D is the average distance between end-attached PEO chains, and R is the protein radius. These authors emphasize that these results evolve from the trade-offs between steric repulsion and the assumed weak hydrophobic interaction between the PEO layer and the protein.

Some surfaces have been precoated to attain desired characteristics. Albumin is the most abundant plasma protein. Albumin molecules in the native state are well known not to be included in thrombus formation and platelet aggregation. Therefore, good thromboresistance of a polymer could be achieved by the selective adsorption of albumin onto a polymeric surface. Sato *et al.* (1987) noted that albumin precoating on controlled porous glass minimized thrombin inactivation. These authors suggest that albumin protects the thrombin from self-hydrolysis.

It has been shown by researchers (Absolom *et al.*, 1987; Lee and Kim, 1974) that the surface properties and more specifically the surface tension of various potential cardiovascular implant materials is related to the protein adsorption to those surfaces. Absolom *et al.* (1987) utilized the sedimentation volume method (V_{sed}) to characterize the surface tension of protein-precoated polymer particles. A maximum in V_{sed} occurred when the surface tension of the suspending liquid was equal to that of the particles. The position of the V_{sed} maxima and hence the surface tension, γ_{PV}, of the particles was found to depend on polymer surface tension as well as on the type and bulk concentration of the coating protein solution. Experiments were performed with the powders polytetrafluoroethylene (PTFE), polyvinyl chloride (PVC), and nylon-6,6. The serum proteins used for precoating were human serum albumin (HSA), human serum immunoglobulin G (IgG), and fibrinogen. These authors indicate that at high bulk concentrations the nature of the underlying substrate materials are entirely masked. The advantage of this method is that it is inexpensive and versatile. It can be applied to investigations involving, in general, any type of particulate material. It does not rely on the use of sophisticated optical equipment that has to be calibrated, and the results are then carefully analyzed. Also, the sedimentation volume method does not require the use of substrates having any specific properties. The only disadvantage or limitation of this method is that it can be applied only to surfaces that can be obtained in particulate form. By considering the advantages and the simplicity in the usage of this method it is anticipated that this will be used more frequently in the future by different workers, especially if one is looking for a nonoptical method of surface characterization.

Lahav (1987) investigated the adsorption of thrombospondin, fibrinogen, and fibronectin. Two of these were in solution and one was surface adsorbed. All binding assays were performed in 7-mm diameter wells made of polystyrene. All these proteins form part of the blood coagulation process (Lahav *et al.*, 1982; Leung, 1984), and have been shown to interact with each other when one of them is attached to the surface (Leung, 1984; Leung and Nachman, 1982). Lahav (1987) concludes that, in general, multicomponent systems in which multiple binding can take place could show a complex pattern of interactions. Because this study (1987) presents a more realistic picture of (blood) protein interactions at surfaces, such studies should be more emphasized in the future. As indicated by Lahav (1987) and as is to be anticipated, these studies will be difficult due to the complex interactions involved. Nevertheless, they are necessary if one wants appropriate physical insights into blood protein adsorption on surfaces under actual circumstances. As a matter of fact, there are a large number of proteins in blood that would significantly affect the interactions among themselves and with the surface during the adsorption process. The complex pattern of protein interactions with the interface should lead to an increasing heterogeneity in adsorption.

The surfaces for protein adsorption and the amount of protein adsorbed need to be better characterized. The heterogeneity of the surface will significantly influence adsorption and subsequent reactions occurring on the surface. The nature of the amount of protein adsorbed is also of significance. Does a monolayer of protein-adsorbed molecules exist or do we also have a second layer of protein adsorption? What is the structure of the adsorbed protein molecules on the surface? The next couple of sections begin to address this problem.

B. Monolayer Adsorption

Proteins are intrinsically surface active and tend to concentrate at surfaces. The chemical composition of the surface plays a major role in protein adsorption. Norde and Lyklema (1979) indicate that the mechanism of protein adsorption to surfaces is rather complex. This involves the attachment of different amino acid residues (segments) of one and the same protein molecule to the sorbent surface so that the molar Gibbs energy of adsorption attains large values. Shirahama *et al.* (1990) indicate that in a solution containing a mixture of proteins the interface will initially accommodate the protein molecules that have the largest diffusion rate coefficient and are most abundantly present in solution. These initially adsorbed protein molecules may be displaced by other protein molecules that have a higher affinity. These authors emphasize that the final composition of the adsorbed layer at a given interface is determined by the concentrations of the various kinds of proteins in solution, the intrinsic adsorption affinities, and the possibilities that the proteins have to desorb.

Researchers in the past have successfully modeled the adsorption behavior of proteins in many cases, using the Langmuir model (which was originally developed for gases) even though it does not conform to theory. Rudzinski *et al.* (1983) indicate that other appropriate liquid counterparts of the gaseous empirical isotherm equations have been developed. These include counterparts of the Freundlich (Rudzinski and coworkers, 1973, 1974, 1983; Dabrowski and Jaroniec 1979a,b, 1980a,b), the Dubinin-Radushkevich (Oscik *et al.*, 1986; Jaroniec and coworkers, 1978, 1981, 1983), and the Toth (Jaroniec and Derylo, 1981) empirical equations. Rudzinski *et al.* (1983) emphasize that the application of these empirical equations to correlate experimental data in solution adsorption has not been accompanied by a sufficient care about the limitations in their applicability. For example, these empirical equations do not reduce correctly to Henry's law at sufficiently low concentrations of one of the components of a liquid mixture. This contradicts rigorous thermodynamic predictions. Nevertheless, these studies with these known constraints provide some restricted physical insights into the protein solution adsorption process on different surfaces.

Rudzinski *et al.* (1983) indicate that in many adsorption systems the structure of the solid-solution interface-system is adequately represented by a monolayer adsorption system. In these types of systems the equilibrium bulk solution exhibits small or moderate departures from an ideal solution behavior. These authors emphasize the short-ranged nature of the solid-adsorbate forces with the second and higher layers being formed primarily by interactions between the molecules of the liquid mixture themselves. They further indicate that these interactions between the molecules of the liquid mixture will be even smaller near the vicinity of the solid surface or liquid–solid interface.

Investigators increasingly pay attention to the fact that structural variations might be taking place (Jaroniec *et al.*, 1983; Cuypers *et al.*, 1987). Lundstrom (1985) utilized the Freundlich isotherm to describe a protein adsorption model that includes more than one orientation or conformation of the adsorbed protein. Their study allowed the determination of important parameters that influence protein adsorption on biomaterials. It is, therfore, of importance to study not only the quantitative, but also the qualitative aspects of protein adsorption to surfaces. The following study of comparative protein adsorption in model systems is a good example.

Example 7.3

Protein adsorption on surfaces indicates quantitative as well as qualitative features (Shirahama *et al.*, 1990).

Solution

Shirahama *et al.* (1990) have studied the adsorption of lysozyme, ribonuclease, and α-lactalbumin on hydrophobic silica and on hydrophobic polystyrene-coated silica, which are both negatively charged. These authors monitored the adsorption process by reflectometry and by streaming potential measurements. Reflectometry provided a quantitative measure of the protein adsorbed, and streaming potential measurements provided qualitative information of the protein adsorbed (composition of the adsorbed layer). They emphasized that both sequential and competitive adsorption from flowing solutions never led to adsorbed amounts that exceeded values corresponding to monolayer coverage (i.e., 1-2 ng/m^2). This *flowing* condition prevents association between two proteins that then constrains the adsorption to monolayer coverage. These proteins have similar molar mass, (globular) size and therefore diffusion coefficient. Thus, the effects of molecular size and diffusion coefficient on the adsorption preference are practically negligible. The proteins do, however, differ in their isoelectric point, hydrophobicity, and stability.

At the hydrophilic surface (SiO_2) the adsorption is largely determined by electrostatic interaction (Shirahama *et al.*, 1990). This is because: (1) the protein amount adsorbed from single-protein solution increases with increasing charge contrast between the protein and the adsorbed surface; (2) sequential adsorption occurs only if the second protein has a more favorable electrostatic interaction with the adsorbent surface; (3) the final composition of the adsorbed layer essentially consists of the protein that has the most favorable electrostatic interaction with the adsorbent; and (4) the authors emphasize a remarkable feature of their studies for all three proteins is that the initial adsorption rates are not significantly affected by the nature of the surface, whereas at the later stages of the process the curves for $\Gamma(t)$ at the hydrophobic and hydrophilic surfaces do differ markedly. At the later stages of the adsorption process, the surface becomes crowded with protein molecules. Further variations in $\Gamma(t)$ are due primarily to orientation and conformational effects of the preadsorbed molecules. This would lead to an increasing heterogeneity of the adsorbed protein on the surface. These authors also indicate that at the hydrophobic surface (PS-SiO_2) electrostatic interactions have some effect, but they definitely do not dominate the adsorption process.

More studies like the analysis by Shirahama *et al.* (1990) are required that provide information on both the quantitative as well as the qualitative aspects of protein adsorption on surfaces. Although not much information on the qualitative characterization of protein adsorption is available in the literature, the quantitative aspects have frequently been studied.

McNaly and Graf (1990) have proposed a model in which the molecules can exist in three regions: (1) the bulk solution; (2) the surface; and (3) the subsurface, a region several molecular diameters below the surface. The authors utilized the diffusion-controlled model

$$\frac{\Gamma}{c} = 2\left(\frac{Dt}{\Pi}\right)^{1/2} \tag{7.1}$$

to describe the initial adsorption kinetics of hydroxypropyl cellulose (HPC) and hydroxyethyl cellulose (HEC). Here Γ is the concentration of the adsorbate molecules, C is the concentration of the molecules in solution, D is the diffusion coefficient of the molecules in solution, and t is time. The initial adsorption is described by a surface depleted of adsorbed solute molecules in which molecules will instantly move from the subsurface to the surface, thus leaving a zero concentration at the subsurface. This causes a diffusion-controlled gradient between the bulk solution and the subsurface. This rate of diffusion should govern the overall kinetics of adsorption (Langmuir and Schaefer, 1937). The preceeding diffusion-controlled equation implies that a good straight-line relationship can be obtained for the initial time period. After a while, the plots become more nonlinear, caused by the buildup of an energy barrier to adsorption on the surface as well as to bulk diffusion.

Beissinger and Leonard (1980) proposed a model for albumin adsorption in which the adsorbed molecule can exist in one of two adsorbed states, Θ_1 and Θ_2. Albumin molecules in solution near the surface are assumed to be able to adsorb in state 1. In state 1 they can either desorb or enter state 2. At state 2 only desorption of the albumin molecule is possible, and state 2 is only accessible from state 1. The parameters in the model do have physical significance. For example, the rate constant for desorption from state 1 is larger than the rate constant for desorption from state 2. Thus, the molecules adsorbed in state 2 are much more tightly held than those desorbed in state 1. The model formulation gives an indication of heterogeneity of adsorption sites on the surface. These authors extended the model to also include the competitive adsorption of albumin and γ-globulin.

An exchange reaction often takes place in protein adsorption wherein the protein molecules may desorb from the surface in the presence of other molecules. In an exchange reaction, a second molecule gradually replaces the originally adsorbed one (Jennisen, 1978, 1981), and the original bonds are broken one by one. If the number of binding sites to the surface is large, an exchange reaction is an improbable process. If the exchanged molecules are of the same kind as the originally adsorbed ones, the total free energy has not changed after the completed exchange reaction; however, for molecules of different kinds, a lower total free energy of the second molecule when it binds to the surface may be a thermodynamic force for the exchange reaction. This exchange reaction would also contribute toward the heterogeneity of the protein in the adsorbed layer.

Another interesting observation is that when a protein molecule resides on a surface long enough, it forms all its possible bonds with a surface, and this may be the reason for the conformation change of the molecule. This might lead to the stronger bonds formed with the surface with increasing time. This makes the exchange reaction and also desorption more difficult. An increased residence time of adsorbed molecules on the surface would also increase heterogeneity of the adsorbed molecules on the surface. It would be of interest to develop a parameter of heterogeneity like a standard deviation and relate it to the residence time effect. This should shed physical insights into the time-dependent conformational changes occurring at the interface.

Lundstrom and Elwing (1990) presented a simple model where some kinetic parameters of interest were defined for the understanding of protein exchange reactions on a solid surface. They also considered the residence time effect. The authors emphasize that when a protein molecule is reversibly adsorbed on a surface, it can be exchanged by another molecule of the same or another kind with two major phenomena occurring. There is a time-dependent composition change of the adsorbed protein layer and the possible occurrence of conformationally changed molecules in solution. These authors also noted the existence of four different states on the surface. Heterogeneity should also be explicitly incorporated in their model to provide better physical insights into time-dependent conformational changes occurring at the surface.

These authors indicate the existence of three types of surfaces for protein adsorption. The first type is when the exchange reaction and reversible adsorption takes place with small conformational changes because of a short residence time of the protein on the surface. Heterogeneity in this case would be relatively small. The second type is when only the exchange reaction occurs. The residence time is longer than the first type that leads to conformational changes of the adsorbed molecule. The surface will keep releasing conformationally changed molecules into the solution. These conformationally changed molecules in solution may also adsorb on the surface. This would contribute to increasing heterogeneity of molecules both in solution and on the surface. Such a surface may also cause both surface-oriented biological phenomena and unwanted effects away from the surface. The third type is when at least one kind of protein molecule is irreversibly adsorbed on the surface. In this case, the surface is constantly covered with adsorbed protein that undergoes time-dependent conformational changes on the surface. The parameters of heterogeneity (i.e., standard deviation) may or may not be related for the protein molecules in solution and at the surface.

Tan and Martic (1990) have noted that because of their multifunctionalities, protein molecules can exist in several conformational states. The free energy change required to go from one structure to another is several kilocalories per gmole, which corresponds to the dissociation of a few hydrogen bonds. Similarly, due to the flexibile and dynamic fluctuations of the protein molecules similar rearrangements are possible on the surface. Thus, the driving force for adsorption is isentropic resulting from dehydration owing to the hydrophobic interaction between proteins and the surface. Also, the unfolding of the protein as it adapts to its new environment must be considered.

Finally, some comments should also be made about the desorption-exchange process for proteins adsorbed on a surface. Because the protein molecule attaches itself to the adsorbent surface by different amino acid residues (segments), the molar Gibbs free energy of adsorption may attain large values. Thus, adsorbed proteins are difficult to remove even by diluting the solution. However, Shirahama *et al.* (1990) point out that if the solution contains a displacer or other protein whose molecules have an affinity for the adsorbent, then any desorbing segment can be replaced by another. Simply speaking, desorption of the molecule is now virtually an exchange process; and because $\Delta G_{\text{exchange}} \ll \Delta G_{\text{desorption}}$, these authors indicate that this process is much more likely. Because thermodynamics plays an important role in protein adsorption, it is analyzed to some extent in the next section.

C. Thermodynamics

Mesteri *et al.* (1984) and Partyka *et al.* (1986) utilized the calorimetric method to analyze the adsorption mechanism and the structure of nonionic surfactant films on a hydrophilic silica surface by using the differential molar enthalpies of adsorption (ΔH_d^a). ΔH_d^a is exothermic at lower degrees of coverage, θ, and at a higher $\theta = 0.5$, ΔH_d^a becomes endothermic. The molar enthalpy of adsorption reaches an endothermic minimum around $\theta \approx 0.5$ and subsequently when the coverage ratio approaches unity, the enthalpy of adsorption comes close to zero. These authors (Mesteri *et al.*, 1984; Partyka *et al.*, 1986) noted that the longer the polar chain, the greater the endothermic values. Also, if the protein chain increases, ΔH_d^a changes from exothermic to endothermic at higher degrees of coverage. Mesteri *et al.* (1984) indicate that an increase in adsorption with decreasing temperature is in agreement with an exothermic process. Furthermore, the simultaneous decrease in the enthalpies (in absolute values) is certainly due to a heterogeneous surface site distribution, the most energetic sites being the first to be occupied.

Moacanin *et al.* (1977) suggested that a material's thrombogenic response to blood is influenced partially by the polar and dispersive components of a material's surface free energy. The surface free energy of the interface between the liquid and a biomaterial surface, γ_{s1} is a measure of the imbalance of forces at the interface. The larger the value of γ_{s1}, the greater the imbalance in forces.

Young *et al.* (1988) analyzed the adsorption of α_2-macroglobulin on polyether urethane urea, polyethylene, silicon rubber, and plasticized polyvinyl rubber at different bulk protein concentrations. The binding strength of the four biomaterials for protein decreases in the order silicon rubber > polyethylene > plasticized polyvinyl chloride > polyether urethane urea. The polyethylene and silicon rubber were the most hydrophobic, and the polyether urethane urea was the least hydrophobic biomaterial. Protein affinity was found to be the highest for silicon rubber and polyethylene and the lowest for the polyethylene urethane urea. Note that the biomaterial surface-water free energy also decreases in this same order. This supports the theory that a material with a high dispersion component and a low polar component of the surface energy (that is, the hydrophobic material) adsorbs protein films more strongly than a biomaterial with a lower dispersion component (Moacanin and Kaelble, 1977). Finally, the lowest binding strength between the biomaterial and the protein is because the polar and the dispersive components of the biomaterial exactly match those of the protein.

MacRitchie (1978, 1987) analyzed the thermodynamics of protein adsorption at interfaces. Proteins in solution diffuse to the interface. This is thermodynamically favorable because some of the conformational and hydration energy of the protein is lost at the air–water interface (MacRitchie, 1978). Proteins on adsorption at the air–water interface undergo a change from their globular configuration in solution to an extended chain structure. On energetic grounds, it is expected that the polypeptide backbone lies in the plane of the surface with the polar and nonpolar side chains directed toward and away from the aqueous phase, respectively. This author further indicates that when a protein molecule adsorbs, interfaces of low free energy replace an area of high surface free energy. The polar side chains are in water and the nonpolar side chains are in air. The lowering of the free energy is the driving force and gives rise to the unfolding of the molecule at the surface.

Norde and co-workers (1979, 1986) indicate that the change of entropy on adsorption is an important source of information if the nonconformational and conformational contributions can be separated. On adsorption, a conformational change takes place toward a configuration of higher affinity. With time and structural modifications the protein attaches itself to the surface by different segments. These structural changes, though miniscule, contribute toward the adsorption free energy and increasing degrees of heterogeneity of protein adsorbed at the interface. Norde *et al.* (1986) indicate that desorption requires a higher free energy for initial binding. Thus, the desorption isotherm shows a hysteresis curve and does not follow (or coincides with) the adsorption curve. This degree of hysteresis is lower for molecules with a rigid molecular structure. It is further anticipated that molecules that exhibit a higher degree of hysteresis in the adsorption-desorption curves will exhibit a greater degree of heterogeneity of conformational states at the interface. Also, longer residence times of the protein at the surface would increase the degree of hysteresis for flexible molecular structure proteins.

The adsorption of HPA on hydrophobic and hydrophilic oxide surfaces was analyzed (Norde *et al.*, 1986). These authors indicate for protein adsorption at the hydrophobic oxide surfaces that have the same charge sign as the protein molecules the entropy gain must originate from the protein molecule itself. This can either occur from the dehydration of hydrophobic patches, or structural changes, or from both. The authors assume that the helix content in the adsorbed state is comparable with that calculated from the desorbed material. Then the entropy increase from the loss of α-helix content largely compensates for the positive heat of adsorption, ΔH_{ads}. Protein adsorption on a hydrophilic surface having the same charge sign as the protein proceeds simultaneously by virtue of structure changes in the protein molecules.

Lee and Ruckenstein (1988) studied the adsorption of bovine serum albumin onto polymeric surfaces of different hydrophilicities. These authors proposed an improved explanation concerning the thermodynamic driving force for protein adsorption. There are two positive entropic contributions; (1) an entropy gain due to dehydration of the protein surface, and (2) an entropy gain due to adsorption. There are also two enthalpic effects: (1) a positive one associated with dehydration, and (2) a negative one due to interactions with the solid. The total entropic effect dominates and therefore protein adsorption is entropically driven.

D. Adsorption Parameters

There are various parameters that influence protein adsorption. These include electrostatic interactions, isoelectric point, pH, negatively charged surfaces, surface charge, coadsorption of low molecular ions, intermolecular forces between adsorbed molecules, solute–solvent interactions, strength of functional group bonds, chemistry of the solid surface, morphology, and topology. The effects of some of these parameters on protein adsorption follow.

Elgersma *et al.* (1990) studied the effect of electrostatic contributions on the adsorption of monomeric bovine serum albumin (BSA) on polystyrene lattices. They investigated the influence of surface charge on the latex. These authors showed that BSA adsorption occurs spontaneously even when the protein has the same charge sign as the sorbent. The isoelectric point of BSA is 4.7–5.0, and for both the negatively charged lattices the initial slopes decrease with increasing pH. These authors are unclear as to why the negatively charged BSA molecule has a higher affinity for the negatively charged polystyrene surface. They do indicate that analyzing this particular problem on the basis of electrostatic interactions alone is not enough.

Researchers have observed (Bagchi and Birnbaum, 1981; Sonderquist and Walton, 1980; Morrissey and Stomberg, 1974) a maximum in the amount of protein adsorbed and indicate that it is due to the decrease in conformational stability of the protein with increasing net charge on the molecule. This results in a greater tendency for structural rearrangements of the adsorbing molecules that create a larger surface area per molecule and cause a small amount of protein to be adsorbed. These structural rearrangements on the surface would contribute to the microheterogeneity of proteins adsorbed on the surface. Furthermore, at pH values away from the isoelectric point of the protein, there is an increased electrostatic repulsion between adsorbed molecules that leads to a smaller amount of adsorbed protein. This increased electrostatic repulsion would also increase the microheterogeneity of the adsorbed protein molecules. Elgersma *et al.* (1990) indicate that maximum adsorption around the isoelectric point is found with BSA adsorbed on negatively charged lattices. Furthermore, maximum protein adsorption around the isoelectric point has been reported for albumin, immuno-γ-globulins, fibrinogen, hemoglobin, and gelatin; however, for conformationally stable proteins like cytochrome c and RNase no such maximum in adsorption is observed. One may anticipate that lower degrees of microheterogeneity are to be observed for conformationally stable proteins than for proteins that do not exhibit this conformational stability.

Clark *et al.* (1988) and Poole *et al.* (1984) analyzed the adsorption of BSA at the air–water interface. They noted that the addition of polycationic lysozyme to polyanionic BSA at neutral pH extends the range of solution conditions under which stable foams are produced with individual proteins. Electrostatic interactions stabilized the multiprotein complex at the interface. Clark *et al.* (1988) further showed that extensive aggregation of the protein (presumably of electrostatic or hydrophobic origin) is occurring at the two film surfaces, resulting in the formation of a gel-like network. Any such process would presumably be facilitated by the partial unfolding of BSA that occurs following adsorption at the air–water interface. This partial unfolding and aggregation of the protein would lead to an increased microheterogeneity of the adsorbed protein at the interface.

Abramson (1942) noted that horse serum albumin adsorbed on negatively charged quartz and colloidal particles at its isoelectric point. At a pH of 4.8 a maximum of adsorption occurs. Norde and Lyklema (1978a,b) noted that HSA exhibits a maximum in adsorption on negatively charged polystyrene lattices. Norde (1988) further noted that in adsorbed BSA the average position of the carboxyl group is relatively close to the sorbent, probably because of nonelectrostatic interactions. Positive ions from the solution may be incorporated in the contact region between the protein and the surface to prevent an accumulation of net negative charge (van Dulm *et al.*, 1981). Hlady and Furedi-Milhofer (1979) have indicated that the interactions at an HSA-calcium hydroxyapatite interface depend on the surface charge rather than the electrokinetic charge.

It has been found that maximum protein adsorption as a function of pH is not determined by the electrostatic potential of the protein but rather the protein and the particle taken together. This is shown for the albumin-polystyrene latex system, for the immunoglobulin on polystyrene system, and also for other systems with similar properties (Elgersma *et al.*, 1990).

Elgersma *et al.* (1990) indicate that the adsorption of protein involves coadsorption of low-molecular weight ions to screen any excess potentials that may develop in the contact region between the protein and the charged latex surface, due to the tendency of the protein to expose certain groups to the latex. Therefore, the analysis of coadsorption of low-molecular weight ions is important in studying the protein adsorption process.

Moyer and Govin (1940) studied the competitive adsorption behavior of albumin and γ-globulin on the surfaces of quartz and colloidan particles. These authors noted that these proteins hardly adsorbed on each other after a sample had been coated with one protein and then exposed to another, although one protein may replace another at the surface. Also, the nature of the surface influences the adsorption process in which the hydrophilic protein adsorbed more readily to more hydrophilic surfaces and vice versa. One may reasonably anticipate the competitive behavior to increase the microheterogeneity of the adsorbed protein on the surface.

Kochwa *et al.* (1949) studied the sequential and simultaneous adsorption of albumin, γ-globulin, and fibrinogen on artificial surfaces. These authors noted that when a polyurethane surface is first exposed to unlabeled protein, the prior exposure is always found to decrease the uptake of the labeled protein over that observed for labeled protein on a virgin surface. γ-Globulin blocked the sequential application of labeled albumin by 27%, and albumin blocked labeled γ-globulin by 46%. In the sequential studies, the surface was exposed first to one single-protein solution containing only the other protein. In the simultaneous studies, a surface was exposed to a solution containing a mixture of both proteins. These authors observed that for the sequential experiments, the amount and kind of adsorption depended on the sequence of exposure. The main observation for the simultaneous (competitive) adsorption process is that there is a large reduction in the adsorbed amount of one component in the presence of the other as compared with the single-component adsorption of either.

Some of the parameters that influence protein adsorption on different surfaces have been analyzed. Rudzinski *et al.* (1983) indicate that the protein adsorption model is complicated by surface heterogeneity or energetic heterogeneity of the surface sites. This fact has often been brought out in the literature (Zchuchovitzky, 1938; Delmas and Patterson, 1960; Siskova and Erdos, 1960a,b; Coltharp and Hackerman, 1973a,b; Everett, 1964, 1965). Initially, Rudzinski and co-workers and others (Rudzinski *et al.*, 1973, 1974; Oscik *et al.*, 1986; Dabrowski 1983; Dabrowski and Jaroniec, 1979a,b, 1980a,b) attempted a quantitative description of solution adsorption on solid surfaces. These authors applied the method of the Stieltjes transform, utilized earlier by Sips (1948), to describe the gaseous adsorption on actual heterogeneous solid surfaces. They noted that Sips' theoretical results on gaseous adsorption could be easily modified to solution adsorption by a simple transfomation of variables. However, later on, Rudzinski *et al.* (1983) indicated that the method of

Stieltjes transform cannot be applied in the case when molecules of the liquid mixture may have different cross-sectional areas (or heterogeneity) on the solid surface. This surface heterogeneity of the adsorbed protein molecules (or, in a more general sense, the biological macromolecule) may arise due to either the energetic heterogeneity of the surface sites or a heterogeneity of the molecule in question in solution, or due to a combination of the two or other reasons. The influence of heterogeneity on protein adsorption and on reactions on the surface is examined in the following section.

III. HETEROGENEITY IN PROTEIN ADSORPTION

The surfaces for protein adsorption need to be better characterized. The herogeneity of the surface will significantly influence adsorption and subsequent reactions occurring on the surface. The heterogeneity of the surface will also influence the rate and extent of protein denaturation on the surface. It would be of interest to develop a measure of heterogeneity of the surface and then be able to relate it to the extent of protein denaturation or conformational changes that occur on protein adsorption at the surface. Such studies exhibit the potential to provide novel physical insights into the nature of protein adsorption on different surfaces. Norde and Lyklema (1979) have emphasized that a detailed analysis of structural rearrangements of proteins adsorbed on surfaces has eluded investigators. They estimated and analyzed the structural contributions of the adsorbed proteins to thermodynamic functions. In this section we will examine the influence on protein adsorption at interfaces of: (1) heterogeneity in solute, (2) heterogeneity on surfaces, (3) models incorporating heterogeneity, and (4) implications of this heterogeneity on protein adsorption and mediation of further reactions on the surface.

Proteins are known to be heterogeneous. Besides, the energies for protein adsorption on a surface need not necessarily be homogeneous; in fact, it is reasonable to assume a distribution in energies for protein adsorption. The application of mathematical distributions of proteins adsorbed on surfaces is a complex problem. However, the application of mathematical distributions of proteins adsorbed on surfaces is necessary, because it is a more realistic approach to the actual situation. This also presents a novel technique to gain valuable physical insights into the protein adsorption process and into the influence on subsequent adsorbed protein-mediated reactions occurring on the surface. The approach using distributions would provide a knowledge of the time-dependent compositions and conformational changes of proteins in the adsorbed layer on the surface.

In a relevant though not similar study Malhotra and Sadana (1987a and b) assumed a continuous normal distribution of thermal activation energy for deactivation, and by using this they developed a simple mathematical model to find the activity-time trajectories for a microheterogeneous enzyme. By using this model, these authors were able to show a time-dependent change in the composition of the enzyme. This composition change was revealed as a change in the width and in the mean of the distribution of the activation energies of deactivation for the enzyme. Malhotra and Sadana (1989) further analyzed the influence of intraparticle diffusion on the deactivation characteristics of microheterogeneous enzymes. These authors noted that intraparticle diffusion effects alleviated the influence of microheterogeneity on the deactivation characteristics of an enzyme exhibiting first-order kinetics of deactivation. Lundstrom and Elwing (1990) in their analysis of simple kinetic models for protein exchange reactions on solid surfaces also noted that it would be of interest to analyze the possible influence of diffusional limitations on the initial coverage of molecules in the different states. Finally, Malhotra and Sadana (1990) analyzed the role of the initial state distribution on first-order deactivation of microheterogeneous enzyme samples. Their analysis primarily showed that detailed deactivation data are necessary to distinguish between different distributions of activation energies of deactivation in enzymes.

There is apparently not much information available in the literature concerning the heterogeneity of protein adsorption on different surfaces. This is a complex process, especially when differences in molecular sizes between the components of a solution exist. Jaroniec *et al.* (1983) presented a model of multisolute adsorption from dilute aqueous solutions involving energetic heterogeneity of the solid and differences in the molecular sizes of the solutes. These authors determined the parameters characterizing energetic heterogeneity of the solid and the ratio of the molecular sizes of the two arbitrary solutes. The authors also assumed negligible effects of association and dissociation in both phases. Jaroniec (1981) also proposed an equilibrium constant for protein adsorption that involved a symmetrical quasi-distribution of adsorption sites and inequality of molecular sizes of both solutes. Using the model proposed by Jaroniec (1981), Jaroniec *et al.* (1983) were able to show that for some systems the effects connected with differences in molecular sizes of solutes play a more important role than the heterogeneity effect. However, for systems where the molecular size ratio of the two solutes is close to one, then the heterogeneity effects are dominant.

The most advanced treatments of heterogeneous adsorption from solutions composed of molecules of different sizes have been by Jaroniec *et al.* (1983), Dabrowski (1983), and Rudzinski *et al.* (1983). Jaroniec *et al.* (1983) and Dabrowski (1983) adopted a rather kinetic approach for the derivation of the adsorption isotherm, while the Rudzinski *et al.* (1983) isotherm was derived by means of the condensation method. Other studies on the adsorption on energetically heterogeneous surfaces are also available (Borowko and Jaroniec, 1983; Nikitas, 1985).

Nikitas (1989) developed a simple mathematical method that makes possible the development of isotherms for adsorption from dilute solutions composed of molecules with different sizes starting from isotherms based on the equality of the molecular sizes of the components. The treatment was restricted to random heterogeneous surfaces. This method was able to extend the Temkin and Langmuir-Freundlich isotherms to include size effects. This author utilized three distribution functions of partial surface coverage on sites, V_i, with adsorption energy, U_i. A uniform distribution generated the generalized Temkin isotherm valid for solvent and solute molecules of equal size. A heterogeneity factor, $\gamma = U_0/(kT)$ was defined where U_0 is the mean adsorption energy, k is a constant, and T is temperature. This factor describes the width of the adsorption energy distribution, and thus, it increases with increasing heterogeneity of the adsorbent. The extension of the Temkin isotherm to include size effects resulted in a complicated expression for the isotherm. The exponential distribution generates, as a first approximation, the Freundlich isotherm. Finally, a quasi-distribution was related to the Langmuir-Freundlich isotherm. Here the degree of surface heterogeneity is expressed by the parameter c that ranges from 0 to 1 as one passes from heterogeneous to homogeneous surfaces.

Models for adsorption for solute in solid-solution adsorption systems are simple but are complicated by the energetic heterogeneity of the solid surface sites. Everett (1964) and Delmas and Patterson (1960) brought attention to the importance of surface heterogeneity in solution adsorption on solid surfaces. As indicated earlier, Rudzinski *et al.* (1973) tried to quantitatively describe solution adsorption on hetreogeneous solid surfaces by applying the method of Stieltjes transform used by Sips (1948) to describe gaseous adsorption on an actual heterogeneous surface. They showed that Sips' theoretical results on gaseous adsorption can easily be applied to protein adsorption from solution by a simple transformation of variables. The technique does not reduce correctly to Henry's law at significantly low concentration of one of the components in the liquid mixture. Also, the method of Stieltjes transform cannot be used when the molecules of the liquid mixture have different cross-sectional areas on the solid surface. In their earlier studies, Rudzinski *et al.* (1973) did recognize the limitations of their analysis and did try to remove these two limitations. Later, Rudzinski *et al.* (1983) developed a general isotherm that approached the problem of surface heterogeneity in adsorption from a binary liquid mixture on an actual solid surface. This isotherm correctly showed the transition from the Dubinin-Radushkevich and Freundlich isotherm equations to Henry's law. It is generalized by taking into account the different cross-sectional areas of the adsorbed molecules.

The Nikitas method (1989) is now utilized to develop isotherms for heterogeneous adsorption from dilute solutions involving differences in molecular sizes of components. Nikitas (1989) developed new adsorption isotherms from dilute solutions consisting of different size molecules, starting from isotherms based on the equality of the molecular sizes of the components.

The partial adsorption isotherm for protein or other biological macromolecular adsorption on a random heterogeneous surface may be expressed as

$$\frac{\alpha_i}{1-\alpha_i} = K \exp\left(\frac{E_{ads}}{kT}\right) \tag{7.2a}$$

if size effects could be neglected, and:

$$K = \frac{\gamma_s \delta x_A}{\gamma_a}. \tag{7.2b}$$

Besides, the solution is dilute enough so that the activity coefficient of the protein molecule in solution is unity. α_i is the partial surface coverage on sites with adsorption energy, E_{ads}; δ is the adsorption equilibrium constant; x_A is the molar fraction of the adsorbate in the bulk solution; γ_s and γ_a are the surface activity coefficients of the adsorbate (protein or other biological macromolecule) and the solvent, respectively. This author emphasizes that the activity coefficients γ_a and γ_s depend on the surface coverage α over the whole adsorption layer, while they are independent of the adsorption energy, $E_{ads.}$

Protein (or other solute molecules) size effects may be included in Eq. (7.2a) by modifying it to yield

$$\frac{\alpha_i}{(1 - \alpha_i)^m} = K \exp\left(\frac{E_{ads}}{kT}\right), \tag{7.3}$$

where m is a size ratio parameter equal to the ratio σ_a/σ_s of the areas covered at the adsorption layer by a solute s_a and a solvent, s_s molecule.

In this case

$$K = \frac{\gamma_s^m \, \delta x_A}{\gamma_a}. \tag{7.4}$$

Next, Nikitas (1989) considers the expression:

$$\frac{\alpha_i}{b - \alpha_i} = K \exp\left(\frac{E_{ads}}{kT}\right), \tag{7.5}$$

which bears physical meaning, when the parameter b equals unity. The heterogeneity is incorporated in the analysis by including a distribution function $\chi(E_{ads})$, and the total surface coverage is obtained by a weighted-average expression. The equilibrium constant is given by

$$K = \Phi\left(\frac{\alpha_i}{b}\right), \tag{7.6a}$$

where

$$\int_{-\infty}^{\infty} \frac{\alpha_i}{b - \alpha_i} \chi(E_{ads}) \, dE_{ads} = c\Phi(\alpha_i/b) \tag{7.6b}$$

and c is a constant.

The Nikitas (1989) expression of an isotherm for adsorption from dilute solutions composed of different size protein (or other biological macromolecules) molecules and same-size solvent molecules on a random heterogeneous surface is given by

$$c_1 x_A = \frac{\gamma_A}{\gamma_s^m} \left. \frac{d^n \phi\left(\frac{\alpha}{b}\right)}{db^n} \right|_{b = 1}, \tag{7.7}$$

where c_1 is a constant.

Example 7.4

Provide applications for heterogeneous adsorption of solutes from dilute solutions (Nikitas, 1989).

Solution

Case One. Nikitas (1989) indicates that the analytical expressions obtained from Eq. (7.7) are dependent on the distribution function $\chi(E_{ads})$. Let us consider a uniform distribution function and a quasi-Gaussian distribution (Sips, 1948).

The uniform distribution function is given by (Nikitas, 1989)

$$\chi(E_{\mathrm{ads}}) = \begin{cases} 0 & E_{\mathrm{ads}} < E_{\mathrm{ads,m}} - E_{\mathrm{ads,0}},\ E_{\mathrm{ads}} > E_{\mathrm{ads,m}} + E_0 \\ 1/2\ E_{\mathrm{ads,0}} & E_{\mathrm{ads,m}} - E_{\mathrm{ads,0}} \leq E_{\mathrm{ads}} \leq E_{\mathrm{ads,m}} + E_0 \end{cases}. \tag{7.8}$$

Young and Cromwell (1962) and Nikitas (1988) indicate that the preceding above uniform distribution function generates a generalized Temkin isotherm valid for solvent and solute molecules of equal size

$$\alpha = \frac{kT}{2E_{\mathrm{ads,0}}} \ln \frac{1 + K[(E_{\mathrm{ads,m}} + E_{\mathrm{ads,0}})/kT]}{1 + K[(E_{\mathrm{ads,m}} - E_{\mathrm{ads,0}})/kT]}, \tag{7.9}$$

where K is given by Eq. (7.2b).

From Eqs. (7.5) and (7.8) one obtains

$$\Phi\left(\frac{\alpha}{b}\right) = \frac{1 - \exp(2\lambda\alpha/b)}{\exp(2\lambda\alpha/b - \lambda) - \exp(\lambda)} \exp\left(-\frac{E_{\mathrm{ads,m}}}{kT}\right) \tag{7.10}$$

where $\lambda = E_{\mathrm{ads,0}}/kT$ is the heterogeneity factor. Nikitas (1989) indicates that this heterogeneity factor describes the width of the adsorption energy distribution. This factor, thus, increases with an increase in the heterogeneity of the adsorbent.

Equation (7.10) yields the adsorption isotherm

$$c_1 x_{\mathrm{A}} = \frac{2\lambda\exp(2\alpha\lambda)\,[\exp(\lambda) - \exp(-\lambda)]\gamma_{\mathrm{a}}}{[\exp(2\lambda\alpha - \lambda) - \exp(\lambda)]^2\lambda_{\mathrm{s}}^2} \tag{7.11a}$$

and

$$c_1 x_{\mathrm{A}} = \frac{2\lambda\alpha\exp(2\lambda\alpha)\,[(1 - \lambda\alpha)e^{2\lambda\alpha-\lambda} - (1 + \lambda\alpha)\exp(\lambda)]\,[\exp(\lambda) - \exp(-\lambda)]}{[\exp(2\lambda\alpha - \lambda) - \exp(\lambda)]^3} \times \left(\frac{\gamma_{\mathrm{a}}}{\gamma_{\mathrm{s}}}\right)^3 \tag{7.11b}$$

for $m = 2$ and 3, respectively. Note the similarity in the two expressions. Nikitas (1989) emphasizes that the inclusion of size effects utilizing the Temkin isotherm results in a complicated expression. It is of interest to note that as the heterogeneity parameter $\lambda \to 0$, Eqs. (7.11a and b) reduce to the corresponding expression valid for homogeneous surfaces, given by

$$c_1 x_{\mathrm{A}} = \frac{\alpha}{(1 - \alpha)^{\mathrm{m}}} \frac{\gamma_{\mathrm{a}}}{\gamma_{\mathrm{s}}^{\mathrm{m}}}. \tag{7.12}$$

Sips (1948) indicates that the quasi-Gaussian distribution

$$\chi(E_{\mathrm{ads}}) = \frac{1}{kT} \frac{\sin(\pi h)\exp[h(E_{\mathrm{ads,m}} - E)/kT]}{1 + 2\cos(\pi h)\exp[h(E_{\mathrm{ads,m}} - E)/kT] + \exp[2h(E_{\mathrm{ads,m}} - E)/kT]} \tag{7.13}$$

is related to the Langmuir-Freundlich isotherm

$$\left(\frac{\alpha}{1-\alpha}\right)^{1/c} - Kx_A. \tag{7.14}$$

Nikitas (1989) indicates that the degree of surface heterogeneity is given by the parameter c that goes from zero to one as we go from heterogeneous to homogeneous surfaces.

The function Φ (α/b) is given by

$$\phi\left(\frac{\alpha}{b}\right) = c_1\left(\frac{\alpha}{b-\alpha}\right)^{1/c}. \tag{7.15}$$

Then the extension of Eq. (7.14) to include size effects may be written as

$$c_1x_A = \frac{\alpha^{1/c}}{(1-\alpha)^{m-1+1/c}}\frac{\gamma_a}{\gamma_s^m}. \tag{7.16}$$

As expected, Eq. (7.16) reduces to Eq. (7.12) when c = 1 for homogeneous surfaces.

Case Two. Other studies on the effect of heterogeneity are also available. Jaroniec *et al.* (1983) presented a simple equation for multisolute adsorption from dilute aqueous solutions on solids. Their proposed model provided a simple relationship between amount adsorbed and the concentrations of the two arbitrary solutes. From their model, these authors could obtain a parameter that characterized the energetic heterogeneity of the solid and the ratio of molecular sizes of two arbitrary solutes. The assumptions made include (1) monolayer adsorption, (2) differences in molecular sizes of the solutes, (3) ideality in the adsorption space and in the bulk solution, (4) energetically heterogeneous solid, and finally (5) effects of association and dissociation in both phases being neglected.

Jaroniec *et al.* (1983) successfully applied their model to the adsorption of different phenols (2,4-dichlorophenol, etc.) in dilute solution on activated carbon at 293 K. These authors noted that the effects connected with differences in molecular sizes of solutes are considerably greater than heterogeneity effects. It would be of interest to extrapolate these studies to the adsorption of proteins and other biological macromolecules of interest to appropriate surfaces-interfaces.

Case Three. Corsel *et al.* (1986) analyzed the adsorption and desorption of prothrombin, albumin, and fibrinogen to phospholipid bilayers by ellipsometry. Adsorption of proteins to biological membranes is of importance in many physiological processes, and is of significance especially in blood coagulation. In this case, the final product is a blood clot of polymerized fibrin that is formed after the splitting of circulating fibrinogen by thrombin. These authors indicated that there were complications during their measurements of the adsorption, k_{on}, and desorption, k_{off}, rate constants. This was due to the presence of different classes of binding sites. Note that it has been shown that the values of k_{off} and k_{on} are generally dependent on the surface concentration (Γ) of the protein (Cuypers *et al.*, 1987).

Kop *et al.* (1989) indicate that sorption rate constants should therefore preferably be measured at initial values, that is, at values of low surface coverage. Equilibrium measurements of the binding of prothrombin to dioleoylphosphatidylserine (DOPS) demonstrated a biphasic adsorption with sites characterized by an equilibrium constant, $K_d(= k_{off}/k_{on})$ equal to 6×10^{-10} M, $\Gamma_{max} = 0.26$ μg/cm^2, and additional low affinity sites with $K_d = 10^{-7}$ M and $\Gamma_{max} = 0.12$ μg/cm^2. Here Γ_{max} represents the adsorbed quantity of prothrombin (in this case) at maximal surface coverage (Kop *et al.*, 1984). Corsel *et al.* (1986) obtained similar results for the same system except that Γ_{max} in both cases was equal to 0.18 μg/cm^2. They explained this shift by the use of a more physiological calcium concentration of 1.5 mM in their studies compared to 10 mM utilized by Kop *et al.* (1984). Corsel *et al.* (1986) indicate that in the absence of specific biological binding sites protein adsorption to the phospholipid bilayers would include secondary changes in the surface interactions of protein molecules. These secondary changes would then lead to a heterogeneity of the protein adsorbed on the surface.

Example 7.5

There is a paradox between concentration dependent adsorption and lack of desorption in pure buffer (Kop *et al.*, 1989).

Solution

Case One. The preceding phenomenon wherein the radiolabeled adsorbed protein does not show net adsorption after dilution of the protein solution but readily exchanges with unlabeled protein has been observed for albumin (Brash and Samak, 1978; Cheng *et al.*, 1987) and fibrinogen (Chan and Brash, 1981). Several models have been proposed in the literature to explain these observations; some of these include time-dependent structural changes in the adsorbed protein layer and specific models of the exchange of the bound and unbound protein molecules. These structural changes could lead to an increasing heterogeneity of the adsorbed protein on the surface.

Case Two. Kop *et al.* (1989) analyzed the binding of coagulation factor V to planar phospholipid double layers by ellipsometry. At 20°C, coagulation factor V in buffer solution undergoes a rapid (half-life approximately 25 min) spontaneous denaturation. This destroys the binding capacity of this protein to the phospholipid bilayers. Because the dissociation constant, $K_d = k_{off}/k_{on}$, a decrease in k_{on} leads to overestimations of K_d of several orders of magnitude and an apparently reversible binding isotherm for coagulation factor V.

It is of interest to note that in both cases we have time-dependent structural changes-denaturation of the adsorbed protein that lead to an increasing heterogeneity, helping to elucidate the so-called paradox.

The multiple adsorption states exhibited by proteins would yield, in general, a plethora of different structures at the interface exhibiting slightly different functionalities. This multiple state of the protein adsorption at different sites of the interface should exhibit heterogeneous deactivation behavior at the interface. In any realistic model for protein–enzyme inactivation at interfaces this heterogeneity of adsorption and the subsequent heterogeneity in deactivation should be taken into account. In general, a heterogeneity in an enzyme sample leads to an enhanced stabilization when compared with a homogeneous enzyme (Malhotra and Sadana, 1987a and b). This heterogeneity, as indicated later may be denoted by a distribution in the activation energy for deactivation or in the conformational states. It would be of significant interest to characterize this heterogeneity and distribution in adsorption, and to get an estimate or a measure of this heterogeneity. Then an analysis could be performed on the overall effect of this heterogeneity on reactions occurring at the interface, protein stability, and properties at the interface.

Ion exchange chromatography has been used as a standard application for protein purifications (Chase, 1984). Gill *et al.* (1994) indicate that, however, the fundamental aspects of ion-exchange chromatography have not been described in any detail. It would be of interest to analyze theoretically, even briefly, the adsorption of a protein by ion-exchange chromatography.

These authors have analyzed the non-Langmuirian adsorption of recombinant soluble core of rat cytochrome b_5 on a polymeric strong anion exchanger, Mono Q. The non-Langmuirian adsorption isotherms obtained along with concave upward Scatchard plots and values of the Hill coefficient less than unity indicate a heterogeneous adsorption due to: (1) negative lateral interactions between adsorbed molecules, or (2) nonuniform binding affinities of adsorbent sites. They indicate that the low and fractional number of binding interactions between the protein and the adsorbent surface implies a surface heterogeneity of binding sites. The low and fractional number of binding sites represents actually an average number of the distribution of binding sites. Furthermore, Gill *et al.* (1994) emphasize that the charge distribution and three-dimensional shape of the protein are factors that significantly affect the ion exchange.

An analysis of the state of research on protein adsorption at different interfaces indicates that though much is known about the quantitative nature or amount of protein adsorbed, little qualitative information (such as the actual nature of the adsorbed layer or about the functional consequences of the adsorption process) is presently available. Though methods are available to determine the quantity adsorbed, there are few techniques or studies available that delineate the structure or orientation of an adsorbed protein. Techniques are also lacking that describe the relationship of a particular aspect of the adsorption process to its influence on other processes at the interface. Besides, predictive models for any aspect of the adsorption process in terms of specific properties of proteins at the interface are lacking. Also, readily measurable quantities (e.g., constants) that can be collected and confidently compared between laboratories are lacking. Careful and detailed studies are urgently required to describe more clearly the effect or influence of some of the factors described previously on protein adsorption at interfaces, and their subsequent effect on the proteins themselves, and other processes occurring at that interface.

Johnsson *et al.* (1985) compared the adsorption isotherms for IgG and secretory fibronectin (HFN) on silica with two different surface energies by *in situ* ellipsometry. The results were interpreted as time-dependent conformational changes in the adsorbed protein film, where the degree of changes was dependent on the solid surface free energy. These time-dependent conformational changes of the adsorbed protein molecule lead to heterogeneity in adsorption. Also, the shapes of the adsorption isotherms may depend on the heterogeneity of the protein preparations, the interaction between adsorbed molecules, the concentration-dependent structural changes in the adsorbed film, or a heterogeneous surface with several types of adsorption sites (Bagchi and Birnbaum, 1981; Fair and Jamieson, 1988).

Thus it is clear methods are required that both qualitatively and quantitatively analyze the adsorbed state of the protein molecule on different surfaces. It is safe to assume that the adsorbed protein is heterogeneous. Now effective experimental techniques are required that can estimate both the quantity of the protein adsorbed and also the heterogeneity of the protein adsorption. This heterogeneity of protein adsorption is one qualitative measure of protein adsorption. Appropriate models that provide a measure of heterogeneity of protein adsorption are required. From protein adsorption data the experimental technique should also provide a measure of heterogeneity of protein adsorption. Then the measures of heterogeneity obtained by experiment and by modeling may be compared. Such comparisons may consequently lead to better model development or even better experimental techniques that provide more reliable measures of heterogeneity. Clearly, this is an avenue that will provide more physical insights into protein adsorption on different surfaces.

As instrumentation advances and becomes more and more sophisticated, this aspect of protein heterogeneity on adsorption to different surfaces will become more and more important and prominent. This may well play a significant role in the influence of proteins on reactions occurring at the surface. Thus, it is essential to evaluate or estimate the effect of heterogeneity on protein adsorption. Unfortunately, this aspect has been rather neglected. The next section looks at a few experimental techniques that have been utilized to qualitatively characterize protein adsorption on surfaces. The following section then examines the most recent models that have been utilized to describe protein adsorption on different types of surfaces.

IV. TECHNIQUES FOR QUALITATIVE CHARACTERIZATION OF PROTEIN ADSORPTION

Because of their complex (often patchwise) chemical constitution, proteins may adsorb by different mechanisms on different surfaces. Although it is well known that physicochemical properties strongly affect the protein adsorption, such studies on chemically and morphologically well-characterized surfaces are scarce. Few techniques lend themselves to direct study of the structural properties of proteins at interfaces. The ideal approach should produce quantitative real-time data *in situ* concerning the amount, activity, and conformation of proteins at the interface. Most approaches are only approximations of this optimum and are generally restricted in their application. We now analyze some of the techniques that have been used to qualitatively characterize protein adsorption on surfaces. These techniques are ellipsometry, total internal reflection fluorescence (TIRF), spectroscopy, immunogold staining technique, and other methods.

A. Ellipsometry

Ellipsometry is an *in situ* method used to make more quantitative the thickness and refractive indices of adsorbed protein films. In this optical technique the change in state of polarization of light on reflection from a surface is used to characterize the surface. If proteins are allowed to adsorb to that surface, ellipsometry makes it possible to determine the thickness, the refractive index, and the specific amount of adsorbed molecules. Not much information is available on the qualitative nature of protein adsorption. Morrissey *et al.* (1976) studied adsorbed fibrinogen layers as a function of the surface potential by means of ellipsometry. Changes in compactness as calculated from these parameters were interpreted as indications of conformational changes of the protein. Cuypers *et al.* (1987) demonstrated the possibility of different protein orientations of hydrophobic versus hydrophilic chromium substrates. Stoner and Srinivasan (1970) measured the thickness and simultaneously the interfacial capacitance (i.e., surface coverage) for fibrinogen on platinum as a function of the applied potential. It was shown that an attractive electrostatic potential resulted in a flat conformation of the protein adsorbate.

Johnsson *et al.* (1985) compared the adsorption isotherms for IgG and secretory fibronectin on silica with different surface free energies by *in situ* ellipsometry. The isotherms were obtained by either direct-addition or successive-addition of the proteins. A significant difference between the direct- and successive-addition isotherms was found for both proteins on hydrophobic silica, whereas the isotherms essentially coincide for the proteins on hydrophilic silica. These authors interpreted their results as time-dependent conformational changes in the adsorbed protein film where the degree of changes was dependent on the solid surface free energy. These changes were most pronounced on hydrophobic silica. For example, secretory fibronectin adsorbed to hydrophobic silica showed less tendency to undergo surface conformational changes as compared with fibronectin adsorbed to hydrophobic silica. Also, at low surface concentration lack of competition for surface adsorption sites results in a flatter adsorbed conformation, while at high surface concentration intermolecular repulsion causes a more extended conformation with fewer surface attachments.

Golander and Kiss (1988) wanted to correlate the surface functional properties of smooth Electron Spectroscopy Chemical Analysis (ESCA)-characterized polymer films with their adsorption behavior vis-a-vis some well-known proteins as studied by ellipsometry. These authors used ellipsometry to investigate the adsorption of BSA; IgG; fibrinogen; and poly-L-lysine (PLL) to silicon wafers, which were surface modified by attaching PVC, PMA, or PEO films, all of which were characterized by ESCA. They noted that the adsorption of the three plasma proteins and one cationic polyelectrolyte, PLL, is generally lower to the hydrophilic PMA and PEO films than that to the PVC films. This demomstrated the importance of the hydrophobic driving force for protein adsorption. Also, the chemical constitution of the substrate surface has a significant influence on the course of protein adsorption. For example, the protein isotherms obtained on PVC may be explained by assuming dynamic adsorption models with two adsorption modes, that is, native and denatured molecules in equilibrium having different affinities to the surface. This would also lead to a heterogeneity (as indicated in the earlier sections) of the adsorbed protein states. Of course, not only the chemical constitution of the substrate surface is important in adsorption but also the protein that affects the nature of the protein interaction. Finally, these authors also noted that a low degree of protein adsorption, $\Gamma < 0.5$ mg/m^2, was observed for surfaces covered with surface-grafted PEO chains (molecular weight of 1900) that were covalently linked by means of terminal CHO groups to the surface amino groups.

Quantitative analysis of protein adsorption and interaction at a solid–liquid interface is usually made on surfaces with a homogeneous chemical composition. At the outset, it is important to realize that a surface with a homogeneous composition is an idealization because there is no such surface with a homogeneous composition. This aspect and the importance of heterogeneity (either of surface or protein) have been emphasized throughout this chapter. In a detailed quantitative analysis of the influence of a certain surface constituent, this surface constituent should be varied. This, if done properly, indicates that several samples need to be analyzed—a time-consuming and expensive procedure.

Elwing *et al.* (1987) have utilized the wettability gradient method to study protein interactions at the solid–liquid interface. There is a gradient in a surface constituent (in this case, methyl groups) that is formed by diffusion of $Cl_2(CH_3)_2Si$ on a flat silicon surface. The surface so formed has hydroxyl groups at one end and methyl groups at the other. Primary adsorption of human γ-globulin, fibrinogen, and lysozyme was made more quantitative along the gradient using ellipsometry and was related to the degree of wettability determined by an independent method.

The capillary rise method is used for the investigation of wettability gradients. Two glass plates with wettability gradients (there is one hydrophobic end and one hydrophilic end) are put together with a support that separates the plates. The lower edges of the parallel plates are then brought into contact with a trough filled with water. Water moves upward between the plates and the height of the liquid meniscus is determined by the wettability of the surface of the plates. The height of water is higher at the hydrophobic end compared with the hydrophilic end. The contact angles with water on the hydrophobic and the hydrophilic sides of the gradient were determined to be 85 and 10.3, respectively. The adsorbed amount of human fibrinogen was about 0.7 μg/cm^2 at the hydrophobic end and about 0.3 μg/cm^2 at the hydrophilic end. In between was a sigmoidal decrease. Similarly, the adsorbed amount of γ-globulin was around 0.55 μg/cm^2 at the hydrophobic end and 0.3 μg/cm^2 at the hydrophilic end of the gradient.

The desorption induced by the addition of 4 M urea and acid buffer (pH 2.3) was also studied and shown to be maximal at the hydrophilic side of the gradients although there was a considerable amount of protein also desorbed at the hydrophobic parts of the gradient. There were other qualitative differences in the desorption pattern of γ-globulin and fibrinogen that may be partly explained by assuming different degrees of surface-induced conformational changes of the adsorbed protein molecules.

The technique is rather appealing in that it analyzes the influence of a predetermined and controlled heterogeneous surface on protein adsorption. This, to the best of this author's knowledge is the first analysis that examines, albeit unconsciously, the influence of a heterogeneous surface on protein adsorption. More studies of this type are necessary because they more correctly represent the real-life situation. It would, of course, be better, and more true to the real-life situation if the wettability method could be modified and used not just as a gradient method but more as a technique to evaluate the heterogeneity of a surface that is present in a random fashion.

Finally, it should be realized that ellipsometry and the other techniques to be presented (such as TIRF, protein fluorescence, and circular dichroism) are often suggestive in regard to possible conformational alterations on protein adsorption at different surfaces. More direct methods that, for example, measure activity loss of enzymes on adsorption at surfaces would be beneficial. These techniques are now presented in the following sections.

B. Total Internal Reflection Fluorescence

Because in most practical applications protein adsorption takes place from flowing solutions, experiments designed to measure adsorption kinetics must be conducted under well-defined conditions such that mass transfer limitations, if present, can be quantified. The technique of TIRF has been shown to be well suited for such studies (Lok *et al.*, 1983a,b). Relating TIRF fluorescence signals to protein surface concentrations, or calibrating the TIRF results, is a difficult problem requiring the careful consideration of a number of factors. The calibration technique developed by Lok *et al.* (1983a) appears to have overcome the reservations associated with obtaining accurate calibrations for use of TIRF studies.

Cheng *et al.* (1987) utilized a modified form of the Lok *et al.* (1983a) method to examine the initial adsorption, desorption, and exchange kinetics of the protein BSA on six polymer surfaces with widely varying surface properties and functionalities. These authors covalently attached fluorescein isothiocyanate to primary amine groups of BSA. The molar fluorescein:BSA ratio was approximately unity. The results indicate that the fluorescence intensity of adsorbed Fluorescein Isothiocyanate (FITC)-BSA is proportional to the protein surface concentration for each surface. The initial rate of protein adsorption onto a surface is determined by both transport of protein to the surface and the intrinsic kinetics of adsorption at the surface. This has been described by a convection-diffusion model with appropriate boundary conditions for the channel geometry of the TIRF apparatus (Lok *et al.*, 1983b). Relating fluorescence signals to protein surface concentrations or calibrating the TIRF results is a difficult task, and several authors have failed to account for all the factors (Norde *et al.*, 1986; Leveque, 1928; Hsu and Sun, 1988; Langmuir, 1918).

Cheng *et al.* (1987) have shown how bulk solution ionic strength and pH can dramatically affect the fluorescence signal in a TIRF experiment in the absence of any changes in the protein surface concentration. Even the technique and interpretation used by these authors are not entirely flawless. The TIRF detection point is an approximately 1- to 3-mm oval region in the center of the microscopic slide. Cheng *et al.* (1987) acknowledge some spatial variations but basically they assume the adsorption to be essentially homogeneous with the wetted plate. This is not entirely true because we do recognize that protein adsorpton is heterogeneous, and one should factor this into the calculations. This is especially true if the measure of heterogeneity of protein adsorption is significant. The problem is compounded if a small degree of protein adsorption heterogeneity significantly affects the reactions occurring at the surface that are mediated by the protein adsorbed.

The results of this study show that the initial adsorption of BSA on three of the polymeric surfaces is diffusion limited up to wall shear rates of 4000 s^{-1}. The initial adsorption of BSA on another polymer is diffusion limited at shear rates below about 70 s^{-1} but becomes kinetically controlled at higher shear rates. Studies of the kinetically limited BSA adsorption on this last polymer show the adsorption process can be described by a kinetic rate expression that is first order in protein concentration. Also, the desorption of adsorbed proteins on five out of the six polymer surfaces studied is shown to be kinetically limited.

Example 7.6

Briefly describe the competitive adsorption of HSA, IgG, and fibrinogen on silica made hydrophobic by methylation or plasma deposition of hexamethyldisoloxane (HMDSO) using *in situ* ellipsometry and TIRF (Malmsten and Lassen, 1994).

Solution

In many practical applications including biomedical, diagnostic, and bioseparations, protein adsorption occurs from complex mixtures containing proteins that have a wide range of concentrations, shapes, sizes, etc. Malmsten and Lassen (1994) analyzed the influence of the structure of the adsorbed layer in mixed protein systems. These authors studied the competitive and sequential adsorption of proteins at methylated and HMDSO-treated silica surfaces from model binary protein mixtures. Ellipsometry and TIRF were used.

These authors noted in competitive adsorption experiments that IgG and fibrinogen adsorbed preferentially over HSA at the modified silica surfaces. In fact, in 50:50 mixtures of IgG and HSA, the adsorption of IgG was almost complete. Surprisingly though in sequential protein adsorption experiments, preadsorption of HSA resulted in a significant decrease in the adsorption of IgG and fibrinogen. They offer the explanation that there is a surface-induced irreversible conformational change of HSA at hydrophobic surfaces, especially at low surface coverages. Similar results were obtained with both methylated and HMDSO-treated surfaces using ellipsometry and TIRF. These authors emphasize that these types of results not only find application in biomedical and in bioseparation systems, but also may be used in helping to block sites of nonspecific adsorption in immunoassay techniques. BSA and HSA are used routinely in these applications. Furthermore, Malmsten and Lassen (1994) indicate that the irreversible nature arises due to a side-on adsorption (Uzgiris and Fromageot, 1976; Sonderquist and Walton, 1980). The analysis of Malmsten and Lassen (1994) is of interest both in theory and for practical applications. The analysis provides insights into the structure of the protein layer adsorbed on the surface, for example, the end-on adsorption. The practical interest lies in helping to minimize nonspecific adsorption in immunoassays. This is of particular interest when analyzing mixtures with dilute concentrations of analytes. Such analysis should also significantly improve: (1) the quality of the proteins and different biological products separated during bioseparation processes, and (2) minimize or control protein adsorption in artifical biomedical systems used to enhance the quality of life in humans.

Example 7.7

Briefly describe plasma protein adsorption onto glutathione immobilized on gold (Lestellius *et al.*, 1995).

Solution

It has been shown that thiol-modified gold surfaces with varying functionalities may be used to modulate protein adsorption (Prime and Whitesides, 1991; Tengvall *et al.*, 1992). Lestelius *et al.* (1995) indicate that the combination of noble metals and thiol chemistry is an effective method to modulate protein adsorption. These authors emphasize that evaporated gold films are smooth and optically reflecting. Direct *in situ* techniques like ellipsometry can easily be adapted to follow adsorption kinetics. These authors analyzed the important differences in protein adsorption from plasma onto two molecular monolayers, L-cysteine (l-cys) and glutathione (GSH) immobilzed on gold. (See Figure 7.1.) The charges shown in this figure are expected at a pH around 7.

In situ ellipsometry was utilized to delineate the differences in protein adsorption on pure gold (Au), and on L-cys- and GSH-modified surfaces (Lestelius *et al.*, 1995). These authors estimated the amount adsorbed from

$$\Gamma = d(n_F - n_B)/(dn/dc). \tag{7.17}$$

Here d is the equivalent thickness of the adsorbed layer based on a fixed refractive index ($n_F = 1.465$, equal to that of silicon dioxide). The refractive index, n_F of the protein film is not known, but may vary from $1.35 < n_F < 1.55$. n_B is the refractive index increment of the ambient solution due to introduction of proteins into the solution. De Feijter *et al.* (1978) indicate that dn/dc is around 0.15–0.2 ml/g for most proteins. Lestelius *et al.* (1995) used a

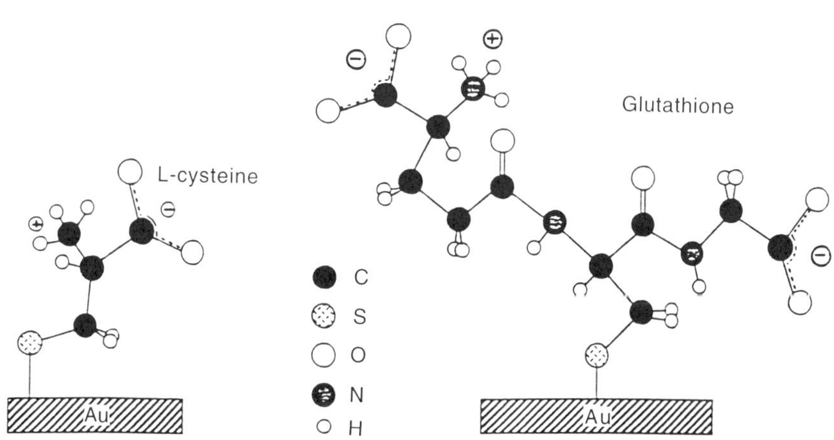

FIGURE 7.1 Structure of L-cysteine and glutathione immobilized on gold. At pH 7 the expected charges are shown. [From Lestelius, M. *et al.* (1995). *J. Colloid Interface Sci.*, *171*, 533.]

value of $dn/dc = 0.18$ and $n_F - n_B = 0.13$ in their calculations. Thus, a protein layer of thickness $d = 0.13$ nm corresponded to approximately 0.7 μg/cm^2. These authors ensured that they had a monolayer of L-cys or GSH deposited on the gold surface.

Lestelius *et al.* (1995) noted that calcium plays an important role in the buildup of the protein layer on GSH and L-cys surfaces, but not on gold. Phosphate-Buffered saline (PBS) and Hank's buffer were used as buffers. The PBS buffer consistently displayed lower protein adsorption. As expected, these authors noted that the adsorption and desorption kinetics differed for the different surfaces and for the two buffers.

The most striking observations were made with the GSH-modified surfaces. Both the GSH and the L-cys molecules bind via their sulfur atoms toward the gold as indicated in Fig. 7.1 Both present a "zwitterionic" group to the protein solution. However, the GSH-modified surface presents a more protein-like appearance than the L-cys-modified surface because it exposes a negatively charged group and two amide linkages. Thus, this proteinlike layer exhibits a different behavior with regard to protein adsorption from solution. These authors indicate that differences in protein adsorption onto GSH-modified and L-cys-modified surfaces can be further explained by calcium binding from the buffer. Lecompte *et al.* (1984) and Ivarsson and Lundstrom (1986) have indicated that Ca^{2+} helps stabilize the adsorbed proteins. The binding of Ca^{2-} on the GSH-modified and L-cys-modified surfaces is qualitatively different due to the structural differences of the molecules. These authors indicate that these qualitative differences in calcium binding to the two surfaces also contribute to the differences in protein adsorption. Finally, Lestelius *et al.* (1995) indicate that GSH is a larger molecule than L-cys. This leads to a surface with larger mobility for GSH. This again would lead to changes in protein adsorption.

The Lestelius *et al.* (1995) analysis of protein adsorption on GSH- and L-cys-modified gold surfaces is of interest because it attempts to explain the differences in protein adsorption on these surfaces. It would be of further interest, as suggested by these authors, to explore whether these favorable properties extend to the actual contacting with tissue or blood.

Though different techniques have been utilized to analyze protein adsorption at interfaces, Cullen and Lowe (1994) indicate that they all lack the ability to generate data with high spatial resolution (i.e., less than 0.5 μm). Thus, many facets of the protein adsorption are not understood. The influence of micro- and even nanoheterogeneity on protein adsorption needs to be carefully examined. These authors indicate that atomic force adsorption (AFM) microscopy is a tool that appears ideal for the study of protein adsorption processes.

Example 7.8

Briefly describe the adsorption of IgG and glucose oxidase (GO_x) to highly oriented pyrolytic graphite (HOPG) as analyzed by AFM (Cullen and Lowe, 1994).

Solution

AFM has been utilized to analyze the adsorption of IgG and GO_x on HOPG (Cullen and Lowe, 1994). These authors indicate that AFM generates a real-space topographic image of a surface with both high lateral and high vertical resolution. Positional information from piezoelectric actuators may be utilized to form a three-dimensional image of the surface. The authors indicate that IgG and GO_x have important commercial application either in immunodiagnostics or in electrochemical biosensors for the clinical measurement of glucose (Wilson and Turner, 1992).

Cullen and Lowe (1994) indicate that the mode of adsorption for IgG and GO_x on HOPG is strikingly different as observed in their AFM studies. IgG adsorption occurred after nucleation at a number of sites. These sites promoted localized binding, which eventually led to a homogeneous binding of the IgG after a long time. In contrast, GO_x binding displayed far fewer nucleation sites on the surface. These nucleation sites, presumably at HOPG surface step defects, promoted localized binding. This binding eventually led to a heterogeneous adsorption of GO_x, that included bare regions of HOPG. Furthermore, these authors suggest that based on their studies the IgG is adsorbed in a native conformation, and the GO_x is adsorbed in a denatured form. In addition, they indicate that the lateral force microscopy of GO_x adsorbed on HOPG supports the interpretation of the topographic image data.

The authors emphasize that their AFM studies yield a spatial distribution of proteins adsorbed to surfaces in real time. This is done with significant conformational changes in the proteins. Moreover, the relative strengths of protein-surface and lateral–lateral interactions compared with the forces applied during the AFM process are of importance so that one may obtain an unperturbed imaging of protein adsorption.

It would be of interest to analyze the influence of additives on protein adsorption to surfaces. This would impact not only bioseparation processes but also some immunodiagnostic applications. Agarose is utilized frequently as an adsorbent during chromatographic separations. A better understanding of the processes that occur on these adsorbents and of what facilitates the adsorption behavior of proteins and other biological macromolecules is of interest, especially if it increases the efficiency of the bioseparation process. Oscarsson *et al.* (1995) have analyzed the protein adsorption behavior on amphiphilic-based agarose adsorbents in the presence of different salts and at different salt concentrations. These authors used serum as a model. The evaluation in this case is more difficult. However, it is more informative because it contains different proteins exhibiting a range of physicochemical characteristics. These authors defined a term *salt-dependent adsorption capacity* (SAC). This is "the percentage of the protein adsorbed that can be released from the adsorbent by omitting the salt from the elution buffer."

They were trying to obtain answers to some fundamental questions, such as, "Are Π,Π-complexation and hydrophobic interactions synergistic in their adsorption effects?" Electrically neutral ligands contain delocalized Π-electrons that interact selectively with Π-electron-rich amino acid side groups (e.g, indolyl) located at the surface. Salt-dependent adsorption capacity was high for pyridyl-S-agarose. Results indicated that proteins adsorbed on phenyl and octyl gels remain after omitting salt from the buffer. A significant weakness in commercially available hydrophobic adsorbents was revealed because a large amount of protein remained on the octyl-sepharose and on phenyl-sepharose even after treatment with 1 M NaOH, especially, where washing with 1 M NaOH is a standardized washing method. Pyridyl-S-agarose eluted with a high efficiency under the same conditions. These authors emphasize that there is a high probability of contamination of fractions on reusing the adsorbent. Furthermore, the reuse of the adsorbent can lead to significant changes in capacity and in selectivity.

These authors express some degree of surprise, correctly so, that the previously mentioned undesirable property of hydrophobic adsorbents has not received more attention. They mention that one may attempt to use less hydrophobic adsorbents, but their capacity is too low. They emphasize that pyridyl-S-agarose is better than the conventional adsorbents used for hydrophobic interaction chromatography in several respects. However, it does differ in its affinity properties. Finally, these authors emphasize that new less hydrophobic adsorbents need to be explored that exhibit a high capacity, along with minimum retention of the protein following a rinsing step. This would significantly minimize contamination when the adsorbent is used for subsequent purifications. This is especially of importance in industrial applications.

The analysis of Oscarsson *et al.* (1995) is of interest because it states clearly that commercially available hydrophobic interaction chromatographic adsorbents are not ideal for downstream processes because they are too hydrophobic. This is because too much protein remains on the column after omitting salt from the elution buffer. One might resort to less hydrophobic adsorbents, but their capacity is too low. These authors emphasize that new types of less hydrophobic adsorbents need to be developed that: (1) exhibit a high SAC, and (2) maintain a high desorption efficiency. They state that the pyridyl-S-agarose is better than the conventional adsorbents used in quite a few respects. It does differ with these conventional adsorbents as far as affinity properties are concerned. Further research is underway to help optimize the adsorbent properties of pyridyl-S-agarose for chromatographic separations. This is being done by using different salts for different proteins.

C. Protein Fluorescence and Circular Dichroism

Thus far we have examined the adsorption of proteins at solid–liquid interfaces—a system that has been most rigorously studied. An example of the adsorption of blood proteins at air–water interfaces is now presented. Clark *et al.* (1988) utilized far-UV circular dichroism and intrinsic protein fluorescence to compare the spectral properties of resolubilized BSA with native BSA and interpreted the results in terms of the conformational properties of the proteins.

Far-UV circular dichroism spectra reveal only minor changes in the protein secondary structure evidenced by a small reduction in helix content after foaming. The biggest differences in conformation appear to be at the tertiary structure level and are readily detected by intrinsic fluorescence. A major irreversible reduction (>30%) in the intensity of tryptophan emission is reproducibly observed in the foamed sample. The change in conformation induced by foaming does not apparently reflect a change in the state of aggregation of the foamed protein. The native and foamed BSA samples used in the experiments contained similar amounts of oligomer as judged by nondenaturing polyacrylamide gel electrophoresis (PAGE). These authors acknowledge that their approach will only allow the observation of irreversible conformational changes that occur as a result of foaming and persist after resolubilization. Nevertheless, they state that their technique has allowed a more thorough study of the nature of these irreversible changes than by fluorescence quenching techniques. They indicate that in the future low-angle X-ray and neutron-scattering techniques may be usefully employed in the investigation of the structural properties of the adsorbed proteins *in situ* at the interface. We agree with them on this, and that in the meantime major compromises must be made if preliminary studies are to be made in this field.

Finally, the characterization of possible structural changes of the protein on surface interaction has been limited to techniques such as those presented previously. Even with these procedures results can often only be suggestive in regard to conformational alterations. Sandwick and Schray (1981) indicate the advantage of employing enzymes as a way of characterizing conformational changes occurring during the protein–solid surface interaction. A change in the conformation of the enzyme molecule is indicated by a resultant loss of the enzyme's activity on the enzyme's interaction with a solid surface.

Mizutani (1980) has reported a loss in enzyme activity when it interacts with a glass surface. Sandwick and Schray (1981) indicate that loss of enzyme activity would occur provided that a sufficient amount of time and adjacent surface area is available for the enzyme to establish a spread conformation. Thus, at relatively high surface area: initial protein concentration ratios inactivation may be observed, while at lower ratios the individual enzyme molecules are restricted by other adjacently adsorbed enzyme molecules in their spreading and therefore their inactivation. Sandwick and Schray (1981) investigated the desorption of four nonblood proteins/enzymes (horse radish peroxidase, alkaline phosphatase, catalase, and β-galactosidase) onto a hydrophobic surface. Their results demonstrate that at low relative initial concentrations the enzyme adsorbs and is subsequently altered in conformation while at relatively high initial enzyme concentrations the enzyme tends to adsorb and remain in its native, active conformation. Thus, according to the authors, proteins can adhere to a surface in either the native or the uncoiled (stretched) conformation. The amount of each form present on the surface at any particular instance will be dependent primarily on enzyme solution concentration, but also on other factors such as surface area available, temperature, and solution characteristics (pH and ionic strength).

The techniques that help assess the qualitative nature of the protein adsorbed at interfaces have been examined. Both of these aspects should be presented in future studies. Really required are predictive models that provide at least estimates of both of these features—the quantitative as well as the qualitative aspects of protein adsorption. The next section describes some typical models that provide some quantitative features of protein adsorption to surfaces. Hopefully in the future as the techniques for the qualitative characterization of protein adsorption on surfaces improve, a parameter or parameters that delineate this qualitative characteristics or heterogeneity may be appropriately defined. Then this heterogeneity factor could be suitably added into the present-day quantitative models for protein adsorption.

V. MODELS FOR PROTEIN ADSORPTION ON SURFACES

Nonflowing as well as flowing systems will be examined. Finally, a probabilistic analysis for protein adsorption is also presented.

A. Nonflowing Cylindrical Flowing System

In a nonflowing cylindrical system protein transport is a diffusional process (assuming no convection due to thermal or concentration gradients) described by (Young *et al.*, 1988)

$$\frac{\partial c}{\partial t} = \frac{1}{r}\frac{\partial}{\partial r}\left(Dr\frac{\partial c}{\partial r}\right), \tag{7.18}$$

where c is the local protein concentration, t is time, D is the diffusivity of the protein, and r is the radial coordinate from the center of the tube radius, R. For an infinitely long tube c is a function of r and t only, that is, $c = c(r, t)$. The boundary conditions are

$$c(r, 0) = c_0, \tag{7.19a}$$

where c_0 is the initial protein concentration. The zero-flux boundary condition at the center of the tube yields

$$\frac{\partial c}{\partial r}(0, t) = 0. \tag{7.19b}$$

The adsorption rate boundary condition at the tubing wall is

$$-D\frac{\partial c}{\partial r}(R, t) = R_{\text{ads}}(c, c_s). \tag{7.19c}$$

Here R_{ads} is the intrinsic adsorption rate constant, and is a function of the solution concentration, c, and the surface concentration, c_s. The intrinsic adsorption rate constant is the adsorption rate in the absence of any diffusional limitations. The first case occurs when the diffusional flux to the surface is much faster than the intrinsic adsorption kinetics. In this case, the adsorption kinetics are not limited by diffusion, and the observed adsorption rate, dc_s/dt, is equal to the intrinsic adsorption rate

$$dc_s/dt = R_{\text{ads}}(c, c_s). \tag{7.20}$$

The other limiting case occurs when the diffusional flux is much slower than the intrinsic kinetics, and the observed adsorption rate is actually the diffusion rate. In this case, each protein molecule that approaches the surface is immediately adsorbed, and the concentraton of soluble protein adjacent to the surface is zero. Equation 7.20 may be replaced by

$$c(R, t) = 0. \tag{7.21}$$

The solution of Eqs. (7.17), (7.18), and (7.20) yields the concentration distribution inside the tube

$$c(r, t) = 2c_0 \sum_{n=1}^{\infty} \frac{J_0(\alpha_n r/R)}{\alpha_n J_1(\alpha_n)} \exp(-\alpha_n^2 Dt/R^2), \tag{7.22a}$$

where J_0 and J_1 are Bessel functions of the first kind of order zero and one, respectively, and α_n is the nth zero of J_0. The surface concentration is the time integral of the flux of the protein to the surface

$$\begin{aligned} c_s(t) &= \int_0^t - D \left.\frac{\partial c}{\partial r}\right|_{r=R} dt \\ &= 2c_0 R \sum_{n=1}^{\infty} \frac{1}{\alpha_n^2} [1 - \exp(-\alpha_n^2 Dt/R^2)]. \end{aligned} \tag{7.22b}$$

A plot of dimensionless surface concentration $X = c_s/(c_0 R)$ against real time may be obtained.

Example 7.9

Briefly describe a macroscopic model for a single-component protein adsorption (Al-Malah *et al.*, 1995).

Solution

Al-Malah *et al.* (1995) proposed a macroscopic model for a single-component protein adsorption at a solid–water interface. These authors indicate that at a solid–water interface, protein adsorption is influenced by molecular size, shape, charge, hydrophobicity, and thermodynamic stability. At a hydrophobic interface, experimental observations suggest that protein hydrophobicity and conformational stability play a significant role. They wanted to incorporate selected model parameters into a macroscopic model to quantitatively predict a single-component protein adsorption at hydrophobic solid–water interfaces.

Their model included the following assumptions:

1. A reversible equilibrium exists between the bulk phase and the interface.
2. Monolayer coverage of protein is the upper limit of adsorption.
3. Figure 7.2 shows the schematic of the proposed adsorption mechanism.

In Fig. 7.2 as the protein enters the interface, it undergoes a change as it adapts to its new environment. After some time (or on the reaction coordinate) a pseudo-equilibrium exists between the bulk phase and the interface. They emphasize that in the absence of electrostatic effects on adsorption and of specific biochemical interactions (e.g., receptor–ligand), the equilibrium state is effectively characterized by the work of adhesion between the protein and the surface. The equilibrium constant, K (= dimensionally adsorbed mass, ΓA_p per dimensionless concentration, $V_p C_{eq}$), at low protein concentration, is given by

$$K = \exp(W_a A_c / RT). \tag{7.23}$$

Here W_a is the work of adhesion between the protein and the surface (J/m^2), A_c is the surface area required by an adsorbing protein molecule to anchor itself on the surface (m^2/mol), C_{eq} is the apparent equilibrium protein concentration

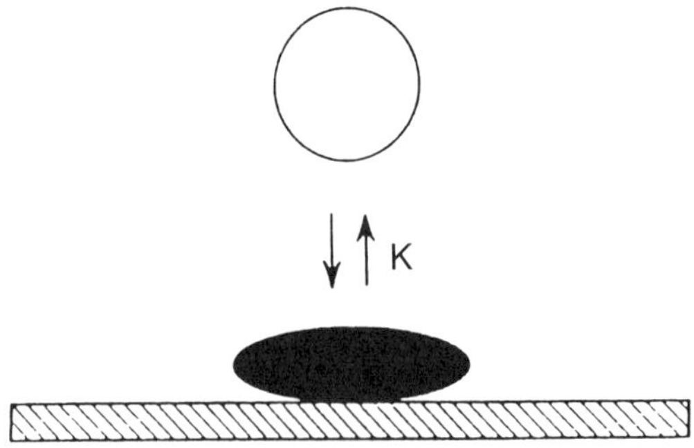

FIGURE 7.2 A pseudo-equilibrium of protein exists between the bulk phase and the interface. [From Al-Malah, K. *et al.* (1995). *J. Colloid Interface Sci., 160,* 261.]

in solution (mol/m^3), R is the gas constant, and T is the temperature. Also, A_p is the partial molar area occupied by the protein at the interface, and V_p is the partial molar volume of the protein solution (m^3/mol).

Al-Malah *et al.* (1995) were able to obtain good agreement between the model and experimentally measured isotherms for the milk proteins α-lactalbumin, β-lactoglobulin, and BSA at hydrophobic silica. These authors noted that the extent of protein adsorption correlated well with molecular size and the strength of hydrophobic interaction between the protein and the solid surface. The larger the molecular weight is, the larger the adsorbed mass is. Also, for proteins of equal size, the stronger the hydrophobic interaction is, the larger the value of K and the larger the adsorbed mass per unit area are.

B. Protein Convection (Desorption Kinetics)

Consider an adsorbing surface with adsorbed protein in equilibrium with a flowing protein solution at concentration c_0. For desorption to occur, the protein solution is replaced by a buffer solution with no protein. At steady state, a concentration boundary layer is established in solution adjacent to the adsorbing surface. Within the region of the concentration boundary layer (where $y \ll b$), a model describing transport-limited desorption at long times is given by (Cheng *et al.*, 1987)

$$\gamma y \frac{\partial c}{\partial x} = D \frac{\partial^2 c}{\partial y^2}. \tag{7.24}$$

The boundary conditions are

$$\begin{aligned} x &= 0, \quad & c &= 0 \text{ for all } y \\ y &\geq \infty, \quad & c &= 0 \text{ for all } x \\ y &= 0, \quad & c &= c_0 \text{ for all } x > 0. \end{aligned} \tag{7.25}$$

c_0 is the solution concentration of protein that is in equilibrium with the surface protein concentration. Solving Eq. (7.24) subject to Eq. (7.25) yields

$$\frac{dc_s}{dt} = \frac{-1}{\Gamma(4/3)9^{1/3}} \left(\frac{\gamma}{Dx}\right)^{1/3} Dc_0. \tag{7.26}$$

Equation (7.24) can be used to calculate the expected transport-limited desorption rate given c_s, the surface protein concentration; and c_0, the solution protein concentration in equilibrium with c_s.

C. A Probabilistic Analysis for Protein Adsorption

The adsorption of proteins to solid surfaces is often modeled by resorting to the assumptions made by Langmuir (1918) in deriving the adsorption equation. This derivation is based on the mean or averaged behavior of the particles in the system, and thus only macroscopic characteristics appear (Boughey *et al.*, 1978; Petersen and Kwei, 1961). A stochastic approach is capable of providing more details about a dynamic system (Stanislaus *et al.*, 1977). Hsu and Sun (1988) adopted a statistical analysis to model the transient behavior of reversible adsorption of small particles on a solid surface. These authors were able to estimate both the mean and the fluctuating characteristic of the adsorption in a straightforward manner. Though they did not give examples of protein adsorption to solid surfaces, their analysis is interesting and should yield valuable insights into protein adsorption on solid surfaces. These authors successfully modeled the deposition of polystyrene particles on nylon fibers (Boughey *et al.*, 1978), the adsorption of CO as a function of time on alumina after preadsorption of water vapor (Stanislaus *et al.*, 1977), and the adsorption of hydrogen by $LaNi_5$ (Tanaka *et al.*, 1977).

The probabilistic method does demonstrate significant potential to provide novel physical insights into the adsorption of proteins and other biological macromolecules on different types of surfaces. These analyses, if performed, should significantly contribute to the understanding of biological macromolecular adsorption at interfaces.

Example 7.10

Develop the equations between flowing blood proteins and an artificial surface (Schaaf and Dejardin, 1987).

Solution

Schaaf and Dejardin (1987) indicate that thermodynamic and structural information may be obtained by the determination of adsorption isotherms and adsorbed layer thicknesses. A dynamic equilibrium is definitely established between the flowing blood and the surface. These authors indicate that at least two aspects need to be considered: (1) the rate at which the proteins (or biological macromolecules) become attached to the solid surface when they are in close proximity to the surface, and (2) the diffusive flux from the bulk solution to the depleted interfacial layer.

D. Diffusion-Controlled Regime

The idealized situation of a surface acting as a perfectly adsorbing barrier was initially considered by Smoluchowski (1916). Herein, any molecule reaching the surface is adsorbed. The appropriate equation and boundary conditions are

$$\frac{\partial c_p(x, t)}{\partial t} = D \frac{\partial^2 c_p(x, t)}{\partial x^2} \tag{7.27}$$

and:

$$c_p(0, t) = 0, t > 0 \tag{7.28a}$$

$$c_p(x, 0) = c_{p0}, x > 0, \tag{7.28b}$$

respectively. Here x is the distance from the interface (or surface), D is the diffusion coefficient of the solute (protein or other biological macromolecule), and c_p is the concentration of the protein (or other biological macromolecule). The concentration of the adsorbed protein molecules is given by

$$c_{p,s}(t) = 2c_{p0}\left[\frac{Dt}{\pi}\right]^{1/2}. \tag{7.29}$$

The boundary condition [Eq. (7.28a)] specifies that the process is completely diffusion controlled.

E. Kinetic-Controlled Regime

In this case the interfacial concentration $c_p(0, t)$ is equal to the bulk concentration. Then

$$c_p(0, t) = c_{p0}. \tag{7.30}$$

The rate of protein adsorption is then controlled by the chemical kinetics at the interface-surface. Consider a Langmuir-type approach. Then the adsorbed protein concentration may be obtained from

$$\frac{dc_{p,s}}{dt} = k_a c_p(0, t)\,(1 - \Gamma' c_{p,s}) - k_d c_{p,s}, \tag{7.31}$$

subject to the boundary condition [Eq. (7.30)]. Here k_a and k_d are the adsorption and desorption rate constants of the protein molecule, and Γ' is the surface occupied by the adsorbed protein molecule. The adsorbed protein concentration is given by

$$c_{p,s} = \frac{k_a c_p(0, t)}{k_d + k_a(0, t)\Gamma}\{1 - \exp[-(k_a c_{p0}\Gamma + k_d)]\}. \tag{7.32}$$

For small time, t, or during the initial rate of adsorption, Eq. (7.32) yields:

$$c_{p,s} = k_a c_{p0} t. \tag{7.33}$$

In general, when experimental data with regard to protein adsorption to surfaces are analyzed, the data apparently do not fit either Eqs. (7.29) or (7.33). Thus, the protein adsorption process seems to neither follow the diffusion-controlled regime nor the kinetically controlled regime.

Other complications may also arise. Heterogeneity of the surface sites or of the solute molecules themselves also needs to be examined to correctly model the more realistic case. Besides, Schaaf and Dejardin (1987) correctly indicate and Collins and Kendall (1949) also point out that the Smoluchowski solution leads to an infinite initial adsorption rate. These authors utilized a simple and discrete model to describe material exchange in the vicinity of the interface, and indicated that the boundary condition, [Eq. (7.28a)] needs to be modified to obtain physically sound results at very short times. It is apparently critical to characterize relative amounts and the kinetics of adsorption of proteins and other biological macromolecules to surfaces to understand the reactions at the surface. However, other parameters too, as expected, will play a significant role. This is clearly demonstrated in the next example, which examines protein adsorption from buffer and plasma onto different copolymers.

Example 7.11

There are some correlations between blood protein adsorption and surface properties (Grainger *et al.*, 1989).

Solution

Grainger *et al.* (1989) have analyzed the influence of substrate hydrophilic-hydrophobic balance on the adsorption of proteins from buffer and plasma using a series of amphiphilic multiblock coploymers composed of PEO and polystyrene (PS). These authors analyzed the adsorption of albumin, fibrinogen, and IgG from single-component buffer; and plasma solution in contact with polymer-coated beads. Initial attempts have been made to correlate protein adsorption and platelet adhesion to polymer surfaces by focusing on the effect of the hydrophobic and hydrophilic balance of constituent chains in amphiphilic surfaces (Yui *et al.*, 1984; Okano *et al.*, 1991; Grainger *et al.*, 1987).

Grainger *et al.* (1989) comment that a myriad of molecular plasma constituents, including more than 200 proteinaceous components, probably compete to differing degrees in the adsorption process occurring at material interfaces. These authors emphasize that the complex interactions between components in the adsorbed state and in bulk solution are further perturbed by exchange–desorption influences and by denaturation–renaturation on the surface and in solution. All these factors and others would increase the overall heterogeneity of the adsorbed protein on the surface, thereby further influencing the subsequent reactions occurring in the solution and on the surface. It would be difficult, but not impossible, to include some of these effects in a more realistic model of protein adsorption. It is therefore not surprising these authors comment that few correlations were obtained between blood platelets *in vivo* and whole blood thrombogenicity. They emphasize the shortcoming of their study by analyzing only three proteins in plasma, whereas dozens of proteins are important in blood–surface interactions. Other factors that increase the heterogeneity of the protein adsorbed, besides just the adsorbed protein amounts and kinetics, are also important. These parameters may include denaturation, degradation, exchange, etc.

Example 7.12

Describe a technique for measuring protein adsorption wherein protein molecules are not modified by the introduction of some extrinsic label that might affect the adsorption kinetics (Norde and Rouwendal, 1990).

Solution

Norde and Rouwendal (1990) developed the streaming potential technique (*in situ* monitoring) to study protein adsorption kinetics on the surface of a flow cell. The streaming potential may be converted into the electrokinetic or ζ-potential or just ζ, of the surface of the flow cell according to Dukhin and Dejardin (1974)

$$V_s = \frac{\epsilon\zeta}{\eta c G}\,\Delta p. \tag{7.34}$$

Here ϵ is the dielectric permittivity, η is the viscosity, c is the cell constant, G is the conductance of the cell filled with the solvent, and Δp is the pressure drop. Norde and Rouwendal (1990) indicate that, in general, the adsorption of a protein (electrically charged) influences the ζ of the surface. In their experimental setup of two parallel glass plates, these authors applied a laminar flow of the protein solution.

The analytic solution of the mass-transport limited equation is the Leveque solution (1928) given by

$$\frac{d\Gamma(y)}{dt} = 0.81\left(\frac{\gamma}{yD}\right)^{1/3} Dc. \tag{7.35}$$

Here D is the diffusion coefficient of the protein molecule in solution, γ is the shear rate at the cell wall $\left(\left.\frac{dv}{dx}\right|_{x=a} = \frac{a\Delta b}{\eta l}\right)$, 2a is the separation distance between the two parallel plates, l is the length of the parallel plates and y is the distance in the direction of flow. Norde and Rouwendal (1990) indicate that Eq. (7.35) is applicable only under steady-state conditions with respect to the concentration boundary layer. Also, the diffusion rate of the protein across this layer must be low relative to the rate of interaction of the protein molecule with the cell wall.

The following values were used in Eq. (7.34) to determine the streaming potential. $\Delta p = 1.7 \times 10^4$ N/m^2, $\epsilon = 78.5$, and η equals 8.9×10^{-4} N/m^2 sec. For the bare glass-buffer solution surface, a ζ-potential of about -48 mV has been derived. These authors analyzed the adsorption isotherms of myoglobulin, ribonuclease, and lysozyme. At pH 7, they concluded that myoglobulin is isoelectric, and ribonuclease and lysozyme are positively charged. The differences in the shapes of the adsorption isotherms, that is the initial slopes (which represents the affinity for adsorption) and the plateau values, may be explained by the differences in the electrical charges between the three proteins that interact with the negatively charged surface.

Norde and Rouwendal (1990) conclude that the initial adsorption rates of lysozyme, ribonuclease, and myoglobulin on the glass surface are transport limited. This is because the observed effects of wall shear rate and of protein concentration in solution (for low concentrations) on the kinetics of protein adsorption from laminarly flowing solutions are in close agreement with the Leveque convective-diffusion model (Leveque, 1928). No information was provided by Norde and Rouwendal (1990) on the conformational changes or heterogeneity of the protein in the adsorbed state. This aspect was not incorporated in the model.

Tilton *et al.* (1990) analyzed the lateral diffusion of BSA adsorbed at the solid-liquid interface (PMMA and PDMS) by a combination of TIRF and fluorescence recovery after photobleaching techniques. These authors indicate that lateral mobility, conformation, orientation, and ordering are probably associated in a complex manner. For instance, a conformational-structural change after adsorption may alter the lateral mobility of a protein. This change in the lateral mobility of the adsorbed protein may alter its ability to interact with the protein's nearest neighbors. They emphasize that the lateral mobility of adsorbed proteins has not been fully characterized, and much of the evidence that supports lateral mobility after adsorption is circumstantial. These authors note that adsorbed proteins do form organized layers, and this may be attributed to lateral mobility (Brash and Lyman, 1969; Dass *et al.*, 1987; Ratner *et al.*, 1981; Fair and Jamieson, 1988).

Variations on the fluorescence bleaching technique has most commonly been used to investigate the slow self-diffusion (Eldridge *et al.*, 1980; Schindler *et al.*, 1980; Thompson and Axelrod, 1980, Tilton *et al.*, 1990). The primary requirement for the technique is that the mobile species bear either an intrinsic fluorescent moiety or a tightly bound extrinsic fluorophore. The rates of molecular transport are determined by creating a gradient of fluorescent and nonfluorescent molecules with a photobleach pulse of high intensity laser illuminaton.

Tilton *et al.* (1990) indicate that the diffusion coefficient is a measure of the dynamics of the adsorbed BSA molecules, and the fractional mobility provides insight into the distribution of dynamic states. The mobile fraction could be verified by an examination of the long-time asymptote of the fluorescence recovery. Fractional mobilities, f, less than unity indicate nonuniformity of the adsorption states of eosin isothiocyanate-labeled bovine serum albumin (EITC-BSA). These authors obtained an f value equal to 0.37 ± 0.05; this indicates that different populations of EITC-BSA characterized by different mobilities prevail on the PMMA surface. They indicate that the coexistence of tightly packed adsorbed protein aggregates and isolated adsorbed proteins leads to a nonuniform lateral mobility. Although a distribution of lateral mobilities may be a consequence of an ordering phenomenon, this lateral mobility in itself may be a prerequisite for the formation of such ordered arrangements. The authors emphasize that this nonuniformity may also be due to a distribution of BSA conformational states. This would then lead to a heterogeneity of the adsorbed BSA on the EITC surface.

Lu and Park (1991) analyzed the influence of surface hydrophobicity on the conformational changes of adsorbed fibrinogen. Such studies are essential because a significant amount of attention has been paid to the conformational changes of protein adsorbed on solid surfaces due to the importance of protein conformation on the activity of the adsorbed proteins (Lenk *et al.*, 1989; Kato *et al.*, 1987). Tomikawa *et al.* (1980) emphasize that the conformational changes of fibrinogen adsorbed on solid surfaces are thought to be reasonsable for the platelet adhesion to the surface, because the intact fibrinogen in solution does not interact with the platelets under the same conditions.

Lu and Park (1991) analyzed the extent of conformational changes of fibrinogen adsorbed on germanium, poly(hydroxyethylmethacrylate), Biomer, and polystyrene surfaces using Fourier transform infrared spectroscopy (FTIR) coupled with attenuated total reflectance (ATR) optics. The authors noted that some α-helical structures were changed into unordered structures and the content of the β-turns was increased on the protein adsorption. Basically, these authors noted that the adsorbed fibrinogen underwent a larger degree of conformational changes as the surface hydrophobicity increased.

These authors underscore that because the analysis of the conformational changes by their weighted-shift method is new, it is difficult to calculate at this point that their method quantitates the absolute magnitude of protein conformational changes. Nevertheless, the sum of the weighted peak shifts is expected to correlate with the relative extent of conformational changes. Furthermore, Iwamoto *et al.* (1985) also found that fibronectin experienced greater conformational changes on a more hydrophobic silica surface. Lu and Park (1991) emphasize that when a protein adsorbs on a solid surface with high hydrophobicity, the hydrophobic core is likely to become exposed to the surface due to the hydrophobic interaction. Therefore, the larger conformational changes on more hydrophobic surfaces would lead to increasing heterogeneities on the surface.

VI. CONCLUSIONS

The causes and influence of heterogeneity on initial protein adsorption, and the mediation of subsequent reactions on the surface presented provide for a more realistic picture of the adsorption of proteins at the interface. A significant amount of evidence presented (qualitative characterization techniques, modeling studies, energetics of surface sites, etc.) indicates that heterogeneity in protein adsorption does exist. Protein adsorption on surfaces-interfaces will lead to differing degrees of conformational changes at the interface. These conformational changes will, in most cases, either decrease or increase the rate of subsequent reactions on the surface. It is worthwhile estimating the conformational changes (or qualitative aspects of protein adsorption) by a suitable heterogeneity parameter. This heterogeneity parameter should initially be defined, estimated, and then evaluated as a time-dependent function. To date, very few models for protein adsorption exist that define an appropriate heterogeneity parameter; models are really required that can relate this heterogeneity parameter to experimental results.

Further effort that appropriately incorporates the influence of heterogeneity in protein adsorption studies, and delineates the influence of heterogeneity (or conformational changes) on the mediation of subsequent reactions at the surface is urgently required to not only shed novel physical insights into the adsorption process but also provide for a more realistic picture of the events occurring at the interface. The introduction of heterogeneity in an analysis of protein adsorption on surfaces, and the collection of such data by different investigators should then provide an initial and useful framework for analyzing subsequent protein adsorption studies. This framework should also help build more predictive techniques to analyze not only the quantitative but also the qualitative aspects of protein adsorption.

REFERENCES

Abramson, H. A. (1942). In *Electrophoresis of Proteins*, Reinhold: New York.

Absolom, D. B., Policova, Z., Bruck, T., Thomson, C., Zingg, W., and Neumann, A. W. (1987). *J. Colloid Interface Sci., 117,* 550.

Al-Malah, K., McGuire, J., and Sproull, R. (1995). *J. Colloid Interface Sci., 170,* 261.

Aptel, J. D., Carroy, A., Dejardin, P., Pefferkorn, E., Schaaf, P., Schmitt, R., Varoqui, R., and Voegel, J. C. (1987). In *Proteins at Interfaces. Physicochemical and Biochemical Studies,* American Chemical Society: Washington, DC.

Baeir, R. E. and Dutton, R. C. (1969). *J. Biomed. Mater. Res., 3,* 191.

Bagchi, P. and Birnbaum, S. M. (1981). *J. Colloid Interface Sci., 83,* 460.

Barnes, D. W. (1984). *Anal. Biochem., 137,* 196.

Beissinger, R. L. and Leonard, E. F. (1980). *ASAIO, 3,* 160.

Borowko, M. and Jaroniec, M. (1983). *Adv. Colloid Interface Sci., 19,* 137.

Boughey, M. T., Duckworth, R. M., Lips, A., and Smith, A. L. (1978). *J. Chem. Soc. Faraday Trans., 74,* 2220.

Brash, J. L. (1981). *Interaction of the Blood with Natural and Artificial Surfaces,* Salzman, E. W., Ed., Dekker: New York (1981).

Brash, J. L. and Lyman, D. J. (1969). *J. Biomed. Res., 3,* 175.

Brash, J. L. and Samak, Q. M. (1978). *J. Colloid Interface Sci., 65,* 495.

Bull, H. B. (1956). *Biochim. Biophys. Acta, 19,* 464.

Chan, B. M. C. and Brash, J. L. (1981). *J. Colloid Interface Sci., 82,* 217.

Chase, H. A. (1984) In *Ion Exchange Technology,* Naden, C. D., and Streat, M., Eds., Horwood: Chichester, p 401.

Cheng, Y. L., Darst, S. A., and Robertson, C. R. (1987). *J. Colloid Interface Sci., 118,* 212.

Clark, D. C., Mackie, A. R., Smith, L. J., and Wilson, P. R. (1988). *Food Hydrocolloids, 2,* 1209.

Coltharp, M. T. and Hackerman, N. (1973a). *J. Colloid Interface Sci., 43,* 176.

Coltharp, M. T. and Hackerman, N. (1973b). *J. Colloid Interface Sci., 43,* 185.

Collins, F. C. and Kimball, G. E. (1949). *J. Colloid Interface Sci., 4,* 425.

Corsel, J. W., Willems, G. M., Kop, J. M. M., Cuypers, P. A., and Hermens, W. T. (1986). *J. Colloid Interface Sci., 111,* 544.

Cullen, D. C. and Lowe, C. R. (1994). *J. Colloid Interf. Sci., 166,* 102.

Cuypers, P. A., Hermens, W. T., and Henker, H. C. (1977). *N.Y. Acad. Sci., 283,* 77.

Dabrowski, A. (1983). *Monatsh. Chem., 114,* 875.

Dabrowski, A. and Jaroniec, M. (1979a). *J. Colloid Interface Sci., 73,* 475.

Dabrowski, A. and Jaroniec M. (1979b). *Acta Chim. Acad. Sci. Hung., 99,* 255.

Dabrowski, A. and Jaroniec, M. (1980a). *J.Colloid Interface Sci., 77,* 571.

Dabrowski, A. and Jaroniec, M. (1980b). *Z. Phys. Chem. Leipzig, 261,* 359.

Dass, D. V., van Enckevort, H. J., and Langdon, A. G. (1987). *J. Colloid Interface Sci., 116,* 523.

Davis, S. S., Illum, L., McVie, J. G., and Tomlinson, E. (1984) Eds., *Microspheres and Drug Therapy. Pharmaceutical, Immunological, and Medical Aspects,* Elsevier Science: Amsterdam.

De Feijter, J. A., Benjamins, J., and Veer, F. A. (1978). *Biopolymers, 17,* 1759.

Delmas, G. and Patterson, R. (1960). *J. Phys. Chem. 64,* 1827.

Dukhin, S. S. and Dejardin, B. V. (1974). In *Surface and Colloid Science,* Matijevic, E., Ed., Vol. 7, John Wiley & Sons: New York.

Dunnill, P. (1983). *Process Biochem., 18*(10), 9.

Eldridge, C. A., Elson, E. L., and Webb, W. W. (1980). *Biochemistry, 19,* 2075.

Elgersma, A. V., Zsom, R. L. J., Norde, W., and Lyklema, J. (1990). *J. Colloid Interface Sci., 138,* 145.

Elwing, H., Welin, S., Askendal, A., Nilsson, U., and Lundstrom, I. (1987). *J. Colloid Interface Sci., 119,* 203.

Everett, D. H. (1964). *Trans. Faraday Soc., 60,* 1803.

Everett, D. H. (1965). *Trans. Faraday Soc., 61,* 2478.

Fair, B. D. and Jamieson, A. M. (1988). *J. Colloid Interface Sci., 121,* 240.

Gill, D. S., Roush, D. J., and Wilson, R. C. (1994). *J. Colloid Interface Sci., 167,* 1.

Golander, C. G. and Kiss, E. (1988). *J. Colloid Interface Sci., 127,* 240.

Graham, D. E. and Phillips, M. C. (1979). *J. Colloid Interface Sci., 75,* 403.

Grainger, D. W., Okano, T., and Kim, S. W. (1987). In *Advances in Biomedical Polymers,* Gebelein, C. G., Ed., Plenum: New York, p 229.

Grainger, D. W., Okano, T., and Kim, S. W. (1989). *J. Colloid Interface Sci., 132,* 161.

Gribnau, T. C., Leuvering, J. H. W., and van Hell, H. (1986). *J. Chromatogr., 376,* 175.

Grinell, F. and Feld, M. K. (1982). *J. Biol. Chem., 257,* 4888.

Haynes, C. A., Sliwinsky, E., and Norde, W. (1994). *J. Colloid Interface Sci., 164,* 394.

Hlady, V. and Furedi-Milhofer, H. (1979). *J. Colloid Interface Sci., 69,* 460.

Hlady, V. and Andrade, J. D. (1986). *Adv. Polym. Sci., 79,* 1.

Hoffman, A. S., (1982). *ACS Series,* Vol. *119,* 3.

Hsu, J. P. and Sun, S. S. (1988). *J. Colloid Interface Sci., 122,* 73.

Hummel, J. P. and Anderson, B. S. (1965). *Arch. Biochem. Biophys., 112,* 443.

Ihlenfeld, J. V., Mathis, T. R., Riddle, L. M., and Cooper, S. L. (1979). *Thromb. Res., 14,* 953.

Ivarsson, B. and Lundstrom, I. (1986). *CRC Crit. Rev. Biocompatability,* 2(1).

Iwamoto, G. K., Winterton, L. C., Stoker, R. S., van Wagenen, R. A., Andrade, J. D., and Mosher, D. F. (1985). *J. Colloid Interface Sci., 106,* 459.

Jaroniec, M. (1981). *Thin Solid Films, 81,* 97.

Jaroniec, M., Derylo, A., and Marczewski, A. W. (1983). *Chem. Eng. Sci., 38,* 307.

Jaroniec, M. and Derylo, A. (1981). *J. Colloid Interface Sci., 84,* 191.

Jaroniec, M. and Derylo, A. (1981). *Chem. Eng. Sci., 36,* 1017.

Jaroniec, M., Narkiewicz, J., and Rudzinski, W. (1978). *J. Colloid Interface Sci., 65,* 9.

Jennisen, H. P. (1978). *J. Chromatogr., 159,* 71.

Jennisen, H. P. (1981). *Adv. Enzyme Regul., 19,* 377.

Jeon, S. J. and Andrade, J. D. (1991). *J. Colloid Interface Sci., 142,* 159.

Jeon, S. I., Lee, J. H., Andrade, J. D., and De Gennes, P.G. (1991). *J. Colloid Interface Sci., 142,* 149.

Johnsson, U., Malmquist, M., and Ronnberg, I. (1985). *J. Colloid Interface Sci., 103,* 360.

Joly, M. (1965). *A Physicochemical Approach to the Denaturation of Proteins,* Academic: London, p 15.

Kato, K., Matsui, T., and Tanaka, S. (1987). *Appl. Spectrosc., 41,* 861.

Kochwa *et al.* (1949). *Kinetics of Chemical Change,* Hinshelwood, C. N., Ed., Oxford University Press: London.

Kop, J. M. M., Cuypers, P. A., Lindhout, T., Hemker, H. C., and Hermens, W. Th. (1984). *J. Biol. Chem., 259,* 1393.

Kop, J. M. M., Willems, G. M., and Hermens, W. T. (1989). *J. Colloid Interface Sci., 133,* 369.

Lahav, J., Schwartz, M. A., and Hynes, R. O. (1982). *Cell, 31,* 253.

Lahav, J., Lawler, J., and Grimbone, M. A. (1984). *Eur. J. Biochem., 145,* 151.

Lahav, J. (1987). *J. Colloid Interface Sci., 119,* 262.

Langmuir, I. (1918). *J. Am. Chem. Soc., 40,* 1361.

Langmuir, I. and Schaefer, V. J. (1937). *J. Am. Chem. Soc., 59,* 2400.

Lecompte, M. F., Clavallier, J.,. Dode, C., Elion, J., and Miller, I. R. (1984). *J. Electroanal. Chem., 163,* 345.

Lee, R. G. and Kim, S. W. (1974). *J. Biomed. Mater. Res., 8,* 251.

Lee, S. H. and Ruckenstein, E. (1988). *J. Colloid Interface Sci., 125,* 365.

Lenk, T. J., Ratner, B. D., Gendreau, R. M., and Chittur, K. K. (1989). *J. Biomed. Res., 23,* 549.

Leonard, E. F., Turitto, V. T., and Vroman, L. (1987). *Annals N.Y. Acad. Sci., 516.*

Leung, L. L. K. (1984). *J. Clin. Invest., 74,* 1764.

Leung, L. L. K. and Nachman, R. L. (1982). *J. Clin. Invest., 70,* 542.

Lestelius, M., Tengrall, P., and Lundstrom, I. (1995). *J. Colloid Interface Sci., 171,* 533.

Leveque, A. (1928). *Ann. Mines, 13,* 201.

Lok, B. K., Cheng, Y. L., and Robertson, C. R. (1983a). *J. Colloid Interface Sci., 91,* 104.

Lok, B. K., Cheng, Y. L., and Robertson, C. R. (1983b) *J. Colloid Interface Sci., 91,* 87.

Lu, D. R. and Park, K. (1991). *J. Colloid Interface Sci., 144,* 271.

Lundstrom, I., Ivarsson, B., Johnson, B., and Elwing, H. (1987). In *Polymer Surfaces And Interfaces,* Feast, W. D. and Munro, H. S., Eds., John Wiley & Sons: New York, p. 201.

Lundstrom, I. (1985). *Progr. Colloid Polym. Sci., 70,* 76.

Lundstrom, I. and Elwing, H. (1990). *J. Colloid Interf. Sci., 116.*

MacRitchie, F. (1978). *Adv. Protein Chem., 32,* 283.

MacRitchie, F. (1987). In *Proteins at Interfaces. Physicochemical and Biochemical Studies*, ACS Symposium Series, 343, Brash, J. L. and Horbett, T. A., Eds., Americn Chemical Society: Washington, DC.

MacRitchie, F. (1978). *Adv. Protein Chem.*, *32*, 283.

Malhotra, A. and Sadana, A. (1987a). *Biotechnol. Bioeng.*, *30*, 108.

Malhotra, A. and Sadana, A. (1987b). *Biotechnol. Bioeng.*, *30*, 1041.

Malhotra, A. and Sadana, A. (1989). *Biotechnol. Bioeng.*, *34*(5), 725.

Malhotra, A. and Sadana, A. (1990). *J. Theor. Biol.*, *145*, 143.

Malmsten, M. and Lassen, B. (1994). *J. Colloid Interface Sci.*, *166*, 490.

McNaly, E. J. and Graf, G. (1990). *J. Colloid Interface Sci.*, *138*, (1990).

Mesteri, J., Partyka, S., and Brun, B. (1984). *Cal. Anal. Therm.*, *15*, 155.

Mizutani, T. (1980). *J. Pharm. Sci.*, *69*, 279.

Moacanin, J. and Kaelble, D. H. (1977). *Polymer*, *18*, 475.

Morrissey, B. W. and Han, C. C. (1978). *J. Colloid Interface Sci.*, *65*, 423.

Morrissey, B. W. and Stromberg, R. R. (1974). *J. Colloid Interface Sci.*, *46*, 152.

Morrissey, B. W., Smith, L. E., Stomberg, R. R., and Fenstermaker, C. A. (1976). *J. Colloid Interface Sci.*, *56*, 557.

Moyer, L. S. and Govin, M. H. (1940). *J. Biol. Chem.*, *133*, 605.

Narendarnathan, T. J. and Dunnill, P. (1982). *Biotechnol. Bioeng.*, *24*, 2103.

Nikitas, P. (1985). *J. Chem. Soc. Faraday Trans.*, *181*, 1767.

Nikitas, P. (1988). *Electrochim. Acta*, *33*, 647.

Nikitas, P. (1989). *J. Colloid Interf. Sci.*, *129*, 579.

Norde, W. and Lyklema, J. (1978a). *J. Colloid Interface Sci.*, *66*, 277.

Norde, W. and Lyklema, J. (1978b). *J. Colloid Interface Sci.*, *66*, 266.

Norde, W. (1983). *Croat. Chem. Acta*, *56*(4), 705.

Norde, W., MacRitchie, F., Nowicka, G., and Lyklema, J. (1986) *J. Colloid Interface Sci.*, *112*, 447.

Norde, W. and Lyklema, J. (1979). *J. Colloid Interface Sci.*, *71*, 350.

Norde, W. (1986). *Adv. Colloid Interf. Sci.*, *25*, 267.

Norde, W. (1988). In *Surfactants in Solution*, Mittal, K. L. and Bothorel, P., Eds., Vol. 5, Plenum: New York, p 1027.

Norde, W. and Rouwendal, E. (1990). *J. Colloid Interface Sci.*, *139*(1), 169.

Okano, T., Shimada, M., Aoyagi, T., Shinohara, I., Kataoka, K., Prime, K. L., and Whitesides, G. M. (1991). *Science*, *252*, 1164.

Oscarsson, S., Angulo-Tatis, D., Chaga, G., and Porath, J. (1995). *J. Chromatogr. A*, *689*, 3.

Oscik, J., Dabrowski, A., Jaroniec, M., and Rudzinski, W. (1986). *J. Colloid Interf. Sci.*, *56*, 403.

Partyka, S., Lindheimer, M., Zaini, S., Keh, E., and Brun, B. (1986). *Langmuir*, *2*, 101.

Petersen, C. and Kwei, T. K. (1961) *J. Phys. Chem.* *65*, 1330.

Poole, S., West, S. I., and Waters, C. L. (1984). *J. Sci. Food Agric.*, *35*, 701.

Prime, K. L. and Whitesides, G. M. (1991). *Science*, *252*, 1164.

Ratner, B. D., Horbett, T. A., Shuttleworth, D., and Thomas, H. R. (1981). *J. Colloid Interf. Sci.*, *83*, 630.

Rechnitz, G. A. (1987). *J. Clin. Annu.*, *1*, 308.

Rosselin, G., Assan, R., Yallow, R. S., and Berson, S.A. (1966). *Nature (London)*, *212*, 355.

Rudzinski, W., Oscik, J., and Dabrowski, A. (1973). *Chem. Phys. Lett.*, *20*, 5.

Rudzinski, W., Waksmundzki, A., Jaroniec, M., and Sokolowski, S. (1974). *Ann. Soc. Chim. Pol.*, *48*, 1985.

Rudzinski, W., Lattar, L., Zajac, J., Wofram, E., and Paszli, J. (1983). *J. Colloid Interface Sci.*, *96*, 339.

Sakurai, Y. (1986). *J. Biomed. Mater. Res.*, *20*, 1035.

Sandwick, R. K. and Schray, K. J. (1981). *J. Colloid Interface Sci.*, *121*, 1.

Sato, H., Tomiyama, T., Morimoto, H., and Nakajima, A. (1987). In *Proteins at Interfaces. Current Issues and Future Prospects*, Brash, J. L. and Horbett, T. A., Eds., ACS Symposium Series, 343, American Chemical Society: Washington, DC.

Schaaf, P. and Dejardin, P. (1987). *Colloids Surfaces*, *24*, 239.

Schindler, M., Osborn, M. J., and Koppel, D. E. (1980). *Nature (London)*, *283*, 346.

Sharma, C. P., Kallyankrishnan, V., and Valiathan, M. S. (1982). *Polym. Plast. Technol. Eng.*, *18*, 233.

Shirahama H. and Suzawa, T. (1988). *J. Colloid Interface Sci., 126,* 269.

Shirahama, H., Suzuki, K., and Suzawa, T. (1989). *J. Colloid Interface Sci., 129,* 483.

Shirahama, H., Lyklema, J., and Norde, W. (1990). *J. Colloid Interface Sci., 139,* 177.

Silman, I. H. and Katchalski, E. (1966). *Annu. Rev. Biochem., 35,* 873.

Sips, R. (1948). *J. Chem. Phys., 16,* 490.

Siskova, M. and Erdos, E. (1960a). *Collect. Czech. Chem. Commun., 25,* 1729.

Siskova, M. and Erdos, E. (1960b). *Collect. Czech. Chem. Commun., 25,* 2599.

Sonderquist, M. E. and Walton, A. G. (1980). *J. Colloid Interface Sci., 75,* 386.

Stanislaus, A., Evans, M. J. B., and Mann, R. F. (1977). *J. Phys. Chem., 81,* 1684.

Stoner, J. and Srinivasan, S. (1970). *J. Phys. Chem., 74,* 1088.

Tan, J. S. and Martic, P. A. (1990). *J. Colloid Interface Sci., 136,* 415.

Tanaka, S., Clewly, J. D., and Flanagan, T. E. (1971). *J. Phys. Chem., 81,* 1684.

Tengvall, P., Lestelius, M., Liedberg, B., and Lundstrom, I. (1992). *Langmuir, 8,* 1236.

Thompson, N. L. and Axelrod, D. (1980). *Biochim. Biophys. Acta, 597,* 155.

Tilton, R. D., Robertson, C. R., and Gast, A. P. (1990). *J. Colloid Interface Sci., 137,* 192.

Tomikawa, M., Iwamoto, M., Olsson, P., Soderman, S., and Blomback, B. (1980). *Thromb. Res., 19,* 869.

Uzgiris, E. E. and Fromageot, H. P. M. (1976). *Biopolymers, 15,* 257.

Virkar, P. D., Hoare, M., Chan, M. Y. Y., Dunnill, P., Humphrey, A. E., and Lilly, M. D. (1979). "Enzyme and Fermentation Technology," John Wiley and Sons: New York, Chapter 12.

Vroman, L. and Adams, A. L. (1969). *J. Biomed. Mater. Res., 3,* 43.

van Dulm, P., Norde, W., and Lyklema, J. (1981). *J. Colloid Interface Sci., 82,* 77.

Wilson, R. and Turner, A. P. F. (1992). *Biosens. Bioelectron., 7,* 165.

Young, D. and Crowell, A. (1962). *Physical Adsorption of Gases,* Butterworths: London.

Young, B. R., Lambrecht, L. K., Cooper, S. L., and Mosher, D. F. (1982). In *Biomaterials : Interfacial Phenomena and Applications,* Cooper, S. L. and Peppas, N. A., Eds., *Chem. Ser.,* Vol. 199, American Chemical Society: Washington, DC.

Young, B.R. (1984). Ph.D. Thesis, University of Wisconsin, Madison.

Young, B. R., Pitt, W. G., and Cooper, S. L. (1988). *J. Colloid Interface Sci., 124,* 28.

Yui, N., Tanaka, J., Sanui, K., Ogata, N., Kataoka, K., Okano, T., and Sakurai, Y. (1984). *Polym. J., 16,* 119.

Zchuchovitzky, A. (1938). *Acta Physicochem. USSR, 8,* 531.

Zucker, M. B. and Broman, N. (1969). *Proc. Soc. Exp. Biol. Med., 131,* 318.

8 APPLICATIONS AND ECONOMICS OF BIOSEPARATION

I. INTRODUCTION

The United States, Germany, France, and England are the leading countries for the sales of biotechnological products, with the United States being the leading player (Burrill and Roberts, 1992). It is estimated that by the year 2000 the worldwide sales of biotechnological products will be $100 billion (Burrill and Roberts, 1992). Bioseparation and downstream processing equipment constitute a large fraction, 50%, if not more (Spalding, 1991; van Brunt, 1985) of the cost of preparing a drug, protein, or biological product suitable for market consumption. Ronsohoff *et al.* (1990) indicated that the purification and recovery costs can account for as much as 80% of the total manufacturing cost in the large-scale production of recombinant protein products. Rosen and Datar (1983) emphasized that the ratio of recovery to fermentation costs for an enzyme is 2.0 compared with 1.0 for penicillin and 0.16 for ethanol. It is anticipated that this ratio will be significantly higher if one is to use these products for pharmacological use. During processing, for example, one may have to purify products at 99.9% levels with virtually complete removal of DNA, viruses, and endotoxins.

As expected, the worldwide annual market for downstream processing equipment will grow rapidly [about 20% per year (Spalding, 1991)] from $1.0 billion in 1991 to about $5.2 billion by the year 2000. Because the key to cutting production costs is emphasizing improvements in bioseparation-downstream processing equipment, it behooves one to attempt to better analyze and understand the different facets involved in downstream processing. Better physical insights into downstream processes will help pave the way to cutting costs wherever possible so that one may improve the overall production costs and stay competitive. Upstream processes are well understood and significant improvements have been made. Genetic engineering has facilitated the large-scale production of new proteins and peptides (Paul, 1981). Datar (1986) emphasized that the bottlenecks in bringing recombinant DNA (rDNA) products to large-scale production are a result of lack in sound understanding of downstream processing operations (Atkinson and Mavituna, 1983; Atkinson and Sainter, 1980; Rosen and Datar, 1983). The key is in understanding downstream processing, integrating it with upstream processing, and thus providing better insights into and improving the economics of the whole process itself.

Conceivably, even minor changes in protein upstream processes can have significant impact (economic or otherwise) on downstream processes. Naveh (1985, 1990) and others (Fish and Lilly, 1984; Hedman, 1984) emphasized that the designers of the purification process must consider the upstream-fermentation process impact on downstream processing early in the development of a process. With the high stakes that are involved in, for example, getting some drugs to the market, there is bound to be extremely fierce competition. Companies have realized this and, now more than ever, jealously guard their know-how (both of academic and economic value). Estimates to bring a pharmaceutical drug to the market range from \$100 million (van Brunt, 1985) to \$300 million (Raab, 1992). Besides, a minimum of 7 to 10 years is required. The Biotechnology Industry Organization indicates that it takes 10 to 12 years to move a product from bench to bedside. This is twice as long as it took 20 years ago. Also, the cost of a drug has increased by a factor of five to \$360 million (Stone, 1995). Hassler (1995) in a journal editorial indicated that it takes about \$350 to 400 million to develop a drug. There are indications, that due to the very high cost involved, eventually there will be 1 to 3 fully integrated biopharmaceutical companies, 10 to 15 platform companies, and presumably 50–100 successful boutiques. Everybody else will be doing something else in some other capacity.

Thayer (1995), too, indicated the strategy where the discovery efforts of small, research-focused companies are being synergistically combined with the drug development, manufacturing, and marketing resources and experience of the large, well-known corporations. This is exemplified with the creation of a new research and development (R & D) company called Allergan Ligand Retinoid Therapeutics. Allergan with nearly \$1 billion in annual sales is combining forces with Ligand with nearly \$13 million in annual research revenues and a net loss of more than twice as large. Thayer (1995) emphasized that many small companies are content with remaining just drug discoverers rather than drug marketers. There is a surge in such corporate arrangements due primarily to the fact that "in the pharmaceutical industry there is a lack of distinctive new products in pipelines that can command high margins while providing major, cost-effective advances in treatment" (Feinstein, 1995). Table 8.1 shows some of the research partnerships that have occurred in the last 3 years, along with the areas of focus and the net value of these partnerships. Cancer treatment is the most common area of the joint ventures (Thayer, 1995).

Furthermore, Egan *et al.* (1995) indicated that the drug spends another 3 years after process development under the watchful eyes of the Federal Drug

TABLE 8.1 Small Research Firms Partner with Big Drug Companies[a]

Company	Partner	Value[b] ($ millions)	Start-up date	Term (years)	Area of focus
Allelix Biopharmaceuticals	Hoechst Roussel	$ 53	1/95	5	Psychiatric disorders
Arris Pharmaceutical	Bayer	70	11/94	5	Inflammation
Cadus Pharmaceutical	Bristol-Myers Squibb	45[c]	7/94	3–5	Proprietary
ICOS	Abbott Laboratories	na[d]	4/95	na	Cancer
Ligand Pharmaceuticals	Allergan	100	6/95	Joint venture	Cancer
	SmithKline Beecham	22	2/95	3–5	Hematopoiesis
	American Home Products	44	9/94	3–5	Women's health
	Abbott Laboratories	26	7/94	3–5	Inflammation
	Glaxo	20	9/92	5	Cardiovascular disease
	Pfizer	17	5/91	5	Osteoporosis
Millennium Pharmaceuticals	Hoffman-LaRoche	70	3/94	5	Obesity, diabetes
Oncogene Science	Hoechst Roussel	na	4/94	na	Alzheimer's disease
	American Home Products	na	1/94	3	Diabetes, asthma, immune system, osteoporosis
	Ciba-Geigy	na	8/93	na	Wound healing
	Hoechst	na	4/93	na	Inflammation, arthritis, metabolic disease
	Marion Merrell Dow	17	1/93	5	Cardiovascular disease
	Pfizer	16	4/91	5	Cancer
Onyx Pharmaceuticals	Warner Lambert	25	5/95	3	Cancer
	Eli Lilly	na	5/95	na	Cancer
	Bayer	38.5[c]	5/94	5	Cancer
Sugen	Zeneca	17.5[e]	1/95	5	Cancer
Synaptic Pharmaceutical	Eli Lilly	na	3/95	4	Nervous system disorders
	Ciba-Geigy	na	1994	3	Cardiovascular disease
	Merck	20	1993	na	Neuroreceptors
Vertex Pharmaceuticals	Wellcome	42	12/93	5	AIDS
	Roussel Uclaf	30	9/93	5	Inflammation

[a] **Source:** Company data.

[b] Value of collaboration includes equity investment, research funding, cash, license fees, and potential milestone payments, but excludes any estimates of potential royalties or shared profits.

[c] Excludes possible milestone payments.

[d] na = Not available.

[e] Excludes possible milestone payments and research funding.

Administration (FDA) (Hassler, 1994). Quite often, as the drug is taken through the regulatory gauntlet it experiences clinical trial problems. This can have a devastating influence on the stock performance of a company. Egan *et al.* (1995) suggested a strategy to manage a crisis in clinical trails.

Bienz-Tadmor and Brown (1994) compared the development times of biopharmaceuticals with biologics on approval data available from the FDA between 1982 and 1991. Biopharmaceuticals are drugs derived through biotechnology, for example, rDNA products and monoclonal antibodies excluding recombinant vaccines. The mean and the range of the development times for the biopharmaceuticals anti-CD3, erythropoietin (EPO), granulocyte colony-stimulating factor (G-CSF), granulocyte macrophage colony-stimulating factor (GM-CSF), human growth hormone (hGH), interleukin-2 (IL-2), interferon-A (INF-A), interferon G (INF-G), insulin, and tissue plasminogen activator (tPA) were ~ 7 and 4 to 10 years, respectively. Biologics are therapeutically important drugs purified from natural sources. Similarly, the mean and range of the development times for the biologics alglucerase, α-proteinase inhibitor, anistreplase, antithrombin III, factor VIII, factor IX, and pegademase were ~ 12 and 7 to 23 years, respectively. These authors emphasized that biotechnology-derived drug development is risky, in spite of the approval of the drugs mentioned previously. All participants in the developments of these types of drugs should be aware of the risks, the time, and the financial support required to bring a drug to the market; this understanding helps minimize disappointment.

By considering the time and effort it takes to bring a drug to the market, it is of interest to see a list of drugs that have been successful for the different companies, and their sales value. Table 8.2 provides a list of U.S. companies along with the sales of their major biotechnology products. Thayer (1995) indicated that in 1993 Amgen was the leader with revenues of \$1.4 billion (24 % change, an increase from 1992), and net revenues rose by 16% to \$356 million. However, Genentech leads in the number of products it sells (five). Five others are also involved including Monsanto's bovine growth factor. Genentech's revenues increased by 19% to \$650 million, and its earnings increased by 183% to \$59 million. Thayer (1994), however, cautioned that few biotechnology companies were profitable in 1993.

Rubinfeld (1995) indicated that one of the pitfalls of the biotechnology industry has been that considerable effort has been spent on developing credibility, and appealing to investors to attract capital. Very little effort is sometimes spent on consumer needs and on markets. This author cautioned that not all companies (especially small ones) can emulate Amgen or Genentech, which are fully integrated pharmaceutical companies. Rubinfeld (1995) emphasized that if today's biotechnology companies, as well as tomorrow's biotechnology companies, want to exist, then they must focus on meeting a demand by providing value products for consumers, and ultimately contribute to the health-care system.

Raab (1992) stressed that maturing biotechnology firms will have to grapple with realistic issues (economic and otherwise). He indicated that the vast resources it takes to develop and bring a drug to the market will force a further shakedown in biotechnological and pharmaceutical companies. The costs of some drugs per dose are high (Table 8.3), and one is understandably reticent

TABLE 8.2 Sales of Major Biotechnology Products Grew in 1993

		Sales ($ millions)	
Company	**Products**	**1993**	**1992**
Amgen	Erythropoietin	$586.9	$506.4
	Granulocyte colony-stimulating factor (G-CSF)	719.4	544.6
Biogen[a]	α-, β-, and γ-Interferons; hepatitis B vaccines and diagnostics	136.4	121.7
Centocor	Antisepsis monoclonal antibody[b], diagnostic products	48.1	58.4
Chiron	β-Interferons[c]	11.8	nm[d]
	Interleukin-2 and other oncology products, ophthalmics	147.9	111.6
Genentech	Human growth hormone	216.8	205.9
	Tissue plasminogen activator (tPA)	236.3	182.1
	γ-Interferon	4.3	2.9
	Human insulin, α-interferon, and Factor VIII	112.9[e]	91.7[e]
Genetics Institute	Factor VIII[f]	41.3	nm
	Erythropoietin, granulocyte macrophage colony-stimulating factor (GM-CSF), factor VIII	26.5[g]	27.3[g]
Genzyme	Therapeutics, fine chemicals, diagnostics products and services	233.9[h]	180.0
Immunex[i]	Oncology products[j]	46.7	nm
	Granulocyte macrophage colony-stimulating factor	42.1	26.3

[a] Predominantly royalty income from $1.5 billion in sales by licensees.
[b] European sales in 1992. Sales halted in January 1993.
[c] Initial sales began in third quarter of 1993.
[d] nm = not meaningful.
[e] Royalties from licensees sales.
[f] Factor VII approved for marketing December 1992; includes sales of product or marketing partner Baxter.
[g] Royalties on overseas sales of erythropoietin and GM-CSF, and on U.S. sales of Factor VIII.
[h] Includes about $124 million in therapeutics sales of glucocerebrosidase.
[i] Merged with oncology business of American Cyanamid's Lederle Laboratories in June 1993.
[j] Sales of certain Lederle products between June 2 and December 31, 1993.

to pay such high prices unless (literally) one's life is at stake. For example, the single dose price for activase is $2200 (Genentech); for eminase it is $1700 (SmithKline Beecham); and for streptokinase (Astra Kabi) it is $200 (Raab, 1992). Thus, so that some life-saving drugs may reach a wider section of the common populace, it behooves biotechnological companies (as well as society as a whole) to minimize the economics of production of these products. This chapter provides some economic information into downstream processes. As expected, this information is difficult, if not rare, to obtain in the open literature.

van Brunt (1985) raised different issues about getting a biotechnological product ready for the market. Does the product conform to FDA standards? It

TABLE 8.3 A Comparison of the Estimated Costs for the Production of Some Drugs, Pharmaceuticals, Proteins, and Bioproducts (Inclusive of Downstream Processing Costs)

Drug/protein/ pharmaceutical	Company	Cost	Ref.
Activase	Genentech	\$2200/dose	Raab, 1992
Eminase	SmithKline Beecham	\$1700/dose	Raab, 1992
Streptokinase	Astra Kabi	\$200/dose	Raab, 1992
tPA	Genentech	\$2000/dose	Spalding, 1991
Monoclonal antibody	Invitron	\$300–400/g[a]	Duffy *et al.*, 1989
α-Galactosidase	Not applicable	\$300–1000/million EU[b]	Porter and Ladisch, 1992

[a] Assuming total cost is three to four times the downstream processing costs (ion exchange).
[b] Range based on throughput per year. Cost estimates by modeling.

must be safe, potent, and pure. Also, is the product stable in the injectable form? What is its shelf life? Meeting these and other requirements mandates the involvement of a significant number of purification steps. Common sense dictates that a large-scale purification procedure should be designed to minimize the number of steps while maintaining high yields and product purity, quality, and activity. Howell (1985) emphasized that when an engineer dealing with fermentations designs a downstream process, he will typically employ more stages than are strictly necessary and may use methods that are not easy to scale up. On review, the engineer needs to restructure the process so that it is as simple as possible, recovers only those products that are of commercial interest, and is carried out with the smallest scale of equipment possible. The equipment should have a high use factor and be as efficient as possible, because fewer stages reduce the opportunity of product loss and minimize product contamination (Howell, 1985).

Bonnerjea *et al.* (1986) emphasized utilizing the right process at the right time. Pharmacia (1986) emphasizes that in planning a strategy for protein purification it is important to have a stated purpose and definite goals, aims, and set standards by which to measure success. An initial in-depth study of the target protein-pharmaceutical product, its properties, its native environment, and its sources will prevent unnecessary and unexpected losses of activity and assist in the choice of fractionation techniques. Differential solubility techniques are generally employed at the start of the downstream processing train to remove gross impurities. The higher resolution techniques are generally employed in the latter part of the process (Pharmacia, 1986). Best results may be obtained by an appropriate combination of these techniques that are tailor-made to a particular process or recovery of a biotechnological product.

First-generation therapeutics like tPA sell for \$2000 a dose, or \$2000/100 mg. U.S. production of this protein is 10 kg or equivalently 100,000 doses (Spalding, 1991). These attractive returns have made companies emphasize beating out competitors to the market rather than focusing on improving the economics of the processes. As competition increases and second-generation therapeutics enter the market, more emphasis needs to be placed on improving the economics of providing these proteins-drugs to the market. Lower production costs, primarily downstream processing costs will significantly play a major role in improving the manufacturing costs of these products.

Detailed production costs of the different biotechnological products of significant use are rarely, if at all, available. Lambert (1989) emphasized that the ready availability of commercial downstream processing equipment combined with significant successes in the laboratory scale has in fact hindered the development of large-scale production. Many academic and laboratory efforts are focused on the isolation of a particular product; scant, if any, attention is paid to the economics of the process. Consideration of operational longevity, and activity and stability of the product are generally of no importance. Besides, operational data from laboratory and pilot plant are rarely reported. Also, practical guidelines for controlled assembly and operations are few, are frequently of commercial origin, and often need to be modified for one's particular process of interest (Lambert, 1989). Operational data for downstream processing equipment should be made available. Especially, factors that cause a decline in the performance (such as activity-stability of the product) should be either clearly delineated carefully studied if not available. These factors can significantly contribute to the economic viability of different downstream processing trains.

The proprietary and the undisclosed nature of this type of information is perfectly understandable. Nevertheless, in this chapter we will "piece together" the different bits of information available in the open literature. The intent of presenting this information together is to provide a picture, although incomplete in some respects, to different researchers, industrial workers, and entrepreneurs-venture capitalists that have a significant stake in the development of biotechnological processes-products. Some scale-up and down-scaling procedures and strategies will be presented. These will be followed by different examples of bioproduct production where economic data in the open literature either scarce or (in significant amounts) are available.

II. SCALE-UP PROCEDURES

van Brunt (1985) defined scale-up as the transition from a procedure (a well-established means of doing something) to a process (a series of operations, which in our case produces the biotechnological product). Those companies that succeed in this transition will be the most successful. This author indicated that not all the variables are completely known as the process is scaled-up. For example, the final dose of a pharmaceutical is not known until the end of the clinical trials. Furthermore, van Brunt (1985) added that the number of doses per liter is a critical number and often significantly affects the process design-planning of a process. There should thus be enough flexibility in the process to take care of these unexpected needs. Many of these extra-unexpected requirements are often satisfied by improvements due to research or further familiarization of the process.

Mahar (1993) indicated that, in biotechnology, centrifugation is effectively used to separate mixtures that exhibit very small differences in specific gravity

(Lavanchy, 1979; Svarovsky, 1977; Tiller, 1974). He indicated that compactly designed centrifuges have efficient economies of scale. Centrifugation may be used to separate many biological cell separations such as bacteria, yeast, and mycelia. Also, cell debris recovery from lysate addition, and protein solubilization of inclusion bodies (IBs) are possible with centrifugation. Centrifugation also finds application in liquid–liquid extraction steps. Mahar (1993) emphasized that for many biological cell separations, high g-forces may be required. Then in this case one has to compromise by sacrificing continuous capabilities or the ability to handle high concentrations of solids. The following example provides some economic considerations for the application of the centrifugation process for biotechnological separations.

Example 8.1

Provide a brief economic analysis for utilizing centrifuges for single- and multiuse facilities (Mahar, 1993).

Solution

Economic considerations are a key factor in utilizing centrifuges for single- and multiuse facilities (Mahar, 1993). Mahar broke down the costs into the following four categories: (1) capital costs, (2) operating costs, (3) maintenance costs, and (4) nontangible costs.

Capital costs can be quickly estimated by assuming a straight-line, 15-year depreciation. The capital costs are a significant proportion of the total costs. If a single-use facility is planned, then the costs are assigned to the facility. If, however, a multiuse facility is planned, then the costs may be suitably proportioned. This author indicated that capital cost for centrifuges runs from \$50,000 to \$500,000 depending on the size and configuration selected. Often the cost is strongly dependent on the specifications required of the centrifuges.

Operating costs include utility, chemical, and labor requirements. Batch operations, especially those that require periodic shutdowns are labor intensive. As expected, product recovery specifications significantly influence the operating costs. Mahar (1993) indicated that if additional product recovery is required, this will entail additional utility, chemical, and possibly labor costs.

The author indicated that maintenace costs run from 5 to 10% of the total labor costs. Furthermore, nontangible costs are often neglected when purchasing a centrifuge system. For example, compact centrifuges may save or make available floor space on the shop floor. This is especially true if one is comparing applications using cross-flow membranes or vacuum filters. Besides, the effect on other downstream equipment is also a nontangible cost. For example, microbial cells have a disposal cost associated with them. The drier the cake is, the less is the disposal cost. Mahar's (1993) analysis looks at the "complete approach" to downstream processing. This approach is correct and should be emphasized.

Example 8.2

Describe briefly the changes made by Genentech as the dosage requirements for tPA increased from 1 to 100 mg during clinical trials (Spalding, 1991).

Solution

When Genentech started its tPA process in 1982, the expected dose was 1 mg (Spalding, 1991). As the drug progressed through clinical trials its dosage increased by two orders of magnitude to 100 mg. This necessitated improvements in the process such as: (1) switching host cells from *Escherichia coli* to Chinese hamster cells, and (2) changing the roller-bottle process to a suspension-culture process. The leader of the development process for Genentech, Stuart Builder, indicated that fortunately the progress in the development process kept up with the increase in the required dosage. In other words, the cost per gram fell by about 100-fold (Spalding, 1991). This was a critical improvement, otherwise the drug would have been priced too high to be of any significant economic advantage to Genentech.

In designing a process, the objectives should be clearly stated beforehand, for example, what product is going to be made, what its specifications or quality is (e.g., stability-activity), how much should be made (quantity), and when should it be ready for the market. Many of the process variables change as one takes the process from the laboratory scale to the pilot plant scale. As results are obtained from a pilot plant or higher scale equipment, these may be used as a feedback mechanism to improve the performance in the pilot plant scale so that the stated objectives can be met utilizing this iterative procedure. As the stated objectives are met, then the process may be transferred to the manufacturing department.

The time of transferring the purification process from the laboratory scale to the pilot plant scale is important. Naveh (1990) emphasized that the production host, location, and the physical form of the protein product determine the selection as well as the sequence of the purification stages. The author emphasized that only chromatography offers the high resolution required to obtain a very pure product. Initial volume reduction steps may require clarification. In some cases a chromatographic step may be required up front. The order of purification steps is largely dependent on the product to be purified. Usually a polishing step is required, for example, to remove pyrogens or to exchange into buffer systems suitable for formulation. Gel filtration remains a suitable polishing step in the purification train. The author emphasized that scale-up should be attempted only when: (1) a suitable processing strategy in accord with the product to be purified has been developed, and (2) the process has been optimized on the laboratory scale. This procedure attempts to take out or minimize the surprises that are inherently present in scale-up. A thorough familiarization of the purification strategy at the laboratory scale should be of considerable value not only during scale-up but also while the process is being run on the pilot plant scale.

van Brunt (1985) emphasized that scaling-up from the laboratory scale to the pilot plant is simply not a case of a multiplicative factor. The parameters that were perhaps easily controllable at the laboratory scale may be different and unpredictable at the pilot plant scale. Adjustments are often required. For example, the microorganism selected must remain stable during the fermentation process. Antibiotics by selective pressure prevent the organisms from reverting. The costs at a low volumetric level may not be too much. At the pilot plant level the costs of the antibiotics to exert this selective pressure may be prohibitive. This is where the knowledge of the engineer and the other members of the team comes into play where modifications need to be made continuously to help optimize the process. These modifications may be such that they may be expensive. Thus, (1) enough slack should be built into the initial estimates of the process to be able to take care of these types of cost overruns, or (2) process optimization needs to be carried out continuously to improve the economics of the modified process.

Example 8.3

Describe briefly the modificatons made by Hoffmann-LaRoche during the large-scale processing of α-interferon A.

Solution

Hoffmann-LaRoche biochemical engineers–scientists noted problems during the scale-up of α-interferon A separated from recombinant *E. coli*. The organism was grown in 400-liter fermentation tanks. The crude extract was purified by the classical method of immunochromatography that was used in the bench-scale studies. Problems occurred when the agarose-based columns used in the bench scale were used for large-scale processing. Apparently, at the larger scale there was radial compression of particles in the column that decreased flow rate and caused significant back pressure. A more rigid support was required that also allowed the higher flow rates at the large scale. Silica was found to be suitable to permit the high flow rates and minimize the radial compression experienced by the agarose particles. High flow rates at this step are critical to the economics of the process because they significantly influence the yield.

Chase (1984) analyzed affinity separations utilizing monoclonal antibodies. He developed some scale-up procedures. The author indicated that the performance of small-scale immunoaffinity separation systems should, in general, form the basis of the design of large-scale systems. For batch systems the quantities of adsorbent required should be increased in proportion to the increase in the volume of the fluid to be processed (Chase, 1984). For fixed-bed systems the author recommends that the volumetric flow rate be increased by the factor v_2/v_1. Here v_1 is the volumetric flow rate where the fixed-bed reactor is giving satisfactory performance. v_2 is the new volumetric flow rate. The cross-sectional area of the fixed bed should be increased by the preceding factor. The height of the column should be kept the same.

Furthermore, if monoclonal immunoglobulin G (IgG) molecules can be immobilized and covalently or noncovalently attached to the support without loss of activity, then the total capacity of the immunoadsorbent will be utilized in immunoadsorbent separations (Chase, 1984). In this case, the maximum amount of antigen that can be isolated by 1 mol of immobilized antibody per cycle of operation is 2 mol. In practice, heterogeneity effects, difficulties due to adsorption, steric effects, orientation and others (Velander, 1992) will prevent the preceding theoretical binding capacity to be utilized to its full extent. Eveleigh and Levy (1977) reported that the highest specific activity that they could achieve was 1.25 when the support was highly activated and the resultant density of immobilized protein was less than 1 mg/ml of settled adsorbent.

The following three equations were suggested to estimate the minimum quantity of antibody that will be required to be produced for large-scale immunoseparation (Chase, 1984). The purification of g grams of antigen, g_{Ag} of molecular weight M_{Ag}, requires at least the following grams of antibody, Ab in grams g_{Ab}

$$g_{\mathrm{Ab}} = \frac{(75{,}000)(g_{\mathrm{Ag}})}{M_{\mathrm{Ag}}}. \tag{8.1}$$

Let the density of immobilized antibody on the support be ρ_{Ab}. Then the minimum amount of adsorbent, V, required to isolate g_{Ag} per cycle is given by:

$$V = \frac{(75{,}000)(g_{\mathrm{Ag}})}{(M_{\mathrm{Ag}})(\rho_{\mathrm{Ab}})}. \tag{8.2}$$

Assume that (1) the annual required production of antigen is G_{Ag}, and (2) the adsorbent can be utilized for n cycles. Then the minimum amount of total antibody required to produce the annual target of antigen is given by

$$G_{\mathrm{Ab}} = \frac{(75{,}000)(G_{\mathrm{Ag}})}{(M_{\mathrm{Ag}})(n)}. \tag{8.3}$$

The minimum total cost of the antibody will then be (G_{Ab})(unit cost of the antibody per gram). There are basically two ways by which to minimize the total cost based on Eq. (8.3), and the unit cost of the antibody per gram. Increase the longevity of the large-scale operating column. Simply speaking increase n. Optimization schemes may be set up (1) to get the most out of each column, and (2) to enhance the longevity without sacrificing the activity or quality of the product. The author correctly pointed out that the extent to which the technique will be used commercially will be significantly dependent on the cost of producing the antibody. The accelerated research in this area is bound to drive down the price of producing antibodies. This should significantly influence the economics of these types of commercially operated immunoadsorbent separation systems utilizing antibodies. A better characterization of antibodies on immunoadsorbent surfaces would also be of considerable assistance (Lin *et al.*, 1988). Research into the covalent–noncovalent attachment of the antibodies to the surface, along with a better understanding of diffusional constraints and steric hindrances, should significantly assist in improving the economics of large-scale immunoadsorbent systems. This will particularly be true if such results are available in the open literature.

Large-scale recovery of bioproducts by liquid chromatography is an area of increasing commercial importance. The direct scale-up of conventional, low-pressure, liquid chromatography has been successful. Nevertheless, Fulton *et al.* (1992) suggested that other approaches are required to process ton-scale bioproducts such as recombinant blood proteins or animal growth hormones. Bioseparation processes need to be developed for biologicals such as viral vaccines, non-antibody immunoregulators, monoclonal antibodies, peptide growth factors, hormones, viral insecticides, tumor-specific antigens, and animal cells as a product, etc. (Mizrahi, 1986). Fulton *et al.* (1992) emphasized that on a commercial basis liquid chromatography has a number of limitations. The most restrictive one is the low flow rate that results because of poor diffusive mass transport in porous media. This low flow rate leads to long cycle times. Therefore, the processes are run typically by using large columns. These columns are not only uneconomical but also are inflexible to changes in the overall plant output. This inflexibility puts an unnecessary constraint on the optimization of the performance, especially with respect to changes elsewhere in the plant.

In scaling-up of a chromatographic column Kelley *et al.* (1986) initially suggested maintaining a constant column aspect (length-to-diameter) ratio. Naveh (1990) emphasized that utilizing this approach may result in extremely long columns. For example a 150-ml column with a 1.6-cm diameter and a 75-cm height scales up to an 80-liter column with a 17-cm diameter and a 6-m bed height. At a linear velocity of 0.2 cm/min, this increases the elution time from 6.25 to 51 h. This is definitely impractical. Another practiced approach is to scale up by increasing column diameter while holding the bed height constant. Also, the elution buffers and the volumetric ratios of the feed are held constant. By utilizing this approach, the scale-up equation for reverse-phase chromatography and hydrophobic interaction chromatography is given by (Naveh, 1990)

$$SF = \frac{(V_{\text{bed}})_2}{(V_{\text{bed}})_1} = \frac{V_2}{V_1} = \frac{A_2}{A_1}. \tag{8.4}$$

Here SF is the scale-up factor; and is the ratio of the column bed volumes, (V_{bed}), the volumetric flow rates (V), and the cross-sectional areas (A). The author emphasized that when only the bed diameter is increased, in this multiplicative fashion, significant differences between working with the process in the laboratory scale and in large-scale columns are observed and should be anticipated. This, along with other aspects, is the "uncertainty factor" observed during scale-up. This uncertainity factor may or may not play a significant part in the economics of scale-up; in most cases, it just might. Therefore, the need for better and improved methods of scale-up.

During scale-up it is essential to validate the goodness of the column packing. This author adds that because more band broadening occurs in large-scale systems compared with that of smaller scale systems during residence time distribution (RTD) testing, these RTD profiles are better at the smaller scale. This is one of the reasons that scaling-up (by increasing only the diameter) yields different results than when the process is carried out at the smaller scale. The author suggested using the "novel" down-scaling approach to the scale-up of processes.

Briefly, the steps involved during scale-up using the down-scaling approach are (Naveh, 1990):

1. The separation should be carried out at the laboratory scale using the mass-volume loadings and linear velocities anticipated at the higher scale. Determine the plate number of the column using the RTD test.
2. For the large-scale column determine the diameter keeping the bed height constant. Carry out the RTD test. Compare the plate numbers obtained at the laboratory scale and at the higher scale. Add backmixing to the laboratory system if its plate number is more than that of the large-scale system. Repeat this processs until the height equivalent to a theoretical plate (HETP) for both systems is within $\pm 5\%$.

3. Obtain the resolution of the product from the degraded laboratory system. If there is no difference, then the separation is plate insensitive. More often than not, additional bed height will have to be added to improve the resolution. Redo step (2) again with this new bed height.

4. If results are satisfactory, then one may commit valuable biotechnological materials for the higher scale production.

The preceding technique suggested by the author not only is practical but also exhibits economic characteristics. One does not commit valuable biotechnological materials to the separation process until one has obtained satisfactory results from the down-scaling approach. The suitability of this technique needs to be tested further to see if it can be applied effectively to other biotechnological products of interest. It would be a tremendous asset in the repertoire of the engineer as far as the economics of the process is concerned if this down scaling approach could be applied successfully for the scale-up of a wide variety of biotechnological products. We next present an example of separation by the gel filtration technique where this down-scaling approach has been utilized.

Example 8.4

Demonstrate the applicability of the down-scaling approach for the gel filtration of a polymeric protein mixture that has a molecular weight-size distribution between 30 × 10^6 and 80 × 10^3 Da and a mass average molecular weight of 3.98 × 10^6 (Naveh, 1990).

Solution

A measure of resolution was obtained by examining the polydispersity in any given fraction of the column tested (Naveh, 1990). Lower resolution in a sample is indicated by a greater degree of polydispersity, in other words, more species. Higher resolution results in monodisperse fractions. Fractions consisting of 0.06 ml were collected by this author and injected into another gel filtration column. An increase in peak width obtained on the analytic column indicated that less resolution was obtained in the first (or test) column. Thus, a plot could be obtained for peak variance on the analytic column compared with the different variables (or characteristics) for the test column. Columns with lower variances for the same retention time are more highly resolving.

The author conducted an RTD test of a 150 ml, 75 cm column packed with Sepharose CL-4B. The number of plates was 3400. A similar RTD test conducted on a pilot-scale 80-liter chromatographic column of 75-cm bed height yielded a plate number of 2200. It was expected that a laboratory-scale 75-cm column of 2000 plate number would predict the performance of the 80-liter large-scale chromatographic column. The 2000 plate column was obtained at the laboratory scale by adding backmixing to the original 3400 plate column. Naveh (1990) noted that the performance of the large-scale chromatographic column was closely predicted by the 2000 plate laboratory column. This technique of characterizing the performance of the large scale column by the laboratory scale column demonstrates significant economic advantages. The separation and scale-up of more bioproducts should be attempted by this technique to further validate the down-scaling approach to scale-up.

Let us now examine the economics of bioseparation of different bioproducts by different techniques. This should place at least some of the analysis presented previously in proper perspective and provide an appropriate framework for comparison for different researchers and industrial personnel.

III. ECONOMICS OF BIOSEPARATION

Only a few examples such as those presented later are apparently available in the open literature. Because apparently only chromatographic separations have been applied successfully at the commercial level, there is considerable emphasis on chromatographic separations in the examples presented. This should not convey the incorrect impression that other bioseparation techniques are not of importance and may not be applied successfully at the commercial level. The analysis presented is based on the data available. Besides, Fulton *et al.* (1992) emphasized that large-scale chromatographic systems used in product recovery, purification, and polishing are of particular importance. Nevertheless, chromatographic separations have one major drawback, low flow rates. This is due to the restricted diffusive mass transport in porous media. Spalding (1991) indicated that perfusion chromatography is 1 to 10 times faster than conventional liquid chromatography, including high-pressure liquid chromatography (HPLC).

Perfusion chromatography uses flow through particles to overcome the mass-transport restrictions of conventional liquid chromatography (Afeyan *et al.*, 1989, 1990, 1991). Spalding (1991) emphasized that the key to perfusion chromatography is the structure of the particles. The particles contain two classes of pores. Thoroughpores of 6000 to 8000 Å size are big enough for convective flow through the particle. Smaller diffusive pores (with a significant adsorption surface area) line the interior of the throughpores where the diffusion path lengths are less than 1 μm. This combination permits a rapid transport of chromatographic sample molecules. This may be contrasted to conventional liquid chromatography where the sample molecules are transported to the exterior of the packing molecules by convection. Thereafter, the molecules must diffuse into the interior of the particles where the significant adsorption area lies. Molecular diffusion by its very nature is a very slow process. Chromatographic separations could benefit using perfusion chromatography where run times are of the order of minutes, compared with about an hour for HPLC (Spalding, 1991). This author also indicated that there is a decrease in the time of process development by an order of magnitude on using perfusion chromatography compared with that of conventional chromatography. Thus, there are quite a few benefits in time and money in utilizing perfusion chromatography.

Example 8.5

Provide economic data for the separation of tPA, monoclonal antibodies, and animal growth factors utilizing perfusion chromatography. Present three different strategies for operating chromatographic columns (Fulton *et al.*, 1992).

Solution

Three different strategies were presented for the separation of the preceding three bioproducts utilizing perfusion chromatography (Fulton *et al.*, 1992):

1. One-cycle one-batch is the conventional process design where the column is sized so that it has enough capacity for all the binding material in the batch. Use linear scale-up. Short cycle times could also be used in a cycling mode.
2. Offset cycling refers to all the cycles of one chromatographic stage that are completed and the material pooled before the cycles of the second stage are started.
3. Staggered cycling means the stages are run in parallel. The material purified in the first cycle of the first stage is immediately applied to the first cycle of the second stage. This is run concurrent with the second cycle of the first stage. The authors emphasized that cycling permits tremendous flexibility as far as operating the plant is concerned. Cycling also minimizes the risk of failure of a particular run.

In the cost comparison that follows care was taken to see that the operating loading capacity of the feed streams on the columns was the same for conventional and perfusive supports. The same relative volumes are required for washing, elution, etc. Also, the quantity-yield of the product is the same for all three processes. However, the quality of the product separated was not specified. The time required for processing is also set to be the same. The offset cycling required 10 cycles per batch. The staggered cycling used 30 cycles per batch. Table 8.4 compares the different cost aspects of separating monoclonal antibodies (medium scale hybridoma cell culture), tPA (large-scale mammalian cell culture), and animal growth factors (large-scale fermentations).

Note that the reduction in equipment-media costs using perfusion chromatography compared with that of conventional liquid chromatography is due to the smaller equipment required. Capital costs similarly decrease due to a

TABLE 8.4 A Comparison of the Estimated Costs for Separating Monoclonal Antibodies, Tissue Plasminogen Activator (tPA), and Animal Growth Hormones[a]

Item	Process	Monoclonal antibody	Tissue plasminogen activator	Animal growth hormone
Initial feed volume, liters		4000	10,000	40,000
Equipment cost 10^6\$	Conventional	1.6	1.9	4.8
	offset cycling	0.80	1.1	2.2
	staggered cycling	0.51	0.70	1.1
Total capital cost 10^6\$	Conventional	23	27	45
	offset cycling	22	26	39
	staggered cycling	22	25	38
Total operating cost 10^6\$	Conventional	14	17	23
	offset cycling	13	16	21
	staggered cycling	13	15	20

[a] From Fulton, S. P. *et al.* (1992). *Biotechnology, 10*, 635–639, with permission.

decrease in the floor space required. Because not many steps are required in perfusion chromatography compared with those of conventional chromatography, there is also a reduction in the operating costs. The cost comparison presented in Table 8.4 is of significant value because (1) it provides a cost comparison utilizing three different strategies for bioseparation, and (2) it provides data on three different bioproducts–pharmaceuticals of interest (Fulton *et al.*, 1992). More such data-cost comparisons are required that help facilitate the choice of the process for effective separation of these and other valuable pharmaceuticals. It would be excellent if some numbers on the quality of the product separated could be provided. This could also significantly influence the decision in the selection of the process. Considering the proprietary nature of these types of data and the expected fierce competition in this area, these data will be difficult to get from industrial sources. Nevertheless, if available, they would be of considerable value.

Example 8.6

Provide some reasons why other bioseparation techniques have not been applied on a commercial scale. Consider a particular case, for example, two-phase aqueous systems (Huddleston *et al.*, 1992).

Solution

The status of two-phase aqueous partition systems has been reviewed (Huddleston *et al.*, 1992). These authors indicated that two-phase aqueous systems have primarily been employed as a primary separation processing step. Even though two-phase aqueous systems have proved to be better economically than centrifugation and cross-flow filtration for the separation of about 1000 kg of biomass (Kroner *et al.*, 1984), Huddleston *et al.* (1992) emphasized that this technique is not ready as yet for application at the commercial level.

Much more detailed information is required concerning what drives these partition-types of systems, the molecular interactions between the protein surface and the two-phase aqueous system with emphasis at the interface, and a better understanding of the major physical and chemical reactions occurring in the system (Huddleston *et al.*, 1992). It is only with a better and more complete understanding of the different parameters involved will it be possible to attempt to scale up two-phase aqueous systems with some reasonable measure of success. Scale-up in itself is a difficult process. A lack of understanding of the major variables involved will significantly hinder and complicate any reasonable scale-up attempts. Besides, interfaces are known to cause the loss of structure or denaturation of proteins and other compounds. Thus, even if the required bioproducts may be separated, it is of tremendous interest to separate these bioproducts in an active form. A more complete understanding of the interactions involved at the interface and elsewhere in the system is essential to help minimize this denaturation or conformational changes that lead to a deleterious bioproduct.

Antibodies that have traditional therapeutic values, are finding increasing use in affinity separations, and have lately found considerable application in biosensor applications (Nygren and Stenberg 1985; Stenberg and Nygren, 1982; Stenberg *et al.*, 1986; Sadana and Sii, 1992a,b). Not only is there bound to be an increasing demand for monoclonal antibodies but also the purity levels of these separated antibodies have to be very high. This is the nature of their applications. This is especially the case when they are to be used for therapeutic applications as they must meet ever-increasingly stringent FDA requirements. The ever-increasing demand for very high purity levels of antibody production places a considerable strain on the bioseparation process that needs to be utilized to separate these antibodies from a fermentation medium.

Example 8.7

Provide some economic data on a technique that effectively separates relatively large amounts of monoclonal antibodies (Duffy *et al.*, 1989).

Solution

The economic costs of separating therapeutic monoclonal antibodies using ion-exchange chromatography and protein A chromatography were compared (Duffy *et al.*, 1989). For ion-exchange chromatography these authors used a 35-liter industrial column. The column had a cross-sectional area of approximately 1500 cm^2, and S-Sepharose packing was used. Recombinant protein A was coupled to CNBr-activated Sepharose for the protein A chromatography separation.

Because feed material greater than 1000 liters was to be treated the authors utilized a "pre-concentration" step so that the chromatography step or steps to follow could be more economical. This preconcentration step needed to be "gentle" to minimize the denaturation of the antibodies. They selected an ultrafiltration system with a large membrane surface area. This ultrafiltration system permitted approximately a 50- to 100-fold concentration of the feed material in about 3 to 4 h. No information on the amount of antibody denaturation was given. After pretreatment by the ultrafiltration unit the load of the antibody to the column was close to 100 mg.

They noted the following advantages when comparing the separation of antibodies for therapeutic usage by ion-exchange chromatography and by protein A chromatography:

1. For 10 cycles of use the cost of separation by ion-exchange and by protein A chromatography was \$53 and 217 per gram, respectively. The cost by ion-exchange chromatography is about a quarter of the cost of removal by protein A chromatography.
2. Ion-exchange chromatography does not copurify other immunoglobulins as a result of the nature of separation (by inherent charge properties) of the ion exchange process. Protein A chromatography, however, is unsuccessful in removing some of the contaminating proteins.

Thus, not only is the ion-exchange chromatography technique cheaper than the protein A chromatography technique, but also it is relatively free of contaminating immunoglobulins. This is critical for therapeutic usage where the contaminating immunoglobulins may cause undesirable reactions. Could the possible presence of undesirable contaminants be one of the reasons why one generic drug manufactured by one company is cheaper than the regular drug? No comments were made by the authors on the level of purity of the antibodies separated. It may be presumed that the purity levels were presumably high enough for therapeutic use. Needed are details on the additional processing steps required, and more importantly the cost of these steps if required levels of purity are not initially obtained. It is reasonable to anticipate that obtaining that extra increment in purity level will be rather cost intensive if the pharmaceutical-antibody is already at high purity levels. There is apparently no such information available in the open literature. This sort of information is necessary to comply with the ever-increasing standards proposed by different government agencies on drug quality–purity.

Finally, one may reasonably ask, why then try to compare protein A chromatography with ion-exchange chromatography? Duffy *et al.* (1989) pointed out that ion-exchange chromatography is a very specific method, and for each application an ion-exchange process needs to be developed. Protein A chromatography, however, is a more generic method and can be applied to the separation of a wide variety of antibodies. Besides, as the authors indicated protein A chromatography is fast, requires a shorter process development time, and provides a higher purification than the ion exchange in a single step (Lindmark *et al.,* 1983).

The scale up of the different bioseparation processes is a vexing problem. This is particularly true for immunoadsorption where contaminants may be copurified. In that case Desai (1990) indicated that additional steps are required. The author further emphasizes that immunoadsorption has been successful, especially in the biomedical area. For example, immunoadsorption has successfully removed different substances from blood in an extracorporeal manner (Sato *et al.,* 1989; Somnier *et al.,* 1989). It would be of significant interest to scale up immunoadsorption columns for biomedical and other uses. Prior to being able to do this it is essential to analyze the cost structure of immunoadsorbent columns.

Example 8.8

Analyze briefly some of the major cost elements in designing immunosorbent columns on a large scale (Desai, 1990).

Solution

The separation of biochemicals for therapeutic applications by immunoadsorption has been reviewed (Desai, 1990). The author emphasized that a major cost involved is the preparation and stability of the immunoadsorbent. Compared with amino acid ligands antibodies are more specific, but they are more costly and less stable. Besides, the surface has to be activated; these activating agents are particularly expensive, especially for large-scale use. Also, during large-scale use the author indicated that bed compression is a major hindrance, because it decreases the rate of material processed. As mentioned earlier, Hoffmann LaRoche engineers also noted a similar problem during the scale-up of α-interferon A (van Brunt, 1985). Pharmacia (1986) utilized multiple-stacked columns to get around this problem. Also, Desai (1990) emphasized the importance of elution conditions. These, if not carefully chosen, may deleteriously affect not only the activity–stability of the bioproduct separated but also the useful operating life of the column.

In addition to the actual process problems, more information needs to be developed or made available concerning the kinetics of adsorption. Diffusional constraints (inevitable in these types of systems), heterogeneity of antibody-affinity ligand on the surface, and flow patterns will significantly affect the performance of immunoadsorbent columns. These parameters will significantly influence the nature of the adsorption process and the eventual structure-activity–stability of the final bioproduct separated. More emphasis needs to be placed on analyzing the influence of flow patterns, diffusional constraints, and heterogeneity of ligand-antibody attachment on the adsorption–desorption process in immunoadsorption systems. A systematic analysis will provide much needed physical insights into the better control of and into the activity–stability of the bioproduct separated.

A simple cost function has been proposed for the affinity purification of protein C (Kang *et al.*, 1992). The cost function accounts for the cost of the activated column per unit volume, labor, maintenance, and utilities, and also includes depreciation. These authors actually set up a profit function. Investment costs (such as for research) were not included. They emphasized that the separation of bioproducts from viscous materials (like blood plasma) is particularly difficult because it places stresses on the gel matrices that result in their significant deformation. This hinders the flow rate and the eventual productivity. Novel designs are required, like that proposed by Pharmacia (1985), to either eliminate or minimize these deformable stresses. The development of nondeformable (stronger) gel matrices would also be of assistance.

At this stage it appears that each immunoadsorbent application has its own set of problems associated with it that hinders the effective scale-up and successful commercial application of that particular process. It would be of considerable assistance to be able to come up with some general guidelines or principles of wider applicability that would assist in the commercial application of immunoadsorbent systems.

Data or numbers to help compare or to note the costs of producing-processing the different drugs, pharmaceuticals, enzymes, proteins, and bioproducts are apparently not easy to come by in the open literature. Some of the numbers available, along with appropriate caveats, are presented in Table 8.3. Some of the drugs are rather expensive. It is hoped that as downstream processing improves the price of some of these life-saving drugs will tend to go down. Market forces as well as severe competition should also assist in driving these forces down. More entries in a table like this would be instructive because it provides an overall view of the processing–production costs of these bioproducts for market consumption. It would, of course, be more instructive to separate the downstream processing costs from the other costs. This could then serve as an appropriate framework for comparison for the different companies who are either in the process of getting a drug-pharmaceutical-bioproduct to the market; or even those who have a market share and want to either hold onto their share or attain a bigger portion of it.

Modeling procedures, if available, would help estimate the costs of bioseparation of the different bioproducts. Granted that the model would be rather specific for presumably a certain type of process, nevertheless, such an analysis would provide significant physical insights into the cost structure of different processes, besides highlighting those variables that are cost sensitive. Industrial workers and others already have a feeling for this. The modeling procedure would place all this on a more quantitative and logical basis. Needless to say, a general model or a model having a wide applicability would be of considerable use. The basic components of a general model should, of course, include the major cost items. These cost items may be premultiplied by appropriate coefficients or constants to take care of the variables involved in a particular process, or even in examining or analyzing different bioseparation processes. This is the general approach taken by Datar (1986). We will now examine three examples. The first example provides some economics on sizing chromatographic columns. In the next two examples some modeling is provided to estimate the separation of proteins-pharmaceuticals, etc.

Example 8.9

Briefly present the costs involved in running chromatographic separations on a large scale (Peskin and Rudge, 1992).

Solution

Because chromatographic separations play a very dominant role in industrial-scale bioseparations, Peskin and Rudge (1992) analyzed the cost structure of the different components involved. These authors concluded that the major costs involved include the operating cost (solvent cost), the capital costs (columns, pumps, etc.), and the column packing costs. They proposed empirical equations for the cost of the resin (based on the diameter of the particle), and the cost of the column (based on the capacity or volume of the column). From a cost analysis of the system these authors arrived at the important result that for scale-up of chromatographic columns a particle size range of 20 to 40 μm is apparently the most economical. This is true for resin particles. For example, these authors noted that the cost of the resins was about 99% of the total cost of the column for particle sizes less than 20 μm, and about 50% for particle sizes greater than 60 μm. This analysis is of considerable value because it presents a general result of presumably wide applicability. This result should be validated by other workers in this area for resin and other types of particles. It would be of interest to note if a similar result could be obtained for other types of particles.

Example 8.10

Describe briefly the qualitative features of the Porter-Ladsich model (Porter and Ladisch, 1992) for the cost estimation of separation of α-galactosidase from soybean seeds. In other words comment on the relative costs for each purification step.

Solution

The purification of α-galactosidase from soybean seeds was analyzed (Porter and Ladisch, 1992). The purification train involves the following steps: cyroprecipitation, acid precipitation, ammonium sulfate precipitation, and chromatographic steps. For the chromatographic steps the authors utilized: (1) ion-exchange chromatography followed by affinity chromatography, and (2) two affinity chromatography steps in sequence. These authors presented detailed cost functions for each purification step. They were also careful enough to provide realistic and current numbers for their economic comparisons in their model for estimating the costs of separation by the two different processes.

At relatively low levels of production 4 to 10×10^6 EU enzyme units (EU)/year labor costs clearly dominate the cost structure. They are more than an order of magnitude greater than the sum of all other costs. As the throughput increases, chromatography stationary phase costs increase and gradually become a significant fraction of the labor costs. At a production rate of about 1000×10^6 EU/year the stationary phase costs are about the same as the labor costs. These two costs apparently dominate the cost estimates of the chromatographic separation process. Because the preceding two costs dominate the estimates for separating α-galactosidase from soybean seeds, the authors correctly recommended finding ways by which (1) to decrease–minimize the stationary phase costs, and (2) to introduce automation-process control in these systems so that labor may be more effectively utilized.

Stationary phase costs could be decreased by developing cheaper materials that are as effective. This is bound to happen with the significant research that is being undertaken in this area. Also, optimization schemes could be set up that either: (1) permit the operation of the column for longer intervals of time at the same separation level (considering both quality and quantity of the product separated), or (2) increase the rate of processing of feed for the same longevity of the column and the quality of the product separated. Regeneration of the column may also be attempted. One would then need to optimize both the actual operation cycle as well as the regeneration cycle as done for catalyst fouling in fixed-bed reactors (Levenspiel, 1972).

The Porter-Ladisch (1992) model highlights the importance of labor costs in the chromatographic separation of α-galactosidase from soybean seeds. It would be of interest to note whether labor costs dominate the separation costs in other bioseparation processes or are just significant when chromatography is used as the separation process. We next present an economic analysis of the primary separation steps in recovering useful bioproducts from a fermentation effluent stream (Datar, 1986). Recognize that one of the major requirements of the primary separation steps in the bioseparation train is the significant reduction in volume of the fermentation effluent without significant loss or denaturation of the bioproduct.

Example 8.11

Present briefly the economics of separation of bioproducts for an *E. coli* based fermentation process (Datar, 1986).

Solution

The economic costs during the primary separation steps in the recovery of α-galactosidase from *E. coli* fermentations have been analyzed (Datar, 1986). The author wanted to use this as a model process to develop certain general principles, guidelines, and general methodology. The intent was that the general model could be applied to other fermentations, with the general framework being the same. This could be accomplished by including the specifics or the different characteristics of the other processes into the general model. This would then assist in providing overall cost estimates of different aspects of the varied bioseparation processes.

Bridgewater (1973) had initially indicated that the following major costs need to be considered in an economic analysis: raw materials (*RM*), fixed capital investment (*FCI*), labor costs (*LC*), and utilities (*U*). From this one could obtain the equation for operating cost

$$\text{Operating cost} = 1.2\ RM + 0.17\ FCI + 2.54\ LC + 1.2\ U. \qquad (8.5)$$

Datar (1986) emphasized that the preceding coefficients are mean values, and that they do not: (1) provide the flexibility for adjustment for the different processes; (2) account for adjustments, modifications, and improvements for a particular process; and (3) consider varying costs for raw materials, utilities, and unusual occurrences during the operation of a particular process.

Datar (1986) modified Eq. (8.5) to account for these factors, and provided the equation

$$\text{Operating cost} = \frac{\alpha RM + \beta FCI + \gamma LC + \delta U}{X}. \qquad (8.6)$$

Here X represents a step or overall yield. Also, α, β, γ, and δ are coefficients that may change as: (1) one goes from one process to another; and (2) variations, modifications, and improvements are made within a process itself. α, γ, and δ may even be increasing functions of time to account for the increasing Marshall price index (Porter and Ladisch, 1992). This index (annual) was 895.1 in 1989; 915.1 in 1990; 930.6 in 1991; 943.1 in 1992; 964.2 in 1993; 993.4 in 1994; and 1016.6, 1029, and 1031.7 in the first, second, and third quarters, of 1995, respectively (Chemical Engineering, 1995).

On providing realistic cost estimates for the different components (such as cell harvesting, high-speed centrifugation, and flow microfiltration) in the primary separation steps, Datar (1986) was able to break down the costs in increasing order as follows: raw materials (20%), capital related charges (28%), and labor (36%).

The economic analysis presented by this author is of considerable value, because it correctly breaks down the cost structure of the different components during the primary separation stages. Modifications in the sequence or selection of the primary separation process is, of course, critically dependent on the drug, protein, or bioproduct as it passes through the primary, secondary, and polishing steps. The major concerns as Datar (1986) correctly pointed out are the yield, purity, stability of the bioproduct; and ease of scale up. These factors may considerably affect and influence the choice and the sequence of the bioseparation stages involved in the entire bioseparation process. This includes the primary, secondary, and polishing steps that eventually lead to the final product.

Economic analysis utilizing models for the primary, secondary, and polishing steps treated earlier are of considerable value. What would be invaluable, of course, is the development of a model and an economic sensitivity analysis of all stages involved in the bioseparation process. This would then help determine the cost-intensive items. A sensitivity analysis would help determine the major and minor interactions as each component is changed and modified.

Once such a model is developed it could be optimized utilizing an appropriate objective function (the most obvious one being profit). Cost factors are required for enhancing the quality and the quantity of a finished product. These types of numbers may be available to industrial sources, but are apparently scarcely available in the open literature. The analysis presented by Datar (1986), and even the previous analyses by Porter and Ladisch (1992) and by Peskin and Rudge (1992) are steps in the right direction. These types of analyses should be vigorously pursued in the future.

Example 8.12

Present briefly the process design and economics for the production of polygalaturonases from *Kluyveromyces marxianus* (Harsha *et al.*, 1993).

Solution

A one-step purification scheme for the recovery of polygalacturonases from *K. marxianus* is available (Harsha *et al.*, 1993). The fermentation broth was sent directly to a ion-exchange column run at pH of 4.5. A recovery of 90% of the highly purified enzyme was obtained. These authors emphasized that current commercial processes involve centrifugation, precipitation, membrane and gel filtration, ion-exchange and affinity chromatography, or dialysis (Barnby *et al.*, 1990). Such a process is difficult to optimize where maximum protein recovery and purification are required along with minimum capital and operating cost. A simpler scheme as proposed by Harsha *et al.* (1993) is more amenable to optimization.

Figure 8.1 shows the simplified pectinase (polygalacturonase) separation

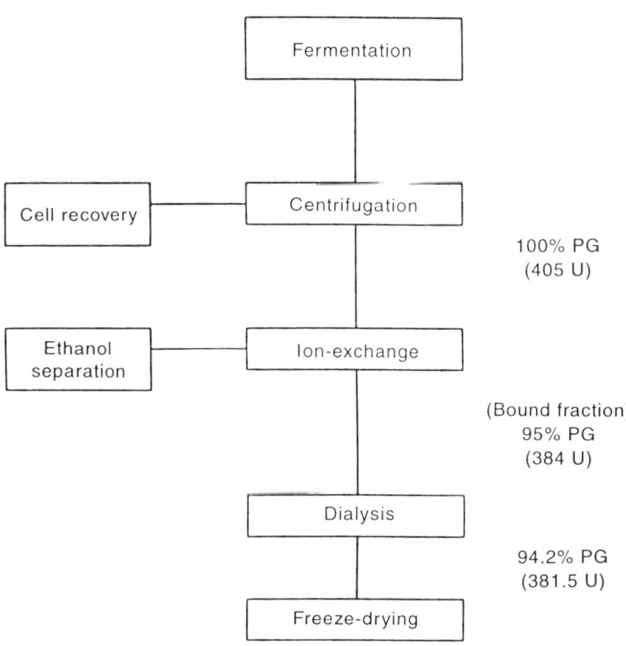

FIGURE 8.1 Simplified bioseparation scheme for pectinase (polygalacturonase). [From Harsha, S. *et al.* (1993). *Process Biochem., 28,* 187–195, with permission.]

scheme. Note that the specific activity is 0.2, 71.5, and 72.6 U/mg in the culture fluid after centrifugation, purification on a CM-Sephadex column, and dialysis, respectively. The enzyme yields (in % w/w) are given in Fig. 8.1. The authors indicated that the centrifugation resulted in a 2% loss in enzyme activity. The ion-exchange separation was conducted at optimum pH (4.5) and temperature (25°C). They emphasized that their simplified process gave enzyme yields over 93% along with a sixfold increase in enzyme concentration. This compares favorably with 55 to 91% yields typically obtained from more expensive processes.

An equation was proposed for the total capital investment and operating cost (Harsha *et al.*, 1993). They based their costs on the production scale (in other words, the fermentation volume, f_v, m^3 per batch). These authors noted that the downstream process costs are almost 60 to 70% of the total investment. Based on 1991 prices, they estimated the following capital investment costs (in \$):

Fermentation system (including cell recovery)	$3.57 \times 10^4 (f_v)^{0.6}$
Ethanol separation	$2.61 \times 10^4 (f_v)^{0.6}$
Total enzyme recovery system	$3.39 \times 10^4 (f_v)^{0.6}$

The estimates for annual operating costs in \$/year follow:

Nutrients	$12{,}000\, f_v$
Gel	$0.0128\, f_v$
Utilities	$754\, f_v$
Maintenance and overheads	(0.12) Investment
Labor	$900{,}000\, (f_v/1{,}000)^{0.25}$
Depreciation	(0.15) Investment

The authors noted that at low scales (e.g., 1 m^3 per batch), the fixed costs are about 95% of the total cost. At a higher scale (e.g., 100 m^3 per batch) the fixed costs are only about 60% of the total cost.

The main advantage of the separation scheme using ion exchange is its simplicity and low cost (Harsha *et al.*, 1993). A comparision of the production costs of their process with current processes is very favorable at scales greater than 10 m^3 per batch. These authors emphasized that at much higher scales (>100 m^3 per batch), the cost of equipment rises, and the cost per kilounit of pectinases decreases very slowly, and there are few economies of scale.

Other factors such as market constraints may also limit the utilization of the process at higher scales (>100 m^3 per batch). The economic analysis by these authors is of interest because it delineates the different aspects that are involved in the economies of scale for a process that includes the fermentation and the purification steps. More such studies are required for other enzymes and other biological products of interest.

IV. CONCLUSIONS

Economic data on bioseparation are scarcely available in the open literature. The analysis presented provides an overall perspective of the economics of bioseparation processes. The few examples presented, especially those where a little modeling is involved, provide some insights into the cost structure and the sensitivity of the economics on a particular component of the entire bioseparation process. The scarcity of the data available considering the high profits and other reasons is understandable. Nevertheless, this is a serious shortcoming and requires urgent attention. More modeling needs to be done on the different components of primary, secondary, and polishing steps taken together. This should facilitate in the optimization of the entire process based on an appropriate objective function. In the long run, the economics will eventually determine the selection as well as the sequence of stages involved in the bioseparation processes.

Hardly any detailed information is available in the literature concerning the effect of selection or sequence of the bioseparation stages on the quantity and quality of the bioproduct produced. Information on the quality of the bioproduct separated is particularly lacking. This sort of information is urgently required to meet not only the market demands but also the ever increasing governmental standards on the bioproducts produced by recombinant and other biotechnological methods. It is hoped that this analysis would highlight this problem in the literature, and also focus on the need to allocate resources to assist in addressing this situation. The availability of an appropriate framework for the economic analysis of a wide variety of bioseparation processes would be of considerable value to the different individuals involved in successfully bringing a bioproduct for market consumption.

REFERENCES

Afeyan, N. B., Fulton, S. P., Gordon, N. F., Mazsaroff, I., Varady, L., and Regnier, F. E. (1989). Perfusion Chromatography: Approach to Purifying Biomolecules, *Biotechnology*, *8*, 203–206.

Afeyan, N. B., Gordon, N. F., Mazsaroff, I., Varady, L., Fulton, S. P., Yang, Y. B., and Regnier, F. E. (1990). Flow-through Particles for the High-Performance Liquid Chromatographic Separation of Biomolecules, *J. Chromatogr.*, *519*, 1–29.

Afeyan, N. B., Fulton, S. P., and Regnier, F. E. (1991). Perfusion Chromatography Packing Materials for Proteins and Peptides, *J. Chromatogr.*, *544*, 267–279.

Atkinson, B. and Mavituna, F. (1983). In *Biochemical Engineering and Biotechnology Handbook*, Macmillan: London, p 890.

Atkinson, B. and Sainter, P. (1980). *DSP: Final Forecast Report*, EEC Fast Project No. FST/C/020/80/UK/H.

Barnby, F. M., Morpetti, F. F., and Pyle, D. L. (1990). *Enzyme Microb. Technol.*, *12*, 891.

Bienz-Tadmor, B. and Brown, J. S. (1994). Biopharmaceuticals and Conventional Drugs: Comparing Development Times, *Biopharm*, *7*(2), 44–49.

Bonnerjea, J., Oh, S., Hoare, M., and Dunnill, P. (1986). Protein Purification: The Right Step at the Right Time, *Biotechnology*, *4*, 954–959.

Bridgewater, A. V. (1973). The Build-Up of Costs, *Chem. Eng.*, *279*, 538-544.

Burrill, G. S. and Roberts, W. J. (1992). Biotechnology and Economic Development: The Winning Formula, *Biotechnology*, *10*, 647–653.

Business Week, (1993). A Star Is Born. How DNase Was Hurled into Combat against Cystic Fibrosis. Science and Technology Section, August 23, pp 66–67.

Chase, H. A. (1984). Affinity Separations Utilizing Immobilized Monoclonal Antibodies. A New Tool for the Biochemical Engineer, *Chem. Eng. Sci.*, *39*, 1099–1125.

Chem. Eng. (1995). *Economic Indicators*, *102*(10), 176.

Datar, R. (1986). Economics of Primary Separation Steps in Relation to Fermentation and Genetic Engineering, *Process Biochem., 21*(1) 19–26.

Datar, R. V., Cartwright, T., and Rosen, C. G. (1993). Process Economics of Animal Cell and Bacterial Fermentations: A Case Study-Analysis of Tissue Plasminogen Activator, *Biotechnology, 11,* 349–357.

Desai, M. A. (1990). *J. Chem. Technol. Biotechnol., 48,* 105.

Duffy, S. A., Moellering, B. J., Prior, G. M., Doyle, K. R., and Prior, C. P. (1989). Recovery of Therapeutic Grade Antibodies: Protein A and Ion-Exchange Chromatography, *Pharm. Technol. Int., 1*(3) 46–52.

Egan, III J. J., Cronan, R. T., and Johnson, III, J. L. (1995). *Biotechnology, 13,* 559–560.

Eveleigh, J. W. and Levy, D. E. (1977). Immunochemical Characteristics and Preparative Immunosorbent Separations, *J. Solid Phase Biochem., 2,* 45–51.

Feinstein, P. (1995). In Thayer, A. M. (1995). *Chem. & Eng. News,* June 5, p 17.

Fish, N. M. and Lilly, M. D. (1984). The Interactions between Fermentation and Protein Recovery. *Biotechnology, 2*(2), 623–628.

Fulton, S. P., Shahidi, A. J., Gordon, N. F., and Afeyan, N. B. C. (1992). Large-Scale Processing and Throughput Perfusion Chromatography, *Biotechnology, 10,* 635–639.

Harsha, S., Zaror, C. A., and Pyle, D. L. (1993). Production of Polygalacturonases from *Kluyveromyces marxianus* Fermentation: Preliminary Process Design and Economics, *Process Biochem., 28,* 187–195.

Hassler, S. (1994). Biotech Goes on Trial, *Biotechnology, 12,* 551–552.

Hassler, S. (1995). Managed Innovation (Editorial), *Biotechnology, 13,* 529.

Hedman, P. (1984). Interfacing Fermentation with Protein Recovery, *Am. Biotechnol. Lab., 1*(7), 29–34.

Howell, J. A. (1985). Downstream Processing, *Pro Bio Tech, Process Biochem. Suppl.,* February, iv–vii.

Huddleston, J., Veide, A., Kohler, K., Flanagan, J., Enfors, S. O., and Lyddiatt, A. (1992). *Trends in Biochem. Sci., 9,* 381.

Kang, K., Ryu, D., Drohan, W. N., and Orthner, C. L. (1992). *Biotechnol. Bioeng., 39,* 1086.

Kelley, T. T., Wang, T. G., and Wang, H. Y. (1986). *Large-Scale Gel Chromatography,* in ACS Symposium Series, 314, Asenjo, J. A. and Hong, J., Eds., American Chemical Society: Washington, DC, pp 193–207.

Kroner, K. H., Hustedt, H., and Kula, M. R. (1984). *Process Biochem., 19,* 170.

Lambert, K. J. (1989). Regulatory Aspects of the Use of Immunoaffinity Reagents, *J. Chem. Technol. Biotechnol., 45,* 45–47.

Lavanchy, A. C. (1979). Centrifugal Separation, In *Kirk-Othmer Encyclopedia of Chemical Technology, 5,* 3rd ed., J. Wiley & Sons: New York, pp 194–233.

Levenspiel, O. (1972). *Chemical Reaction Engineering,* John Wiley & Sons: New York.

Lin, J. N., Herron, J., and Andrade, J. D. (1988). Characterization of Immobilized Antibodies on Silica Surfaces, *IEEE Trans. of Biomed. Eng., 35*(6), 466–471.

Lindmark, R. M., Thoren-Tolling, K., and Sjoequist, J. J. (1983). Binding of Immunoglobulins to Protein A and Immunoglobulin Levels in Mammalian Serums, *Immunol. Methods, 62,* 1–8.

Mahar, J. T. (1993). Scale-Up and Validation of Sedimentation Centrifuges, Part 1: Scale-Up, *Pharm. Technol., 17,* September, 84–96.

McChesney, J. (1993). Costs for Pharmaceutical Processes, Lecture, Chemical Engineering Department, University of Mississippi, October.

Mizrahi, A. (1986) Production of Biologicals from Animal Cells—An Overview, *Process Biochem., 21*(4), 108–112.

Naveh, D. (1985). Scale-Up of Fermentation for Recombinant DNA Products, *Food Technol., 39*(10), 102–109.

Naveh, D. (1990). Industrial-Scale Downstream Processing of Biotechnology Products, *BioPharm, 3,* 28–33.

Nygren, H. A. and Stenberg, M. (1985). Kinetics of Antibody-Binding to Surface-Immobilized Antigen: Influence of Mass Transport on the Enzyme-Linked Immunosorbent Assay (ELISA), *J. Colloid Interface Sci., 107,* 560–566.

Patel, P. (1993). Personal Communication, Doctor of Medicine, Medical Clinic, Coldwater, MS.

Paul, J. K. (1981). *Genetic Engineering Applications for Industry,* Noyes Data Corporation: NJ.

Peskin, A. P. and Rudge, S. R. (1992). Optimization of Large-Scale Chromatography for Biotechnological Applications, *Applied Biochem. & Biotechnol., 34/35*, 49–59.

Pharmacia (1986). Scale-Up to Process Chromatography, *Sep. News, 13*(6).

Porter, J. E. and Ladisch, M. R. (1992). Ion Exchange and Affinity Chromatography Costs in α-Galactosidase Purification, *Biotechnol. Bioeng., 39*, 717–724.

Raab, G. K. (1992). A Natural Selection, Cover Story, *Chief Executive*, May, 35–39.

Ronsohoff, T. C., Murphy, M. K., and Levine, H. L. (1990). Automation of Biopharmaceutical Purification Processes, *Biopharm, 3*(3), 20–26.

Rosen, C. G. and Datar, R. (1983). Primary Separation Steps in Fermentation Processes, *Proceedings in Biotechnology*, Online Conference: Pinner, Middlesex, UK, p 201–209.

Rubinfeld, J. (1995). *Biotechnology, 13*, 934–936.

Sadana, A. and Sii, D. (1992a). The Binding of Antigen by Immobilized Antibody: Influence of a Variable Rate Coefficient on External Diffusion Limited Kinetics, *J. Colloid Interface Sci., 151*(1), 166–177.

Sadana, A. and Sii, D. (1992b). Binding Kinetics of Antigen by Immobilized Antibody: Influence of Reaction Order and External Diffusional Limitations, *Biosens. Bioelectron., 7*, 559-568.

Sato, H., Watanabe, K., Azuma, J., Kidaka, T., and Hori, M. (1989). *J. Immunol. Methods, 118*, 161.

Somnier, F. E. and Landvard, E. (1989). *J. Neuroimmunology, 22*, 123.

Spalding, B. J. (1991). Downstream Processing: Key to Slashing Production Costs 100 Fold, *Biotechnology, 9*, 229–233.

Stenberg, M. and Nygren, H. A. (1982). A Receptor-Ligand Reaction Studied by a Novel Analytical Tool—the Isoscope Ellipsometer, *Anal. Biochem., 127*, 183–192.

Stenberg, M., Stiblert, L., and Nygren, H. A. (1986). External Diffusion Solid-Phase Immunoassay, *J. Theor. Biol., 120*, 129–142.

Stone, R. (1995). Science Scope, *Science, 267*, March 3, 1255.

Svarovsky, L. (1977). Separation by Centrifugal Sedimentation, In *Solid–Liquid Separation*, Ed., Butterworths: Stoneham, MA, pp 125–147.

Thayer, A. M. (1995). *Chem. & Eng. News*, June 5, p 17.

Tiller, F. M. (1974). How to Select Solid–Liquid Separation Equipment, *Chem. Eng., 81*(4), 117–136.

van Brunt, J. (1985). Scale-Up: The Next Hurdle, *Biotechnology, 3*(5), 419–424.

Velander, W. H. (1992). The Use of Fab-Masking Antigens to Enhance the Activity of Immobilized Antibodies, *Biotechnol. Bioeng. 39*(10), 1013–1023.

Weber, J. and Carey, J. (1993). Strong Medicine. The FDA Steps Up a Drug-Industry Crackdown, Even Ordering Some Production Halted, *Business Week*, September 6, 20–21.

9 PROTEIN REFOLDING AND INACTIVATION DURING BIOSEPARATION

I. INTRODUCTION

Bioprocessing of proteins and other biological products of interest requires a careful selection of delicate conditions and sequences of steps so that one may maximize the activity and stability of the product separated. Often the biological product of interest is present in a dilute mixture of similar substances. Besides, one is constrained from using harsh conditions. These and other factors that may apply to a specific process under consideration often tax the ingenuity of chemical engineers, biochemists, process development engineers, and others that are involved in separating biological products of interest. A persistent dogma has existed over the years (and even in this book where one has tried to minimize the inactivation of proteins during bioseparation), which is that once the protein is denatured it is difficult for the protein to regain its activity. This has existed in spite of the work by Anfinsen and coworkers (Epstein *et al.*, 1962; Haber and Anfinsen, 1962) that the information for the protein to adopt its native structure is completely encoded in its primary sequence. The polypeptide chain regains its native and active structure without any assistance from extrinsic factors or input of energy (Anson, 1945; Anfinsen, 1973). Anfinsen and coworkers conclusively demonstrated that by slowly and carefully removing the denaturant the protein could be made to refold to its native and active state. In some cases, for example when irreversible chemical modification of the polypeptide chain has taken place, the protein is said to be irreversibly denatured.

Knuth and Burgess (1987) emphasized that if one can free oneself from the bias of not purposely unfolding a protein during its bioseparation, then one opens powerful avenues to help separate proteins and other biological products of interest (Knuth and Burgess, 1987). Experimenters have allowed proteins and other biological products of interest to be unfolded or become inactive during their bioseparation. Thereafter, the denaturant is slowly and carefully removed under controlled conditions, and the protein is permitted to regain its native state. This is a relatively new and rapidly developing concept that has gained significant importance in recent years, with quite a few groups working in this area.

In principle, the removal of the denaturant should lead to the folding of the protein to its native and active state. However, this does not occur even under controlled conditions, unless one is very careful and adopts the right sequence or protocol. For example, Buchner *et al.* (1991) emphasized that due to improper polypeptide chain interactions or improper disulfide bridge formation, proteins may fold in an incorrect fashion. Nonnative and nonfunctional species may also be formed. These aberrantly formed molecules must than be unfolded and subsequently refolded to the proper form. In essence, the active state is kinetically limited from attaining its native state due to competition with aggregation (Mitraki and King, 1989; Jaenicke, 1987). Marston (1986) emphasized that a large number of proteins are often folded incorrectly when proteins from eukaryotic genes are expressed in bacterial systems. Mendoza *et al.* (1991) emphasized that competing processes may kinetically and chemically trap partially folded protein intermediates in nonnative conformers.

Hagen *et al.* (1990a,b) along with others stated that recombinant DNA technology has presented a proven and viable means by which proteins may be expressed to significant levels. However, often these heterologous proteins are present in an inactive and insoluble state in inclusion bodies (IBs). IBs are intracellular aggregates or refractile bodies whose mechanism of formation is not clearly understood, in spite of the fact that their existence has been known for some time (Rinas *et al.*, 1992). It is of importance to clearly delineate the mechanisms of formation of these IBs. Kane and Hartley (1988) indicated that these IBs consist of densely packed protein molecules that have partial secondary structure. Hagen *et al.* (1990) emphasized that this IB formation severely hinders the activity and stability of the proteins separated, and prevents or limits the success of commercial-scale processes developed using recombinant techniques. The protein molecules must be extracted from these IBs, solubilized, and then renatured under appropriate conditions. The extraction of proteins will be made less difficult if the formation mechanisms of these IBs are better known. In essence, these proteins must be unfolded, and then refolded under contolled conditions.

It is of tremendous importance to be able to refold proteins produced by recombinant technology because they present a major hurdle in the cost-effectiveness of the entire process (Datar *et al.*, 1993). It is also of importance to be able to recover the correctly folded proteins in relatively high concentrations. For example, Weir and Sparks (1987) indicated that the highest recovery of recombinant interleukin-2 is only possible at a concentration of 0.007 mg/ml. This represents a recovery of about 30%. It would be of interest to be able to recover correctly folded forms of the proteins at higher concentrations.

Zardeneta and Horowitz (1994) presented an overview of reactivation of rhodanese (EC 2.8.1.1) by interacting with detergent micelles; liposomes; and chaperonins, GroEL and GroES. Means to control the kinetic limitations in the folding reaction due to aggregation because of hydrophobic surfaces are presented. The control is assisted by the masking of the hydrophobic surfaces. Pain (1994) edited a book in which he describes the molecular mechansims that are responsible to help determine the three-dimensional structure of proteins and their function. He emphasized both the fundamental and (". . . one of the most intriguing intellectual challenges in molecular biology . . .") and the practical (rescuing inactive proteins or aggregates) aspects of protein folding. In this chapter we hope to highlight the significant breakthroughs that are occurring; and to provide a framework and physical insights into the refolding process, especially as it applies to bioprocessing.

Some nonproteinaceous additives have been traditionally used to assist in the folding process. It would be of interest to mention some of these additives.

Example 9.1

Briefly mention some of the nonproteinaceous materials or additives that have been utilized to assist in protein refolding (Zardeneta and Horowitz, 1994).

Solution

Zardeneta and Horowitz (1994) indicated that the following nonproteinaceous additives have been utilized to assist in protein refolding. These include low concentrations of denaturants such as urea (Horowitz and Butler, 1993; Mendoza *et al.*, 1991), osmolytes such as glycerol (Maloney and Ambudkar, 1989) or polyethylene glycol (Cleland *et al.*, 1992), and amphiphilic peptides such as propeptides (Shinde and Inouye, 1993) or peptitergents (Schafmeister *et al.*, 1993).

For example, Schafmeister *et al.* (1993) indicated that integral membrane proteins are of interest in the field of structural biology. These authors suggested that the addition of homogeneous peptides as detergents (peptitergents) leads to a more homogeneous, well-ordered complex for crystallography. They added that when mixed with the peptide, 85% of bacteriorhodopsin and 60% of rhodopsin retained their structure and remained in solution over a peroid of 2 days. The peptitergents sequester the hydrophobic membrane-spanning region of these membrane proteins. They pack around the protein in a rigid, well-ordered, parallel α-helical arrangement. Furthermore, Schafmeister *et al.* (1993) added that these peptitergents can be tailor-made to solubilize particular membrane proteins. Their homogeneity and variation in their properties through sequence variation allow them to be effective small-molecule detergents for the solubilization and crystallization of integral membrane proteins.

II. DIFFERENT PURIFICATION PROTOCOLS FOR RECOVERING PROTEINS IN THE DENATURED STATE

The importance of utilizing techniques that involve the separation of proteins in the denatured state has been emphasized (Knuth and Burgess, 1987). At the first instance this appears contradictory in that we want the protein or other biological products of interest in the native and active state. However, if one can successfully renature the protein recovered in the denatured state, then one has access to powerful techniques for bioseparation that were hitherto inaccessible due to the mind-set that proteins should not be purposely denatured. The next example briefly presents the different techniques that have been utilized to separate proteins in the unfolded or inactive (partially or completely) state.

Example 9.2

Briefly present and analyze the different protein purification strategies (protocols) that have been utilized to separate proteins in the denatured state (Knuth and Burgess, 1987).

Solution

A comprehensive review of the different strategies that may be employed in purifying proteins in the denatured state is available (Knuth and Burgess, 1987). As indicated earlier rather powerful techniques become readily available if one allows the separation of proteins in the denatured state. These techniques are particularly useful when it is possible to gradually remove the denaturant in a predetermined fashion to renature the protein. Some of the denaturants utilized to denature proteins include temperature, chaotropes (agents that increase the disorder of the bulk water), organic solvents, charge effects (pH), ligands and substrates (agents that may bind and alter the conformation of the protein), etc. Some of the unfolding techniques that have been utilized to improve the bioseparation of biological products include the following techniques (Knuth and Burgess, 1987).

Case One. Sodium dodecyl sulfate–polyacrylamide gel electrophoresis (SDS-PAGE) is a method in which SDS is used to dissociate the oligomers to give each protein an equal negative charge density. The separation is obtained based on electrophoretic mobility that is proportional to the molecular weight. The authors indicated that SDS-PAGE has been utilized to effectively separate the σ-subunit of RNA polymerase (Hager and Burgess, 1980), urease (Shaik *et al.*, 1980), and other proteins and biological products of interest.

Case Two. Reversed-phase liquid chromatography (RPLC) is good for the separation of proteins and other biological macromolecules, if the proteins or other biological macromolecules separated can be renatured. The denaturation is a result of the prolonged contact with the hydrophobic packing, the eluant, or other factors. This is a high-resolution technique, and may be used effectively if the proteins or other biological macromolecules have been denatured during a previous step in the bioseparation process.

Case Three. Size-exclusion chromatography (SEC) is, in general, very effective when performed in denaturing solutions (Montelaro *et al.*, 1981). Aggregation effects and protein packing interactions in nondenaturing solutions will lead to skewed elution patterns. Knuth and Burgess (1987) provide a word of caution for SEC separation using SDS, in that SDS will alter the column properties. This restricts the use of a column using SDS to the separation of one particular protein or biological macromolecule.

Case Four. Ion-exchange chromatography (IEC), as indicated by Knuth and Burgess (1987), is often used with denaturants. The denaturants may be used to preserve the solubility of the proteins or of other biological macromolecules. Also, these denaturants will dissociate multimeric proteins. The single subunits of these multimeric proteins may then be separated after isolation (Welling *et al.*, 1983; Bloemendale and Groenewoud, 1981).

Some of the other techniques suggested by Knuth and Burgess (1987) included detergent extraction of proteins, denaturation to provide unusual and predictable properties to proteins, extraction into organic solvents, and unfolding and refolding of rDNA inclusion bodies. The authors also provided a summary of techniques to renature proteins. These include slow dialysis–dilution in physiological buffer; guandine, urea denaturation–renaturation; and renaturation of proteins containing disulfides.

They are realistic by indicating that it is reasonable to expect that in not all cases will the refolding–renaturation of the protein be possible. In some cases, alternate uses may be found, such as in subunit exchange chromatography. Nevertheless, we are in agreement with these authors in that one should free oneself from the bias of purifying proteins in the denatured state; especially because some hitherto unavailable powerful techniques may be utilized to advantage. Often it is just possible that due to some other processing step in the bioprocessing train the protein or other biological product of interest gets denatured. In that case, it may be suggested that the bioseparation of the protein should be considered in its denatured state prior to steps that would facilitate its refolding to the native and active form. In some cases the denaturation of the protein or other biological macromolecule may be a "blessing in disguise" in that powerful techniques may be utilized for the separation. Of course, in these cases the basic premise is that the protein can be renatured–refolded to its native and active form. Thus, we note the importance of: (1) the further development of refolding techniques for a wide variety of applications; and (2) the need for a better understanding of the parameters that influence the different stages in protein refolding especially as applied to obtaining the native structure vis-à-vis the inactive aggregative form.

It is quite possible that as the technique of purposely unfolding a protein prior to its bioseparation becomes popular and more researchers become involved in it, newer avenues may open up that exhibit the potential to increase the efficiency (with regard to the quantity and quality of the product separated) of the bioseparation process. So that one may obtain better insights into the refolding process, the next section analyzes the mechanisms that are involved in the refolding process.

III. *IN VITRO* FOLDING MECHANISMS OF PROTEINS

Protein folding is constrained both by kinetics and by thermodynamics (Jaenicke, 1991). The kinetic nature arises due to the vectorial nature of the protein synthesis, and the thermodynamic nature arises due to the necessity of energy minimization of the different states. The driving force of the three-dimensional protein structure formation is the minimization of the free energy of stabilization. The author commented on the hierarchical nature of the three-dimensional structure formation (Go, 1984; Jaenicke, 1991). In brief, short-ranged interactions lead to secondary structure elements. These secondary elements through the process of gradual combination and reshuffling lead to the formation of subdomains, domains, and subunit assemblies. Note the modular nature of the three-dimensional structure formation.

Jaenicke (1991) emphasized that even though the protein structure is encoded in the amino acid sequence, the mechanism of *in vivo* folding is unknown. This is not surprising because there is an intricate balance between the destabilizing and the stabilizing forces that lead to the three-dimensional active protein structure. Presently, one compromises by analyzing *in vitro* unfolding–folding of different reactions. Kuwajima (1989) indicates that in *in vitro* experiments small sections of secondary structures combine to give supersecondary structures. These supersecondary structures then combine (or collapse) to yield the native or active structure. Also, in general, *in vivo* folding reactions are in the time range of seconds and minutes; *in vitro* folding reactions are inevitably longer. For example, Jaenicke (1991) indicated that the half-life of the pyruvate dehydrogenase complex from *Bacillus stearothermophilus* is approximately 8 h. Also, the half-life of the reshuffling of the Fab fragment of immunoglobulin is approximately 15 h.

There is clear evidence that under incorrect (unbalanced) physiological conditions "wrong" conformers may be formed during *in vivo* folding, and these may interfere with structure formation (Jaenicke, 1991). Mendoza *et al.* (1991) and Mitraki and King (1989) emphasized that *in vitro* folding is often constrained, as indicated earlier, by competing processes (such as aggregation of polypeptide chains) that limit the native and active state. Pitsyn *et al.* (1989) indicated that aggregation may occur due to the association of hydrophobic surfaces that are exposed in folding intermediates. Efforts have been made to minimize these aggregative interactions. Because *in vivo* folding of proteins occurs successfully in a complex milieu of reactions, there were indications that *in vivo* proteins were directing or facilitating protein folding reactions. Ellis (1990) termed these proteins *chaperonins*. Mendoza *et al.* (1991) indicated that chaperonins facilitate the *in vitro* folding of monomeric mitochondrial rhodanese. They do this, in part, by interacting with partially folded intermediates. This minimizes the interactions of hydrophobic surfaces that lead to aggregation.

The folding reaction kinetics may be summarized by a multistep mechanistic scheme (Jaenicke, 1991). This is consistent with the merging of the individual subdomains and hierarchical structure components, and the consecutive folding process. These concepts are also consistent with the simple and elegant cardboard box model for protein unfolding–folding transition described by Goldenberg and Creighton (1985). Jaenicke (1991) has provided some mechanisms that have been utilized to describe unfolding–folding reactions. These mechanisms are presented next.

A. Unfolding and Folding Kinetics of γ-II-Crystallin from Calf Lens (a Two-Domain Protein)

These reactions can be described by the three-state model

$$N \rightleftarrows I \rightleftarrows U. \tag{9.1}$$

Here N is the native and active γ-II-crystallin state, I is the intermediate state, and U is the completely denatured state. Sharma *et al.* (1990) indicated that the preceding mechanism has been corroborated by fragment studies. More intermediate steps may be involved, but then their detection and the kinetic analysis tend to become burdensome. Besides, most series-type mechanisms involving more than a single intermediate can be effectively described by a two-step, three-state mechanistic model (Sadana and Henley, 1986).

B. Oligomeric Protein Association and Aggregation

Jaenicke (1991) indicated that the early-stage folding of oligomeric proteins is expected to be similar to the self-organization of single-chain proteins. The mechanistic sequence may be described by

$$nM \rightarrow nM' \rightarrow \left(\frac{n}{2}\right) D \rightarrow \left(\frac{n}{2}\right) D' \rightarrow \left(\frac{n}{4}\right) T \rightarrow \left(\frac{n}{4}\right) T' \rightarrow \cdots P_n. \tag{9.2}$$

Here n is the number of subunits. M, M', D, D', T, T' are monomers, dimers, and tetramers in different conformation states. P_n is the n-mer state. The author emphasized that the single arrows oversimplify the actual situation. As indicated earlier, present kinetic analysis methods are inadequate to address more detailed and complicated unfolding–folding mechanisms.

Jaenicke (1984) indicated that the reconstitution of oligomeric enzymes after denaturation is a useful model for the folding and association of these enzymes during biosynthesis. Hermann *et al.* (1985) have determined the kinetics of reassociation of tetrameric phosphoglyceromutase (EC 5.4.2.1) from yeast, after denaturation in guanidine hydrochloride. The mechanism could be described by

$$4M \underset{k_{-1}}{\overset{k_1}{\rightleftarrows}} 2D \xrightarrow{k_2} T, \tag{9.3}$$

where M, D, and T represent the monomer, dimer, and tetramer, respectively.

C. Inclusion Body (IB) Formation

Inclusion bodies are intracellular protein aggregates or refractile bodies. Haase-Pettingell and King (1988) proposed that IBs are formed from partially folded intermediates and not from the completely unfolded protein. Jaenicke (1991) indicated that in overexpressing strains of bacteria IB formation may result due to the same mechanism responsible for *in vitro* incorrect aggregate formation. There is a competition between the first-order (correct) folding reaction and the diffusion-controlled higher order aggregation reaction that may be described by

$$U \begin{matrix} \overset{k_1}{\nearrow} N \\ \underset{k_2}{\searrow} A \end{matrix} . \tag{9.4}$$

Here U, N, and A are partially folded intermediates, and the native and aggregated states, respectively. k_1 and k_2 are first-order folding and second-order aggregation rate constants, respectively. Heterogeneity of the initial enzyme (in this case, polypeptide chain state) plays a significant role in enzyme deactivation kinetics (Sadana, 1991). Presumably, heterogeneity of the initial polypeptide state will play a significant part in refolding kinetics. To the best of this author's knowledge no such analysis has been performed. It would be of interest to estimate the effects of heterogeneity on the selectivity of the preceding parallel reaction. Does heterogeneity increase or decrease the aggregative tendencies; and if so, by how much? One may reasonably presume that since heterogeneity tends to make reaction orders higher in enzyme deactivation kinetics (Sadana and Malhotra, 1987), the aggregative reaction will be preferred in refolding kinetics.

IB formation has gained technological significance due to the recovery of recombinant proteins (Mitraki and King, 1989; Rudolph, 1990). This is of tremendous interest because significant losses of protein activity can occur if the IB formation is not carefully investigated to help minimize such losses. Ideally, of course, one would like to eliminate the IB formation under appropriate reaction conditions. Thus, the more recent development of general strategies for downstream processing of recombinant proteins takes IB formation into account. It would be of significant scientific as well as of practical interest to be able to develop general principles for the minimization of IB formation, if they cannot be completely eliminated during downstream processing. This is one of the major thrusts in the increasing efforts to understand the kinetics and mechanisms of IB formation, and how (hopefully active and stable) proteins may be recovered from them effectively.

The fraction of native protein formed compared with the aggregate form decreases with increasing temperature (Jaenicke, 1991). The author further indicated that protein chain synthesis is consistent with the mechanism

$$\begin{array}{l} \text{Translation} \rightarrow [I^{\mathrm{pf}}] \rightleftharpoons [I^{\mathrm{pT}}] \rightleftharpoons [pT] \rightarrow \text{native} \\ \qquad\qquad\qquad\quad \downarrow \\ \qquad\qquad\qquad\ [I^{\mathrm{pf}*}] \\ \qquad\qquad\qquad\quad \downarrow \\ \qquad\qquad\quad\ \text{aggregate.} \end{array} \tag{9.5}$$

Here I^{pf} is a productive intermediate, $I^{\mathrm{pf}*}$ is an intermediate species prone to aggregation, and I^{pT} is the intermediate that associates to the protrimer (pT). The protrimer eventually leads to the native active state.

Example 9.3

Briefly describe the effects of mutations on the aggregation of proteins (Wetzel, 1994).

Solution

The influence of point mutations on the aggregation of proteins has been analyzed (Wetzel, 1994). This author emphasized that *in vitro* aggregation of proteins places limits on protein stability and refolding yields. *In vivo* aggregation of proteins not only leads to inclusion body formation in the bacterial production of proteins, but also leads to amyloid disease and similiar phenomena in animals. Furthermore, the molecular mechanisms of protein aggregation extend beyond biotechnological application into understanding the mechanisms of human disease. Figure 9.1 describes the simple mechanisms involved in protein folding including and excluding the intermediate state. Folding-related aggregation is shown in Figure 9.1(b) (Wetzel, 1992; Mitraki and King, 1989). In this case the intermediate depending on the protein and folding conditions may either convert irreversibly to the aggregate, or exist in equilibrium with the unfolded (U) or native (N) state.

Wetzel's (1994) model for off-pathway aggregation is shown in Fig. 9.2. One notes that a specific mutation may directly influence the ability of an unfolded chain to fold correctly to the native structure. The mutation may decrease the folding stability of the native state, thereby leading to concentration increases in the nonnative state. This state, if inclined toward aggregation, will lead to increased aggregation as its concentration increases.

The author emphasized that further detailed analysis of off-pathway aggregation of a protein is of significant importance. If the reasons of off-pathway aggregation, particularly how the amino acid sequence avoids these off-pathway aggregation processes can be determined, then significant improvements can be made in refolding yields. This would also assist in understanding the molecular mechanisms of human disease.

Example 9.4

Briefly describe the simulation of a folding pathway (Hinds and Levitt, 1995).

Solution

Hinds and Levitt (1995) analyzed protein folding pathways. These authors indicated that simulations of protein folding pathways effectively complement the experimental studies. For example, qualitative features of natural folding pathways can be simulated. Also, different features may be probed that would be inaccessible by present-day experimental techniques. Besides, experimental

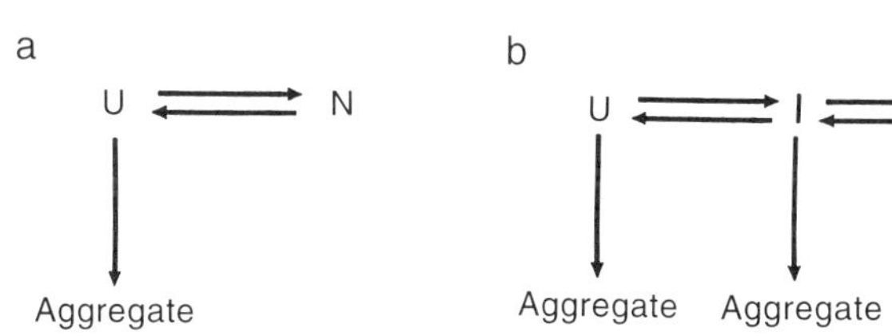

FIGURE 9.1 Protein refolding and aggregation models: (a) aggregation depends on the unfolded state; (b) aggregation depends on both the unfolded state and the intermediate state. [From Wetzel, R. (1994). *TIBTECH, 12,* 193.]

techniques outline the general features and overall framework. The simulations help provide the details that are difficult to study by experimental techniques.

Figure 9.3 shows the free-energy profile of the three distinct stages involved in protein folding. These stages involve all-or-none transitions (Matthews, 1993; Pitsyn, 1994). Basically, the unfolded protein goes to a "molten globule" state that has a compact structure (like the native state) and a relatively stable secondary structure, but a fluid tertiary structure. Hinds and Levitt (1994) emphasized that a large free-energy barrier (see Fig 9.3) separates the molten globule state from the native state. As expected, this is the rate determining step in the folding process.

These authors emphasized as expected, that there are quite a few variations in the preceding simplistic scheme of protein folding. Some proteins follow simple series or sequential steps (Serrano *et al.*, 1992). Other proteins fold by using complex series–parallel schemes (Itzhaki *et al.*, 1994; Radford *et al.*, 1992). Shortle (1993) indicated that the pathway followed may be determined by the heterogeneity or the residual of the unfolded protein.

Hinds and Levitt (1995) emphasized that simulations permit one to manipulate almost any variable or parameter. This level of control is, of course, not possible by experimental techniques. Ideally both modeling and experimental techniques are important in folding pathways. Sometimes the modeling may suggest appropriate experiments along some lines, and sometimes the experiments may guide the modeling developments or procedures.

In spite of the extensive research being carried out in this area, considerable amounts of inactive aggregates or precipitates are formed during the refolding reaction. Efforts continue to improve (minimize) or overcome the formation of these insoluble fractions (De Bernardez-Clark and Georgiou, 1991; Light, 1985; Schein, 1991; Zhu *et al.*, 1989). Nohara *et al.* (1994) suggested that the denaturation step in micrococcal nuclease (Nase) involves a reversible and an irreversible step. They proposed that the irreversible step leads to the aggregates.

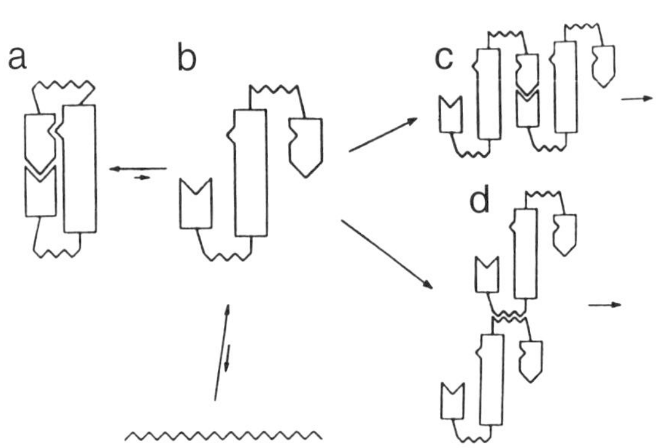

FIGURE 9.2 Mutation-influenced aggregation of a folding intermediate: (a) native state weakened due to mutations leads to increasing concentrations of the folding intermediate; (b) mutations may also affect the nativelike state (c) or nonnative interactions (d), and help stabilize the aggregate state. [From Wetzel, R. (1994). *TIBTECH, 12,* 193.]

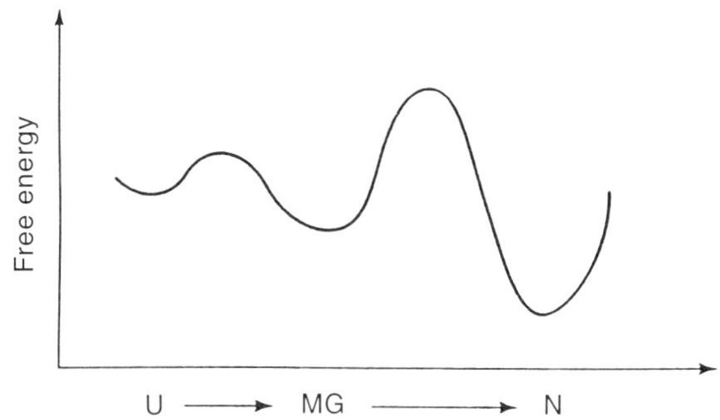

FIGURE 9.3 The three distinct stages and the free-energy profile for protein folding. [From Hinds, D. A. and Levitt, M. (1995). *TIBTECH, 13,* 23.]

Example 9.5

Describe the influence of the reversible and irreversible denaturation of Nase on aggregate formation (Nohara *et al.*, 1994).

Solution

Nohara *et al.* (1994) analyzed the denaturation of Nase with the specific intention of disciminating the reversible step from the irreversible step. Nase is a single-domain protein with a single-stranded polypeptide chain with neither disulfide bond nor cysteine residue (M_r 16,800). Their analysis permitted them to separate the irreversible denaturation of Nase from the reversible step.

Based on their results these authors proposed the mechanism of denaturation

$$X_2 \overset{k_d}{\leftarrow} N \overset{K}{\rightleftarrows} D \overset{k_d}{\rightarrow} X_1. \tag{9.6}$$

They analyzed the influence of sucrose on their reaction, because the presence of polyhydric alcohols increases the stability of proteins in solution. The presence of sucrose affected the reversible step, but did not influence the irreversible step. The irreversible step leads to aggregate formation, and subsequent precipitation.

These authors suggested the following thermodynamic explanation for the increase in stability with increasing sucrose concentration in solution. There are interactions not only between the native protein and the solvent but also between the denatured protein and the solvent. This leads to minor changes in the $\Delta H°$. The authors noted that $\Delta S°$ decreases with an increase in the sucrose concentration. They proposed that in the vicinity of the solvent molecules, the interactions between the solvent molecules and the denatured form are stronger than those between the solvent molecules and the native form (initial state). The addition of sucrose affected the reversible denaturation step, and affected the irreversible step only to a small extent. Thus, overall the addition of sucrose does not signficantly influence the step that involves the refolding of the unfolded Nase.

It is now instructive to briefly analyze a few real-life examples where the refolding of proteins or biological products has been carried out.

Example 9.6

Briefly describe the purification and renaturation of recombinant human interleukin-2 (IL-2) (Weir and Sparks, 1987).

Solution

Weir and Sparks (1987) presented a procedure to purify recombinant IL-2. These authors purified IL-2 partially in denatured form. Thereafter, by using preliminary experiments they identified the correctly folded forms. Finally, the recovery process of these correctly folded forms was systematically improved. Initially, the IL-2 expressed as *Escherichia coli* was isolated as (insoluble) IBs after cell breakage. The IL-2 along with the other contaminants was dissolved in a mixture of 6 M guanidinium chloride and 10 mM dithiothreitol (DTT) at a pH of 8.5. Further purification in the same solvent was performed on the reduced and denatured form by gel permeation chromatography. The product was diluted and refolded by autoxidation. The final purity of the product was 95% as observed by reversed-phase high pressure liquid-chromatography (RP-HPLC). The authors emphasized that to obtain a 30% recovery, concentrations of around 1 μg/ml were necessary. Furthermore, it is essential to maintain the product in reduced form before renaturation and autoxidation. This was done efficiently at pH 8.5 with 1.5 μM $CuSO_4$. Low recovery was primarily due to an aggregation process during the refolding process.

Epstein *et al.* (1962) earlier suggested that protein folding is not just dependent on the sequence of amino acids. Intermolecular reactions could also influence intramolecular reactions (Jaenicke, 1991). This according to Jaenicke (1991) is how chaperones work. They mediate the folding reaction along correct pathways and do not apparently convert incorrect structures or aggregates to the native form. Some common molecular chaperones include protein disulfide isomerase (PDI) (Freedman, 1989), peptidyl-prolyl *cis-trans* isomerases (PPIs), and polypeptide chain binding proteins (PCBs) (Rothman, 1989). Ellis (1990) and Schlesinger (1990) indicated that in a short span of a few years over a dozen such chaperones or helper proteins have been identified. By considering the significant effort that is being directed in this area one may anticipate that quite a few more such molecular chaperones will be identified along with valuable insights into their working mechanism and on how they assist proteins to refold correctly. We now examine some ways by which correct folding may be assisted.

Example 9.7

Briefly describe the *in vitro* folding of glycoprotein hormone chorionic gonadotropin (Huth *et al.*, 1994).

Solution

Huth *et al.* (1994) analyzed the redox conditions to stimulate and enhance the refolding of the glycoprotein hormone chorionic gonadotropin. Human chorionic gonadotropin (hCG) is a member of the glycoprotein family. This hormone stimulates the follicle and thyroid (Ryan *et al.*, 1988). Its activity depends on the appropriate assembly of the α- and β-subunits. Huth *et al.* (1994) indicated that the formation of the disulfide bonds between cysteines 9 and 90, and 23 and 72 is the rate-determining step in the refolding of hCG. The optimum folding conditions obtained by these authors were: (1) 2 mM glutathione buffer, pH 7.4, that contained 1 mg/ml PDI, and (2) 10 mM cysteamine–cystamine, pH 8.7, that contained no PDI. By using these conditions the half-life of the rate-determining folding step was 16 to 20 minutes. This is close to the that obtained in intact cells (4 to 5 min).

These authors suggested that based on their results obtained with hCG, the cysteamine–cystamine redox buffer is an appropriate buffer for the refolding of proteins that contain disulfide bonds. One does, however, need to coordinate the disulfide chemistry along with the conformational changes to assist in the refolding of proteins. They suggested that their buffer is an economic alternative to the more expensive use of large amounts of PDI to facilitate the refolding of proteins. More proteins and other biological products need to refolded using this cysteamine–cystamine buffer to further test its applicability and validate its usefulness and versatility.

D. Chaperones and Chaperonins

Example 9.8

Briefly show the influence of chaperonins and protein disulfide isomerases on the renaturation of single-chain immunotoxin (Buchner *et al.*, 1992).

Solution

Buchner *et al.* (1992) analyzed the renaturation of immunotoxin by chaperonins and by PDI. These helper proteins are known to guide the folding of proteins *in vivo* (Ellis and van der Vies, 1991). Buchner *et al.* (1992) indicated that the *E. coli* proteins DnaK and GroE have been studied extensively. GroE, a complex consisting of the proteins GroEL and GroES, has been shown to facilitate the refolding of several proteins *in vitro*. Skowyra *et al.* (1990) indicated that DnaK has been shown to dissolve the incorrectly folded aggregates of thermally denatured polymerase in an adenosine 5′-triphosphate (ATP)-dependent manner. Bulleid and Freedman (1990) indicated that the depletion of microsomes of PDI (a residual protein of the endoplasmic reticulum) hinders disulfide bond formation. Lower efficiencies of the active protein are also formed due to incorrect disulfide linkage formation (Buchner *et al.*, 1992).

Immunotoxins are complex artificial proteins. B3(Fv)-PE38KDEL, a recombinant immunotoxin, forms inclusion bodies when produced in *E. coli* in a recombinant manner. Brinkmann *et al.* (1991) indicated that B3(Fv)-PE38KDEL specifically kills different carcinoma cells, and causes a complete regression of solid human tumors grown in immunodeficient mice. Buchner *et al.* (1992) indicate that both DnaK and GroE increase the reactivation process. This process depends on ATP. PDI also increases the yield of the immunotoxin. These authors noted that the effects of DnaK and PDI are additive. This permits the possibility of controlling the extent of the refolding reaction to match requirements. Freedman *et al.* (1989) indicated that the formation of native disulfide bonds is catalyzed by PDI. Buchner *et al.* (1992) noted that if equimolar or higher concentrations of PDI were used, then the refolding of the immunotoxin improved. Lower efficiencies of the active protein are also formed due to incorrect disulfide linkage formation.

It is of interest to analyze the influence of chaperonins on the folding of different proteins under different conditions. More information is required on the mechanisms involved in the chaperonin-facilitated refolding of proteins. This information can then be effectively used to manipulate and to control the folding of proteins and other biological macromolecules of interest under different bioprocessing conditions. A particular point of interest to enhance the economics of bioseparation processes would be the reuse of the chaperonins (Buchner *et al.,* 1992), which would also involve the separation of the chaperonins from the reaction solution. The next example briefly looks at the mechanisms involved in the chaperonin-facilitated refolding of proteins.

Example 9.9

Briefly describe the chaperonin-facilitated *in vitro* folding of monomeric mitochondrial rhodanese (Mendoza *et al.,* 1991).

Solution

Mendoza *et al.* (1991) indicated that the *in vitro* folding of the monomeric mitochondrial enzyme, rhodanese (thiosulfate sulfurtransferase; EC 2.8.1.1) is facilitated by the chaperonins 60 (GroEL) and 10 (GroES) in the presence of Mg–ATP and K^+. Westley (1973) initially indicated that rhodanese is found in the matrix of all mammalian mitochondria. Ogata *et al.* (1989) suggested that rhodanese plays a significant role in the formation, maintenance, and control of iron–sulfur-containing electron transport proteins. GroEL or the cpn60 protein of *E. coli* is homologous to a mitochondrial matrix protein, hsp60. hsp60 is a heat shock protein with a molecular weight of 60,000, and Hermann *et al.* (1989) indicated that this is a component of the pathway for the folding of mitochondrial matrix proteins. Mendoza *et al.* (1991) indicated that cpn60 is a tetradecamer (14-mer) of 60-kDa subunits. Pitsyn *et al.* (1989) indicated that this protein facilitates the *in vitro* refolding of the chloroplast protein ribulose biphosphate carboxylase (Rhu-P_2 carboxylase) from unfolded polypeptides. Rhu-P_2 carboxylase refolding requires Mg-ATP, K^+, and cpn10(GroES).

Tandon and Horowitz (1986) had earlier suggested that the nondenaturing detergent lauryl maltoside effectively reduces aggregation. Tandon and Horowitz (1989, 1990) suggested that the refolding of proteins in the presence of lauryl maltoside proceeds along a pathway with detectable intermediates. Apparently the detergents interact with the hydrophobic surfaces on the polypeptide chains, and this minimizes the aggregation. Horwich and Criscimagna (1990) suggested that the chaperonin-facilitated protein refolding also proceeds similarly in this fashion. Mendoza *et al.* (1991) noted that the *E. coli* chaperonins facilitated the refolding of rhodanese, and required K^+ and ATP.

Mendoza *et al.* (1991) noted that chaperonin 60 can combine with the labile intermediate rhodanese-I, which rapidly aggregates at 37°C. This stabilizes the labile intermediate and minimizes the aggregation by preventing the interactions of the hydrophobic surfaces that lead to the aggregation. These authors indicated that the chaperonin forms a complex with the partially unfolded polypeptide chain that looks like a folding intermediate. In essence, the chaperonin arrests the labile intermediate in a non-native configuration and prevents the aggregation of incorrect forms. These complexes are inactive. They emphasized that chaperonin 60 helps guide the folding to the appropriate native form, but it does this at the expense of slowing down the refolding step. Finally, the interaction of chaperonin 10 with ATP leads to a conformational change in chaperonin 60. This step weakens the hydrophobic interactions, permitting the release of the rhodanese and allowing it to complete its final refolding step.

Further detailed analyses like those of Mendoza *et al.* (1991) are required of the chaperonin-assisted refolding steps to help improve the efficiency of each of these steps. Once better physical insights into the different steps are obtained, they can assist in usefully integrating the chaperonin-facilitated refolding of different proteins in the bioprocessing train.

Example 9.10

Compare briefly the refolding of proteins by the use of assistants such as detergents, lipids, and micelles with chaperonin-assisted refolding (Zardeneta and Horowitz, 1994).

Solution

Zardeneta and Horowitz (1994) analyzed the use of assistants such as detergents for the refolding of proteins. These authors emphasized that this is an area in which research has just begun, and that up until now there is no particular system that is the best or superior to others for the refolding of proteins. Nevertheless, there are some required features of these refolding assistants. One of the main drawbacks is the formation of misfolded or incorrectly folded forms or aggregates. These assistants such as detergents, lipids, and micelles apparently promote the refolding of proteins by binding to critical and active sites of the unfolded proteins, thereby minimizing or preventing the formation of misfolded forms. In this sense the mechanistic behavior of these assistants is similiar to that of chaperonin-assisted refolding of proteins.

These authors emphasized that there are advantages of these assistants as compared with using chaperonins for the refolding process. The high cost of chaperonins restricts their utilization, especially for refolding of proteins at higher scale levels. A distinct advantage of using the assistants as compared with using chaperonins is that the assistants are far less expensive than chaperonins. Also, one may use higher concentrations of proteins when one uses the assistants as compared with using the chaperonins. Furthermore, they emphasized that simple purification procedures for separating the nonproteinaceous material from the proteins are available when assistants are used (Zardeneta and Horowitz, 1991; Hagen *et al.*, 1990a). This is a fruitful and hectic area of research, and one may reasonably anticipate that significant effort will be put into using both chaperonins and refolding-assistants to help in the effective refolding of proteins.

There are other methods besides chaperonin-facilitated protein refolding that help in the formation of the native and active state from an unfolded state. Gross *et al.* (1985) indicated that protein properties also help in the partition of protein in the soluble and insoluble (inclusion bodies) fractions. Wetzel *et al.* (1991) indicated that a single amino acid substitution alters the partition of human interferon-γ in the soluble and insoluble fractions. The next example briefly analyzes the influence of a single substitution on the partition of a protein into its soluble and insoluble fractions.

E. Amino Acid Substitution

Example 9.11

Briefly analyze cysteine to serine substitution on basic fibroblast growth factor (bFGF) IB formation during *in vitro* refolding (Rinas *et al.*, 1992).

Solution

Although IB formation has been known for some time, the mechanisms of IB formation are far from understood (Rinas *et al.*, 1992). Some of the factors that influence the amount of protein in the IBs include reduced growth rate of protein (Hart *et al.*, 1990), high temperature during expression (Chalmers *et al.*, 1990), and mutations of the tail-spike protein (King *et al.*, 1990). Besides, Strandberg *et al.* (1991) emphasized that both properties and processing conditions influence protein A–β galactosidase formation in IBs.

The cysteine to serine substitutions do not affect the stability or proteolytic susceptibility of the folded protein (Rinas *et al.*, 1992). However, these substitutions do alter the susceptibility of the folding intermediates to aggregation and proteolytic degradation. For example, these substitutions may alter the folding kinetics by modifying the half-life of the intermediates. More specifically, these authors indicated that a single mutation at position 88 and a double mutation at positions 70 and 88 do not significantly change the bFGF partition characteristics in the insoluble and soluble fractions. However, a single substitution of cysteine 70 by serine decreases the fraction of soluble bFGF considerably. Furthermore, they emphasized that cysteine to serine substitutions affect bFGF proteolytic susceptibility during *in vitro* folding from IBs. Apparently the *in vivo* and *in vitro* folding mechanisms of bFGF must be different. Perhaps chaperones may be involved in the *in vivo* folding of bFGF. More studies like this analysis are required that help analyze the influence of different amino acid substitutions on *in vitro* protein refolding. A framework of data generated would considerably assist in tailor-making IBs to match (active) protein refolding requirements.

There are other methods that may be used to alleviate the aggregation problem in IBs. Cleland and Wang (1990) suggested the addition of folding aids to the dilution buffer. Cleland and Randolph (1992) and Cleland *et al.* (1992) showed that the addition of a cosolvent, polyethylene glycol (PEG), to the dilution buffer increased the refolding of carbonic anhydrase B. Apparently, the aggregation of a folding intermediate is prevented by the binding of the polyethylene glycol (PEG) that leads to a nonaggregating complex. Cleland *et al.* (1992) emphasized that subsequent to polyethylene glycol (PEG) binding all folding reactions occur at the same rate. Thus, PEG only prevents the improper aggregation of the protein.

Example 9.12

Describe PEG-assisted refolding of three recombinant human proteins (Cleland *et al.*, 1992).

Solution

The ability of PEG to enhance the refolding of three recombinant proteins, deoxyribonuclease (rhDNAse), tissue plasminogen activator (rhtPA), and interferon-γ (rhIFN-γ) has been examined (Cleland *et al.*, 1992). These authors indicated that the refolding of carbonic anhydrase B proceeds through an intermediate. Cleland and Wang (1990) emphasized that the aggregation of this intermediate decreases the recovery of this protein. However, PEG binds specifically to the refolding intermediate and forms a nonaggregating complex, and thereby inhibits its aggregation. This nonaggregating complex folds to a second intermediate. The PEG is released and the second intermediate eventually refolds to the native protein.

Refolded rhDNAse was obtained from impure *E. coli*-derived IB with and without using PEG (Cleland *et al.*, 1992). On utilizing a dilution buffer that yields a final PEG to rhDNAse molar ratio of 10 to 1, there is a threefold increase in the recovery of the protein when compared with the case when PEG was not used. Similar increases in recovery were obtained for rhtPA and rhIFN-γ.

These results indicated that PEG binds presumably to specific segments of the proteins analyzed, thereby preventing their aggregation. These authors emphasized that because PEG interacts with these proteins through both hydrophobic and hydrophilic forces, the PEG is only weakly attached to the partially folded intermediate. These weak interactions permit the intermediate to fold to its native protein state by displacing PEG molecules.

The refolding of other proteins should also be attempted using PEG (Cleland *et al.*, 1992). This would shed further physical insights into the PEG-assisted refolding of proteins, and considerably assist in the bioseparation of proteins utilizing recombinant methods and involving IBs.

F. Affinity Ligands

Another way of enhancing protein refolding is by the addition of affinity ligands during the refolding step. For example, Kato and Anfinsen (1969) indicated that the addition of the complementary fragment, S-peptide enhances the refolding of reduced S-protein. Even though the S-peptide does not contain the information to refold this may be induced by another molecule even if it is not linked to the S-protein. More specifically, Carlson and Yarmush (1992) indicated that polyclonal antibodies specific to certain domains in the protein structure may increase the extent of the antigenic structure of the protein. These authors added that antinative antibodies would seem to demonstrate the potential to enhance protein refolding as shown earlier by Chavez and Benjamin (1978). The next example briefly analyzes antibody-assisted protein refolding.

Example 9.13

Briefly analyze the antibody-assisted protein refolding (Carlson and Yarmush, 1992).

Solution

Monoclonal antibodies (MAbs) have been utilized to assist in the refolding of a model protein, S-protein (Carlson and Yarmush, 1992). The S-protein is a proteolytic fragment of bovine pancreatic ribonuclease A, which includes residues 21 through 124, and all eight cysteine groups that participate in the four disulfide bonds (Richards and Vithaythil, 1959). Haber and Anfinsen (1962) indicated that reduced S-protein will only partially refold when it is air-oxidized. Figure 9.4 shows the mechanistic scheme involved in antibody-assisted refolding of S-protein. Carlson and Yarmush (1992) reduced S-protein, purified it, and then mixed it with a predetermined amount of MAb specific to the protein. Enzymatic activity at the end was used as a measure of regaining active protein conformation.

Four antinative MAbs were utilized to enhance the refolding of the S-protein (Carlson and Yarmush, 1992). Out of these four MAbs, only one antinative MAb successfully enhanced the refolding of the S-protein. These authors noted that 54% of the total population was correctly folded, as compared to a 100% folding of the protein, with only 54% attainment of native structure activity. The refolding experiments were carried out for a period of 24 h. They investigated the structure of the refolded protein by size-exclusion HPLC. The native S-protein exhibited a single sharp peak characteristic of a monomer. The S-protein refolded in the presence of the monoclonal antibodies exhibited a broad peak. Two possible reasons were suggested for this. There could be a progressive dissociation of dimers to monomers. Also, there could be a range of molecular sizes in the refolded population. This heterogeneity in the refolded population, if obtained, is not surprising considering the complexities in the reaction and the protocol used.

Because this is a preliminary investigation, Carlson and Yarmush (1992) are further studying the reasons or characteristics that are involved, and particularly those that may enhance antibody-assisted refolding of proteins. Such

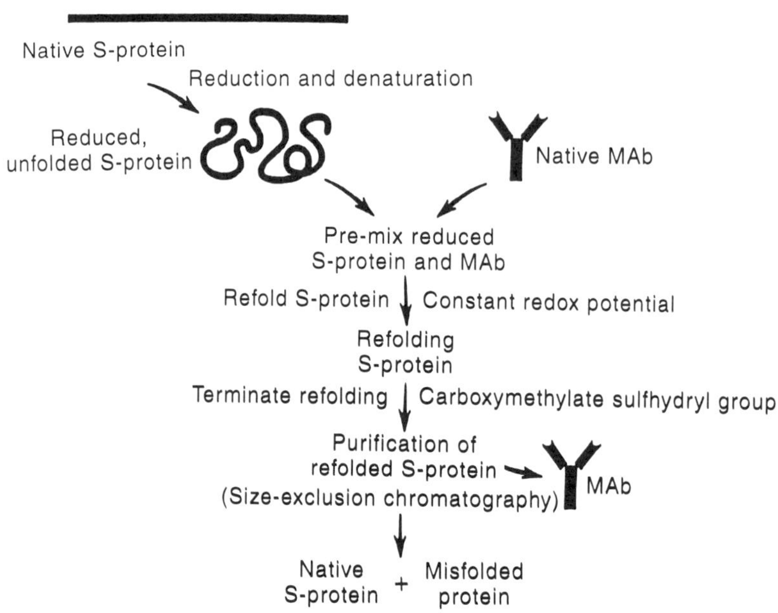

FIGURE 9.4 Proposed mechanism for the antibody-assisted S-protein refolding. [From Carlson, J. D. and Yarmush, M. L. (1992). *Biotechnology, 10,* 86.]

studies exhibit the potential to provide novel physical insights into the assisted (by chaperons or otherwise) refolding of proteins, and perhaps of other biological macromolecules of interest. These studies are highly recommended to be performed to further fully explore and develop the potential of this process.

G. Reverse Micelles

The use of reverse micelles is a possible means to help separate proteins and other biological products of interest. The aggregation of incorrectly refolded proteins is due to the interactions of the hydrophobic patches exposed on the polypeptide chains. Hagen *et al.* (1990a,b) presumed that if the polypeptide chains could be isolated from each other, this would assist in the recovery of activity of these proteins. They suggested isolating these polypeptide chains in reverse micelles. The next example analyzes the refolding of proteins in reverse micelles.

Example 9.14

Briefly analyze protein refolding in reverse micelles (Hagen *et al.,* 1990a).

Solution

Reverse micelles are aqueous-phase droplets in organic solvents that are stabilized by surfactants (Hagen *et al.*, 1990a). The surfactants have their polar heads facing inward toward the water, and their tails are on the outside toward the organic phase. Martinek *et al.* (1981) and Luisi *et al.* (1988) indicated that proteins solubilized in the interior of the reverse micelles retain their activity and their conformation. Hagen *et al.* (1990a) proposed to manipulate the reaction conditions so that only a single protein molecule is solubilized in the interior of the reverse micelle. It is also possible, and as suggested by these authors, that quite a few reverse micelles will go empty. This would tend to reduce the efficiency of the process. An advantage of using reverse micelles is that there is background knowledge on reverse micelles. This should be of considerable assistance. One would thus need to concentrate primarily on the refolding aspects.

The reverse micelles in the Hagen *et al.* (1990a) analysis are water droplets stabilized by bis-(2-ethylhexyl)sodium sulfosuccinate (AOT) (surfactant) in isooctane. Each reverse micelle contains a single protein (albeit not completely folded). Figure 9.5 delineates the protocol suggested by these authors to enhance protein refolding in reverse micelles. Briefly, the following steps were involved: (1) the denatured protein solubilized in guanidine hydrochloride is transferred into the reverse micelles by the phase transfer method; (2) the denaturant concentration is reduced gradually in the reverse micelles; (3) the disulfide bonds in the denatured protein are reoxidized by the addition of a redox agent, and the protein attains its active conformation; and (4) the protein is extracted from the reverse micelles into an aqueous solution.

The effectiveness of the procedure was demonstrated when denatured and reduced ribonuclease A was able to recover almost complete activity within a period of 24 h (Hagen *et al.,* 1990a). These authors emphasized that two problems require further study to fully utilize the effectiveness of this process for

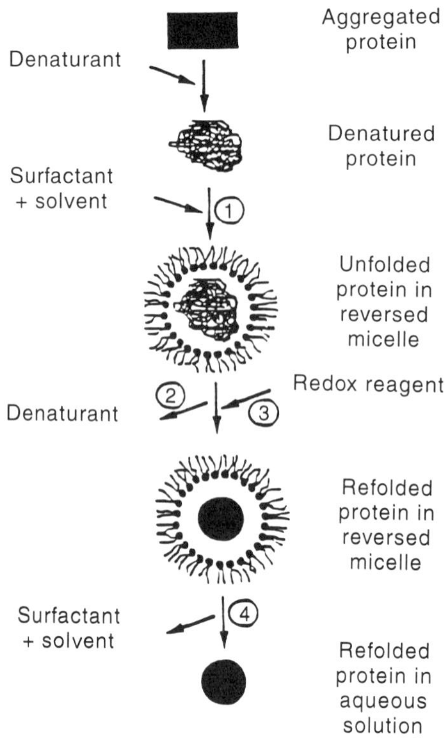

FIGURE 9.5 Proposed protocol for protein refolding in reverse micelles. [From Hagen, A. J. *et al.* (1990a). *Biotechnol. Bioeng., 35,* 955.]

the refolding of proteins and other biological macromolecules of interest. Little information on these aspects is presently known: (1) the behavior of denatured proteins in reverse micelles; and (2) the protein refolding process, more specifically as applied to within the constraints of the reverse micelles. We are in agreement with these authors, and studies in this direction will shed novel physical insights that would be invaluable in developing a commercial process. For example, because surfactants are involved it would be of interest to analyze the interactions of the proteins with the surfactants; and to see how they would influence the protein activity, stability, and entry into–exit from the reverse micelles. In a subsequent study, Hagen *et al.* (1990b) did this, and their analysis is now briefly presented.

Reverse micelles contain different components such as denaturants and detergents. These will affect the conformation of the enclosed "guest" protein or of other biological macromolecules (Hagen *et al.*, 1990b). It is of interest to carefully examine the influence of, for example, detergents on the conformation of the protein in the reverse micelles, because this will significantly affect the refolding process. Lapanje (1978) indicated that some detergents are strong denaturants, and these will significantly affect the conformations of the proteins in the reverse micelles. Hagen *et al.* (1990b) emphasized that the location and the conformational state of the protein in the reverse micelles will significantly affect the refolding process. Luisi *et al.* (1988) emphasized that the hydrophobicity of the guest molecules will significantly influence the detergent–protein interactions. Furthermore, Leotidis and Hatton (1989) indicated that strong interactions exist between hydrophobic amino acids and the micellar interface.

Hagen *et al.* (1990b) indicated that the hydrophobic regions of an unfolded protein will interact with the surfactant layer. For example, γ-interferon contains more hydrophobic regions than ribonuclease. Thus, as expected, its interactions with the surfactant layer will be stronger. This was demonstrated experimentally by the inability of γ-interferon to refold, and its subsequent aggregation on extraction. Ribonuclease, however, overcomes its attraction to the surfactant due to the strong refolding forces that prevail once the denaturant is removed. This ability of RNase to refold was lost by modifying the RNase surface with hydrophobic residues.

H. Environmental Conditions

Example 9.15

Briefly describe the influence of environmental conditions on the refolding selectivity of insulin-like growth factor I (IGF-I) (Hart *et al.*, 1994).

Solution

Blundell *et al.* (1983) initially indicated that IGF-I is a member of the insulin-like family of peptides. Sara and Hall (1990) stated that these are hormones that exhibit a number of metabolic and growth-promoting activities. This depends on the state of the cell. Hart *et al.* (1994) indicated that these IGFs exhibit both structural and functional similarities. Thus, they also exhibit similar behavior during refolding. These authors have analyzed the influence of environmental conditions on the refolding of IGF-I. Studies were conducted on partially purified denatured and reduced recombinant human (IGF-I) obtained from *E. coli* IBs. They indicated that solution polarity and salt concentration strongly influence the refolding characteristics of IGF-I. Also, their effects are interdependent. Furthermore, they emphasized exploring for synergistic effects.

The refolding was performed in gently shaken tubes with an air space to a fluid volume ratio of 2 : 1. This provided enough oxygen to complete the thiol oxidation. These authors indicated that the refolding was carried out for about 3 to 6 h, depending on the conditions. This allowed sufficient time to complete the process. They emphasized that their refolding experiments were performed so that the factors that were potentially interdependent were varied simultaneously. Data were analyzed using the Cochran and Box (1957) method of factorial analysis of variance. This permitted quantitative estimation of factor importance, factor interdependence, and standard deviation.

Hart *et al.* (1994) indicated that solution polarity, salt type and concentration, and chaotrope type and concentration strongly influence the refolding of IGF-I. These authors attempted to explain their results based on current understanding. For example, Cleland and Wang (1990) indicated that association of aggregation-susceptible intermediates tend to follow second-order kinetics, whereas intrachain folding processes tend to follow first-order processes. The

Hart *et al.* (1994) data for the associations leading to mult-IGF-I formation also follow approximately second-order kinetics. These authors emphasized that decreasing the solution polarity decreases the function of mult-IGF-I. Apparently, the decreasing solution polarity reduces the concentration of relatively hydrophobic aggregation-susceptible species or diminishes the tendency to associate. This is consistent with the results of Lustig and Fink (1992) on the thermal denaturation of ribonuclease, where moderate concentrations of methanol help stabilize the hydrophobic intermediates.

Furthermore, Hart *et al.* (1994) indicated that decreasing the solution polarity may help stabilize a specific hydrophobic site on the surface of the folding or folded peptide. This would influence the selectivity of the IGF-I refolding process. Also, these authors suggested that the decrease in the solution polarity would promote the formation of specific structural features. For example, changes in the solution polarity strongly influence changes in structure (Jackson and Mantsch, 1992; Shibata *et al.*, 1992; Zhong and Johnson, 1992). Hua and Weiss (1991) indicated that moderate concentrations of alcohols enhance α-helix content. Hart *et al.* (1994) speculated that reduced solvent polarity may enhance the formation of α-helix content in IGF-I. This leads to an increased production of cor-IGF-I relative to mis-IGF-I.

The analysis of Hart *et al.* (1994) is of significant interest because it attempts to delineate the effects of solution polarity on the refolding characteristics of IGF-I. A possible refolding mechanism based on solution polarity effects is also suggested.

Wolf and Luisi (1979) had initially shown that the hydrophobic regions of a membrane protein will interact with a lipid bilayer. Hagen *et al.* (1990b) indicated that similarly the hydrophobic regions of a protein in a reverse micelle will interact with the micellar surfactant layer. These authors stated that because γ-interferon contains more hydrophobic groups than ribonuclease, it interacts more strongly with the micellar surfactant layer. These interactions prevent γ-interferon from refolding, and the polypeptide chains aggregate on extraction. The surfactant–ribonuclease interactions are milder; and the refolding forces prevail over them, and this then leads to the refolding of the ribonuclease. Presumably, if γ-interferon needs to be refolded in the reverse micelles, then every attempt should be made to minimize or hinder the protein–surfactant interactions at the micellar interface.

Tandon and Horowitz (1988) proposed the concept of "masking" the hydrophobic surfaces to minimize the interaction. Perhaps, the protein can be made to interact with another component, which minimizes surfactant–protein interactions, by a suitable conformational change. The protein can be extracted, and then the protein and the component can be "disengaged." Other suitable schemes (such as using nondenaturing surfactants) can be thought about and presumably tested for their effectiveness. In any case, analyses such as those of Hagen *et al.* (1990a,b) should be aggressively pursued to assist in minimizing protein interactions and enhancing the refolding step. Detailed and precise information will be required not only on the conformational states of the protein at different locations but also on how useful interactions can be promoted and deleterious effects can be minimized. This should considerably assist in manipulating these reactions in desired directions. It would be ideal, of course, to have a regulatory or "corrective or repair" mechanism inherent in these types of systems. These could then automatically take care of aberrantly folded molecules that are bound to occur in the refolding process, in spite of the extreme care that may be taken.

IV. CONCLUSIONS

IB formation during the recombinant production of proteins has been known for quite some time; however, only recently have mechanisms of formation been looked at so carefully. The primary reason is the decrease in the activity and stability of the final product, and the subsequent lowering of the efficiencies of the bioprocess. Because the extraction of the protein from the IB requires an unfolding and a subsequent refolding step, there have been extensive and rather detailed studies on the refolding aspects. Unfolding has been dealt with rather extensively. Mechanistic schemes for the refolding steps are required in different "local" environments for different proteins. Such mechanistic schemes if developed and analyzed should help generate a framework of information that should prove invaluable in assisting and determining the appropriate conditions required to refold a wide variety of proteins–enzymes.

Assisted folding, by chaperones or otherwise, should also be aggressively explored. Such types of studies not only provide quicker and correct pathways to arrive at the native and active structure but also exhibit the potential, in general, to shed novel physical insights into the protein refolding process. Antibody-assisted refolding and the refolding of enzymes in different local environments should also be actively pursued to help determine the optimum medium or conditions under which the refolding process may be performed. Because the refolding process requires the careful balancing of the different destabilizing and stabilizing forces, one is encouraged to try novel and imaginative techniques rather than restricting oneself to only safer and well-tried or seasoned techniques.

REFERENCES

Anfinsen, C. B. (1973). *Science, 181*, 223.

Anson, M. L. (1945). *Adv. Protein Chem., 29*, 205.

Bloemendale, H. and Groenewoud, G. (1981). *Anal. Biochem., 117*, 327.

Blond, S. and Goldberg, M. (1987). *Proc. Natl. Acad. Sci. USA, 84*, 147.

Blundell, T. L., Bedarkar, S., and Humbel, R. E. (1983). *Fed. Proc. Fed. Am. Soc. Exp. Biol., 42*, 2592.

Brinkmann, U., Pai, L. H., Fitzgerald, D. J., Willingham, M., and Pastan, I. (1991) *Proc. Natl. Acad. Sci. USA, 88*, 8616.

Buchner, J., Schmidt, M., Fuchs, M., Jaenicke, R., Rudolph, R., Schmid, F. X., and Kiefhaber, T. (1991). *Biochemistry, 30*, 1587.

Buchner, J., Brinkmann, U., and Pastan, I. (1992). *Biotechnology, 10*, 682.

Bulleid, N. J. and Freedman, R. B. (1990). *Nature (London), 335*, 649.

Carlson, J. D. and Yarmush, M. L. (1992). *Biotechnology, 10*, 86.

Chalmers, J. J. Kim, E. E., Telford, J. N., Wong, E. Y., Tacon, W. C., Shuler, M. L., and Winson, D. B. (1990). *Appl. Environ. Microb., 56,* 104.
Chavez, L. G. and Benjamin, D. C. (1978). *J. Biol. Chem., 253,* 8081.
Cleland, J. L. and Wang, D. I. C. (1990). *Biotechnology, 8,* 1274.
Cleland, J. L. and Randolph, T. W. (1992). *J. Biol. Chem., 267,* 3147.
Cleland, J. L., Hedgepeth, C., and Wang, D. I. C. (1992). *J. Biol. Chem., 267,* 13327.
Cochran, W. G. and Box, G. E. P. (1957). *Experimental Designs,* 2nd ed., Wiley Publications in Statistics, John Wiley & Sons: New York, p 286.
Datar, R. V., Cartwright, T., and Rosen, C. G. (1993). *Biotechnology, 11,* 349–357.
De Bernardez-Clark, E., and Georgiou, G. (1991). *Inclusion Bodies and Recovery of Proteins from the Aggregated States,* ACS Symposium Series, 470, American Chemical Society: Washingotn, DC, p 1020.
Dobson, C. M. (1991). *Clin. Opinions Struct. Biol., 1,* 22.
Ellis, R. J. (1990). *Semin. Cell. Biol., 1,* 1.
Ellis, R. J. and van der Vies, S. M. (1991). *Annu. Rev. Biochem., 60,* 321.
Epstein, C. J., Goldberger, R. F., Young, D. M., and Anfinsen, C. B. (1962). *Arch. Biochem. Biophys. Suppl.,* 223.
Freedman, R. B. (1989). *Cell, 57,* 1069.
Go, N. (1984). *Adv. Biophys., 18,* 149.
Goldenberg, D. P. and Creighton, T. E. (1985). *Biopolymers, 16,* 4014.
Gross, M., Sweet, R. W., Sathe, G., Yokoama, S., Fasano, O., Goldbarg, M., Wigler, M., and Rosenberg, M. (1985). *Mol. Cell. Biol., 5,* 1015.
Haase-Pettingell, C. and King, J. (1988). *J. Biol. Chem., 263,* 4977.
Haber, E. and Anfinsen, C. B. (1962). *J. Biol. Chem., 237,* 1839.
Hager, D. A. and Burgess, R. R. (1980). *Anal. Biochem., 109,* 76.
Hagen, A. J., Hatton, T. A., and Wang, D. I. C. (1990a). *Biotechnol. Bioeng., 35,* 955.
Hagen, A. J., Hatton, T. A., and Wang, D. I. C. (1990b). *Biotechnol. Bioeng., 35,* 966.
Hart, R. A., Rinas, U., and Bailey, J. E. (1990). *J. Biol. Chem., 265,* 12728.
Hart, R. A., Giltinan, D. M., Lester, P. M., Reifsnyder, D., Ogez, J. R., and Builder, S. E. (1994). *Biotechnol. Appl. Biochem., 20,* 217.
Hermann, J., Horwich, A.L., Neupert, W., and Hartl, F.U. (1989). *Nature (London), 341,* 125.
Hermann, R., Jaenicke, R., and Price, N. C. (1985). *Biochemistry, 24,* 1817.
Hinds, D. A. and Levitt, M. (1994). *J. Mol. Biol., 243,* 668.
Hinds, D. A. and Levitt, M. (1995). *TIBTECH, 13,* 23.
Horowitz, P. M. and Butler, M. (1993). *J. Biol. Chem., 268,* 2500.
Horowitz, P. M. and Criscimagna, N. L. (1990). *J. Biol. Chem., 265,* 2576.
Hua, Q. X. and Weiss, M. A. (1991). *Biochim. Biophys. Acta, 1078,* 101.
Huth, J. R., Feng, W., and Ruddon, R. W. (1994). *Biotechnol. Bioeng., 44,* 66.
Itzhaki, L. S., Evans, P. A., Dobson, C. M., and Radford, S. E. (1994). *Biochemistry, 33,* 5212–5220.
Jackson, M. and Mantsch, H. H. (1992). *Biochim. Biophys. Acta, 1118,* 139.
Jaenicke, R. (1984). *Angew. Chem. Int. Ed. Engl., 23,* 395.
Jaenicke, R. (1987). *Prog. Biophys. Mol. Biol., 49,* 117.
Jaenicke, R. (1991). *Biochemistry, 30,* 3147.
Kane, J. F. and Hartley, D. L. (1988). *Trends Biotechnol., 6,* 95.
Kato, I. and Anfinsen, C. A. (1969). *J. Biol. Chem., 244,* 1004.
King, J., Fane, B., Haase-Pettingell, C., Mitraki, A., Villafane, R., and Yu, M. H. (1990). In *Protein Folding: Deciphering the Sceond Half of the Genetic Code,* Gierasch, L.M. and King, J., Eds., American Association of Advances in Science: Washington, DC.
Knuth, M. W. and Burgess, R. B. (1987). *Protein Purification: Micro to Macro,* Burgess, R. R., Ed., Alan R. Liss: New York, p 279.
Kuwajima, K. (1989). *Proteins Struct. Funct. Genet., 6,* 87.
Lapanje, S. (1978). *Physiochemical Aspects of Protein Denaturation,* John Wiley and Sons: New York.
Leotidis, E. B. and Hatton, T. A. (1989). *Langmuir, 5,* 741.
Light, M. (1985). *Biotechnology, 3,* 298.

Luisi, P. L., Giomini, M., Pileni, M. P., and Robinson, B. H. (1988). *Biochim. Biophys. Acta, 947,* 209.

Lustig, B. and Fink, A. L. (1992). *Biochim. Biophys. Acta, 1119,* 205.

Maloney, P. C. and Ambudkar, S. V. (1989). *Arch. Biochem. Biophys., 269,* 1.

Marston, F. A. O. (1986). *Biochem. J., 240,* 1.

Martinek, K., Levashov, A. V., Klyatchko, N. L., Pantin, V. I., and Berezin, I. V. (1981). *Biochim. Biophys. Acta, 657,* 277.

Matthews, C. R. (1993). *Annu. Rev. Biochem., 62,* 653.

Mendoza, J. A., Rogers, E., Lorimer, G. H., and Horowitz, P. M., (1991). *J. Biol. Chem., 266,* 13044.

Mitraki, A. and King, J. (1989). *Biotechnology, 7,* 690.

Montelaro, R. C., West, M., and Issel, C. J. (1981). *Anal. Biochem., 114,* 398.

Nohara, D., Yamada, T., Watanabe, A., and Sakai, T. (1994). *Biotechnol. Bioeng., 44,* 276.

Ogata, K., Dai, X., and Volini, M. (1989). *J. Biol. Chem., 264,* 2718.

Pain, R. H. (1994). *Mechanisms of Protein Folding,* Ed., IRL, Oxford University Press: New York.

Pitsyn, O. B. (1994). *Protein Eng., 7,* 593.

Pitsyn, O. B., Reva, B. A., and Finkelstein, A. V. (1989). *Highlights Mol. Biol., 1,* 11.

Pitsyn, O. B., Pain, R. H., Semisotonov, G. V., Zeronik, E., Radford, S. E., Dobson, C. M., and Evans, P. A. (1992). *Nature (London), 358,* 302.

Razgulyaev, O. I. (1990). *FEBS Lett., 262,* 20.

Richards, F. M. and Vithaythil, P. J. (1959). *J. Biol. Chem., 234,* 1459.

Rinas, U., Tsai, L. B., Lyons, D., Fox, G. M., Stearns, G., Fieschko, J., Fenton, D., and Bailey, J. E. (1992). *Biotechnology, 10,* 435.

Rudolph, R. (1990). In *Modern Methods in Protein and Nucleic Acid Research,* Tschesche, H., Ed., de Gruyter: Berlin.

Ryan, R. J., Charlesworth, M. C., McCormick, D. J., Milius, R. P., and Kentmann, H. T. (1988). *FASEB J., 2,* 2661.

Sadana, A. and Henley, J. P. (1986). *Biotechnol. Bioeng., 28,* 1277.

Sadana, A. and Malhotra, A. (1987). *Biotechnol. Bioeng., 30,* 717.

Sadana, A. (1991). *Biocatalysis: Fundamentals of Enzyme Deactivation Kinetics,* Prentice-Hall: Englewood Cliffs, NJ.

Sara, V. R. and Hall, K. (1990). *Physiol. Rev., 70,* 591.

Schafmeister, C. E., Miercke, L. J. W., and Stroud, R. M. (1993). *Science, 262,* 734.

Schein, C.H. (1991). *Physical Methods and Models for the Study of Protein Aggregation,* ACS Symposium Series, 470, American Chemical Society: Washington, DC, p 21.

Schlesinger, M. J. (1990). *J. Biol. Chem., 265,* 12111.

Serrano, L., Matouschek, A., and Fersht, A. R. (1992). *J. Mol. Biol., 224,* 847.

Shaik, A. K., Guy, A. L., and Pancholi, S. K. (1980). *Anal. Biochem., 103,* 140.

Sharma, A. K., Minke-Gogl, V., Gohl, P., Siebendritt, R., Jaenicke, R., and Rudolph, R. (1990). *Eur. J. Biochem., 194,* 603.

Shibata, A., Yamamoto, M., Yamashita, T., Chiou, J. S., Kamaya, H., and Ueda, I. (1992). *Biochemistry, 31,* 5728.

Shinde, U. and Inouye, M. (1993). *Trends Biochem. Sci., 18,* 442.

Shortle, D. (1993). *Curr. Opinions Struct. Biol., 3,* 66.

Skowyra, D., Georgopoulos, C., and Zylicz, M. (1990). *Cell, 62,* 939.

Strandberg, L. and Enfors, S. O. (1991). *Appl. Environ. Microbiol., 57,* 1669.

Tandon, S. and Horowitz, P. M. (1986). *J. Biol. Chem., 261,* 15615.

Tandon, S. and Horowitz, P. M. (1988). *Biochim. Biophys. Acta, 955,* 19.

Tandon, S. and Horowitz, P. M. (1989). *J. Biol. Chem., 264,* 9859.

Tandon, S. and Horowitz, P. M. (1990). *J. Biol. Chem., 265,* 5967.

Weir, M. P. and Sparks, J. (1987). *Biochem. J., 245,* 85.

Welling, G. N., Groen, G., and Welling-Wester, S. (1983). *J. Chromatogr., 266,* 629.

Westley, J. (1973). *Adv. Enzymol. Relat. Areas Mol. Biol., 39,* 327.

Wetzel, R., Perry, L. J., and Villeux, C. (1991). *Biotechnology, 9,* 731.

Wetzel, R. (1994). *TIBTECH, 12,* 193.

Wetzel, R. (1992). In *Stability of Protein Pharmaceuticals: In vivo Pathways of Degradation and*

Strategies for Protein Stabilization, Ahern, T. J. and Manning, M. C., Eds., Plenum: New York, pp 43–48.
Wolf, R. and Luisi, P. L. (1979). *Biochim. Biophys. Res. Commun., 89,* 209.
Zardeneta, G. and Horowitz, P. M. (1991). *J. Biol. Chem., 267,* 5811.
Zardeneta, G. and Horowitz, P. M. (1994). *Anal. Biochem., 223,* 1.
Zhong, L. and Johnson, W. C. Z. (1992). *Proc. Natl. Acad. Sci. USA, 89,* 4462.
Zhu, X., Ohta, Y., Jordan, F., and Inouye, M. (1989). *Nature (London), 339,* 483.

10 VALIDATION OF THE PRODUCTION OF BIOLOGICAL PRODUCTS

I. INTRODUCTION

Advances in technology have heightened the awareness of potential safety hazards. More stringent purity standards are being set due to advances in measurement and purification technology improvements. Briggs and Panfili (1991) emphasized that therapeutic biopharmaceuticals and *in vivo* diagnostics involve the production of recombinant DNA and monoclonal antibodies harvested from cultures of genetically modified cells. Potentially dangerous impurities and contaminants may be present. Garnick *et al.* (1988) indicated that these contaminants need to be identified by suitable analytic methods. The measurement and interpretation of the testing done for contaminants and impurities are directly linked to the safety of the product for human consumption. Briggs and Panfili (1991) emphasized "the safety, potency, and purity of the injectable product is ultimately the responsibility of the manufacturer and forms the basis of regulatory evaluation." On-going analytic tests are critical in the development of processes, and are a key element of good manufacturing practice and regulatory evaluation.

The Biotechnology Task Force on purification and scale-up in the Parenteral Drug Association (PDA) Report (1992) indicated that process validation is "the assurance that a process when operated within established limits, produces a product of appropriate and consistent quality." The task force members emphasized that careful studies need to be performed so that these parameters can be met on a consistent basis. Formally, process validation is the "assurance that the product quality is derived from a careful attention to a number of factors, including process design, selection, and use of quality parts and materials; and control of the process through appropriate in-process and end-process testing." These authors emphasized that validation should be considered as early as possible during the development of a product. In this way data may be collected during the development studies and during the production of batches for clinical studies.

Akers *et al.* (1994) indicated the importance of validation during each of the three different phases of a product's lifespan: development, pilot scale, and production. These authors emphasized that early consideration of validation requirements and the development of a validation plan can save time, money, and prevent costly regulatory delays. They indicated that one should start thinking about validation requirements as early as the development phase. This is an effective cost-saving strategy, and prevents failed batches in production. For validation to be effective one must consider product characterization, purification design parameters, and current Good Manufacturing Practices (cGMP) requirements.

Example 10.1

Describe briefly some of the considerations that must be examined to set the stage for later validation work (Akers *et al.*, 1994).

Solution

One must initially consider product characterization, purification design parameters, and cGMP requirements as the background material for the development of an effective validation plan (Akers *et al.*, 1994). The characterization of the product by suitable analytic methods is critical in the development of a validation plan. These methods should be sensitive, reproducible, and reliable. Frequently, process variables are changed. These sensitive, and reproducible analytic methods will determine product equivalency. These authors emphasized that these product characterization methods should be employed for the starting materials, at the isolation and concentration steps, during purification, and for the final product.

The requirements for biopharmaceuticals are relatively high, and chromatographic methods are used to obtain the required purity. These authors emphasized that the development scientists should be able to obtain a consistent product quality. Otherwise, one cannot go to the pilot-scale level (the next level). Akers *et al.* (1994) emphasized that when examining purification design parameters, one should also consider the removal of nucleic acids, endotoxins, modified proteins, and host cell proteins (HCPs). Eaton (1995) indicated that the detection of HCP contaminants is, to a large extent, industry driven, and is proprietary in nature. Proprietary reagents and assays must be developed for the quantitation of HCPs that are unique to a novel purification process. Thus, the publication of relatively detailed descriptions of more recent developments is significantly delayed. There should also be cooperation between the fermentation, recovery, and purification scientists. Often fermentation scientists add materials to enhance the fermentation yields, but these may lead to the failure of the purification schemes. Maintenance routines and records of the performance of these routines are essential for a validation plan.

Products to be administered to humans must meet Food and Drug Administration (FDA) cGMP requirements (Akers *et al.*, 1994). In 1978, the FDA began to enforce a set of cGMP guidelines to ensure the adherence to specific validation requirements. Once again, development is a good time to become aware of these cGMP requirements. The compliance with these cGMP regulations means that appropriate documentation must be in place; and standard operating procedures for production, quality assurance, quality control, and facility management operations must be followed. Appropriate cGMP training of personnel must be provided, and records of this training must be present.

The equipment and utilities must also meet with cGMP standards. For example, utility systems for water used in manufacturing, in purified steam, and for compressed gases must be of pharmaceutical grade; and must be designed according to established industry standards. cGMP compliance also requires companies to document the use and cleaning of equipment and associated components.

Finally, these authors emphasized that the introduction of the validation procedure early in the process, though time-consuming, is cost-effective in the long term. It permits for more accurate financial planning and enables a company to enhance the quality of the product. Also, early discussions with the required regulatory officials may be initiated, and costly delays later on may be avoided.

Mahar (1993) defined validation as the "assurance that a process is closely followed during a product's manufacture." This author emphasized that the following industries face increasing validation requirements: biotechnology—parenteral drugs, finished pharmaceuticals, bulk pharmaceuticals, and food products. Furthermore, recombinant products and new technologies place an increasingly greater demand on regulatory agencies to ensure that products are safe and that manufacturing processes are effectively controlled. This author emphasized that validation procedures should be similar from one pharmaceutical manufacturing process to another. Also, validation procedures and requirements for different unit operations should be similar.

Kieffer and Nally (1991) defined validation as "the scientific study of a process to prove that the process is doing consistently what it is supposed to do (that is, the process is under control), to determine the process variables and acceptable limits of the variables, and to set up appropriate in-process controls." Nally and Kieffer (1993) emphasized that the fundamental purpose or value of validation is to increase the knowledge and the understanding of the process. This leads to more effective, more rapid troubleshooting, and better system control; enhances continual improvement of the process, system equipment, etc.; and finally empowers employees to control the processes and continuously improve them. This increased understanding of the causes of the variations in system parameters, and the wide dissemination of this knowledge within an organization lead to total quality (TQ). TQ is the improvement of products and processes, empowerment of all employees, and aggressive pursuit of learning.

Validation should be effectively integrated within the "customer chain," that is, the overall business strategies and the total product delivery process (Nally and Kieffer, 1993). Figure 10.1 shows the customer chain. The overall

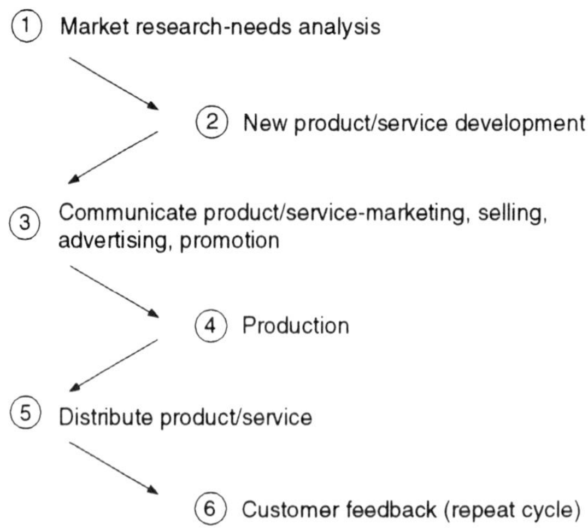

FIGURE 10.1 The customer chain. [From Nally, J. and Kieffer, R. (1993). BioPharm '93, San Francisco, CA, June 13–15, with permission.]

business process starts with identifying customer needs, and ends with supplying to the customer products that satisfy these needs. These authors emphasized that a pharmaceutical manufacturer should merge validation studies into efforts to continuously improve and optimize the overall customer chain and manufacturing process.

Bhote (1991) indicated the importance of the design of experiments that identify, optimize, and reduce the variability of the key variables that affect the process. Nally and Kieffer (1993) emphasized that this identification and reduction of the variability in critical parameters is a first step before validation can be undertaken. This is "making available the established documentation that provides a high degree of assurance that a specific process will consistently produce a product meeting its predetermined specifications and quality attributes."

Brewer (1986) delineated the interactive role of the production department, the analytic group, and the quality assurance group. Each of these should be separate. The production department converts raw materials to value-added products (or drugs). The analytic group provides accurate assays for the raw materials, intermediates in the process, and the bulk drug. This information provides the feedback to the production department to let them know how well their processes are functioning. The author emphasized that drug quality is a combination of the reproducibility of its production, and the sensitivity of the analytic methods utilized to delineate the impurities. Finally, before the drug can be released to the market, the quality assurance group determines if the drug or final product meets the regulatory requirements, and will also review the production department's documents.

In this chapter we will examine the validation of the production of biological products of interest by analyzing the different examples available in the literature. The different categories to be analyzed include the validation of: (1) rDNA processes, (2) column-based separation processes, (3) analytic methods, (4) processing for bulk biopharmaceuticals, (5) clinical monoclonal antibodies, (6) column regeneration studies, and (7) cleaning procedures.

II. VALIDATION OF rDNA PROCESSES

Developments have been analyzed in process control and in analytic methods that permit the drug, β-Urogastrone to meet the required standards and regulations set by governmental organizations (Brewer, 1986). This is a human polypeptide with wound and ulcer healing activities, and is obtained from *Escherichia coli*.

Example 10.2

Briefly describe the procedures involved in the validation of β-Urogastrone (Brewer, 1986).

Solution

Figure 10.2 summarizes the production and purification scheme for β-Urogastrone production. Brewer (1986) emphasized that in-process controls are required to maintain the reproducibility of the different steps. This author indicated that: (1) the maintenance and distribution of the seedstocks, and (2)

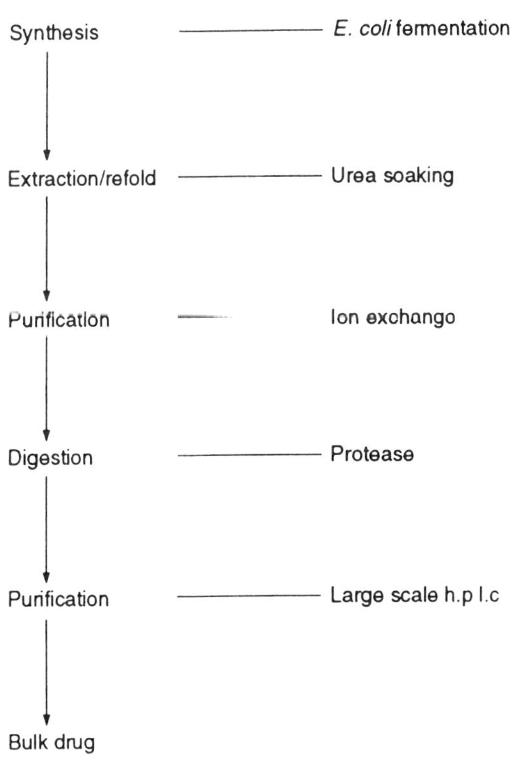

FIGURE 10.2 Production and purification of β-Urogastrone. [From Brewer (1986). *J. Chem. Technol. Biotechnol.*, *37*, 367, with permission.]

the growth of the fermentation inoculum are carefully controlled. Furthermore, there is a significant amount of information available concerning the fermentation conditions, such as the pH, aeration, and feed rates. This permits the control of the biological synthesis in the reactor.

Column chromatography is utilized to achieve the high purity required of this drug (Brewer, 1986). Furthermore, the successful control of the production, extraction, and purification process is possible due to the development of a specific and accurate assay for β-Urogastrone. This author indicated that the rDNA derived protein may be contaminated by endogeneous impurities (originating from the production organism), and by exogeneous impurities (from the culture medium and from the extraction and purification process). Also, heterogeneities in the protein may lead to its existence in different forms.

Low levels of endogeneous impurities, such as *E. coli* proteins, nucleic acids, and endotoxins may contaminate the product. The author indicated that β-Urogastrone has two amino acids, Phe and Thr, which do not appear in the primary sequence. It is unlikely that a contaminating protein would have these same two amino acids missing. Therefore, a sensitive method for the detection of *E. coli*-contaminating proteins is possible (Fig. 10.3).

Nucleic acids and endotoxins are of particular concern. These must be reduced to a level below 1 ng/ml concentration in the final product to avoid pyrogenic effects. Specific assays for endotoxins are available using the limulus lysate (LAL) assay (Garratt *et al.*, 1981). Furthermore, DNA contamination must be extremely low (10 pg of DNA per clinical dose).

Specific assays may be developed for exogeneous impurities such as the chemicals used for production, extraction, and purification of the product (Brewer, 1986). Many salts and buffers will pose only small problems. They can be determined by elemental, ion, and spectroscopic analyses. If bioactive chemicals are used, then specific assays for their detection are required. Penicillin is often used to stabilize plasmids during fermentation. However, this can form immunogenic protein conjugates, and a small fraction of the human population is hypersensitive to them. This was difficult to detect in β-Urogastrone

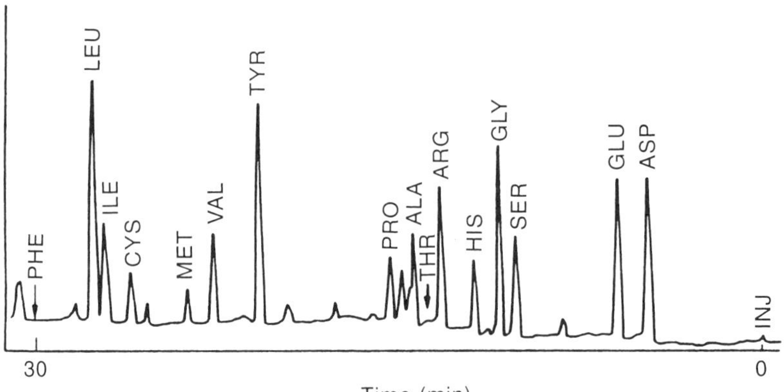

FIGURE 10.3 Amino acid analysis for β-Urogastrone. [From Brewer (1986). *J. Chem. Technol. Biotechnol., 37,* 367, with permission.]

production. Thus, the fermentation was redesigned to allow for the omission of penicillin.

The author noted that proteins, as expected, are not produced with absolute fidelity. Thus, heterogeneities can be expected. Also, modifications may occur due to downstream processing steps. For example, deamination and other chemical modifications will produce charge heterogeneity on protein products. Low percentage contamination of the product by these species may be identified and made quantitative by analytic high-pressure liquid chromatography (HPLC) and by isoelectric focusing.

The analysis presented by Brewer (1986) is of interest because it demonstrates how low levels of impurities and contaminating proteins may be measured by a combination of protein chemical, biochemical, and chemical analysis. All this assists and is critical in the validation process.

Example 10.3

Explain the concern over the removal of DNA and protein impurities in biopharmaceuticals (Briggs and Panfili, 1991).

Solution

There are two causes for concern over DNA and protein contamination in biopharmaceuticals (Briggs and Panfili, 1991). The first deals with good manufacturing practice, where validation and assurance are required that nonrelevant material is removed. Also, the impurities present a theoretical risk to the patient. These authors mentioned that the primary concern with contaminating DNA is that it may: (1) contain an oncogene, (2) cause an oncogene to be activated, or (3) cause a tumor inhibitory gene to be turned off. There is also greater concern relative to DNA contamination in products derived from mammalian cells, because mammalian cell cultures are more likely to harbor a virus that is infectious in humans.

The primary concern over contaminating proteins is the possibility of generating an immune response by the recipient of the biopharmaceutical. The immune response could be either acute (as in an allergic response such as anaphylactic shock) or chronic (such as autoimmune disease). The contaminating protein may also generate a biological response in the recipient. Such biological responses should be anticipated if the contaminating protein is a toxin, hormone, or cytokine with physiological effects in humans. However, the biological effects of contaminating proteins are expected to be transient. Briggs and Panfili (1991) indicated that product proteins differing in immunogenicity and potency may also be considered impure. These could be products with altered amino acids, glycosylation, etc.

The authors were careful to point out that the primary problem with the measurement of trace impurities and contaminants is the presence of a large amount of product protein or biological product. The choice of the analytic method is determined to a large extent by the interference by the product protein in the assay procedure. They concluded by stating that newer drugs will be made available by advancing technologies. However, one should anticipate increasing regulatory controls and concerns with regard to the administration of the therapeutic for human consumption. This leads to the development of more sensitive analytic methods that place greater emphasis on the quantitation of the impurities of significant risk. These developments are essential and critical in the development of a working and suitable validation plan (that matches the regulatory requirements).

Example 10.4

Briefly describe the avoidance of unsafe levels of host cell protein contaminants that might lead to toxic or immunologic reactions (Eaton, 1995).

Solution

The avoidance of unsafe levels of residual HCPs is not a trivial concern, because it may lead to toxic or immunologic reactions (Eaton, 1995). For example, during the early administration of recombinant growth hormone, unacceptable levels of residual *E. coli* HCPs not only eluted anti-HCP antibody, but also resulted in the elicitation of an undesirable antibody against the biopharmaceutical protein itself. This author emphasized improved purification methods that significantly decreased the bacterial HCP content of the hormone alleviated the problem. Thereafter, no specific anti-HCP antibody elicited was observed in the recipients of recombinant growth hormone of mammalian cell origin.

III. VALIDATION OF COLUMN-BASED SEPARATION PROCESSES

The downstream processing of biotechnology products usually includes quite a few column-based purification steps. These column-based separation steps are required to achieve the level of purity essential for therapeutic agents. Column-based separations are of four main types: gel filtration or size exclusion, ion exchange, reverse phase or hydrophobic, and affinity. A PDA Report of the Biotechnology Task Force (1992) indicated that laboratory studies using scaled-down columns and "spiking" experiments can yield valuable validation data. These authors emphasized that clearance studies done with spiking experiments with radiolabeled chemicals, toxic chemicals, or infectious biological agents should be done at small scale for reasons of worker safety and of avoiding contamination of production equipment. They emphasized that validation tests and challenges should be repeated enough times to ensure that reliable and meaningful results are obtained. Furthermore, it needs to be demonstrated that the manufacturing process consistently removes known and potential contaminants at the production scale. This would then eliminate the need for testing every batch for impurities.

Example 10.5

Briefly describe the validation of column-based separation processes (PDA Report, 1992).

Solution

The Biotechnology Task Force in a PDA Report (1992) indicated that process validation of column-based separations usually covers the following four major areas: process chemicals, column packing materials, equipment qualifications, and performance qualification of the process itself. These authors stressed that equipment qualifications may be broken down into installation qualifications (IQ) and operational qualifications (OQ). Basically, these qualifications ensure that the equipment is properly installed, calibrated, and functioning according to specifications.

The PDA Report (1992) indicated that the performance qualification (PQ) of the process will establish that it is effective and reproducible. Furthermore, the product meets with all the established release specifications. These authors emphasized that performance process validation should clearly specify the protocols that are prepared; and the procedures and tests to be conducted, the data to be collected, and the acceptance criteria. One also needs to identify, monitor, and document the important process variables. The PDA Report indicated that in the biopharmaceutical industry, process validation studies are performed by multidisciplinary groups with people from manufacturing, engineering, process development, quality control, and research.

For process chemicals the PDA Report (1992) indicated that chemical reagents, such as buffer salts, used to prepare solutions for column-based separations should be controlled in the same manner as other raw materials. Appropriate raw material sampling plans and specifications should be developed and approved by quality control. Furthermore, these authors emphasized that test procedures and sampling plans should be developed and validated. Finally, the water used in column-based separations should meet predetermined specifications, and the water producing system itself should be properly validated.

The selection of the chromatography media significantly affects the purity, uniformity, and other characteristics of the final product. The PDA report indicated that the column material should be quarantined on receipt. It should be released only after meeting the desired specifications. The media should yield the specified product purity. Once this has been established, it is worthwhile to reach an agreement with the supplier, where he provides a sample for user testing. The required quantity of the same batch is quarantined pending acceptance. It is worth the effort also to discuss the in-process and quality control with the media manufacturer.

Finally, the Biotechnology Task Force in their PDA Report (1992) emphasized that the "most important criterion for validation of a column-based separation is the demonstration that when operated in a specified manner, the overall process, or process step, yields a product of consistent quality which conforms to specifications." It should be clear that the process will not fail when it is operated within the specified ranges of critical process parameters, such as buffer pH and ionic strength, gradient shape, amount of material applied per unit volume of packing material, temperature, flow rate, and system pressure.

The PDA Report of the Biotechnology Task Force (1992) emphasized that there are no explicit rules for the validation of column-based separation processes. Their document is a good starting point. These authors emphasized that column validation along with in-process and quality control of the final product ensures the consistency of the product from batch to batch.

IV. VALIDATION OF ANALYTIC METHODS FOR PHARMACEUTICAL PRODUCT DEVELOPMENT

Hokanson (1994) indicated that guidelines for the validation of analytic methods for the testing of pharmaceutical products have been published in the United States in the *United States Pharmacopeia* (Paul, 1991); in Europe regulatory guidelines are found in *The Rules Governing Medicinal Products in the European Community* (Guidelines on the Quality, Safety, and Efficacy of Medicinal Products for Human Use, 1990); and in Canada regulations are cited in *Drugs Directorate Guidelines* (Health Protection Branch, 1992). This author emphasized that analytic method validation should be considered as a process that continuously provides maximum confidence in the reliability of test procedures, and not just as a regulatory requirement. Futhermore, validation is to be viewed as a dynamic procedure, which is extended further as additional information is obtained and as the test procedures expand to new analysts in different laboratories.

Example 10.6

Briefly describe the life cycle approach to analytic methods during pharmaceutical product development (Hokanson, 1994).

Solution

The life cycle approach to analytic methods during pharmaceutical product development has been analyzed (Hokanson, 1994). This author stated that the life cycle approach is, "The process that is initiated during the development of a new product. Also, the scope of the analytical method that is required is defined and the strategy for their validation must be formalized." Furthermore, the validation process may be divided into those requirements relating primarily to equipment, and to assessing sample and standard specifications (analyst specific). The validation protocol must define the tests necessary to characterize the reliability of the test procedures. Also, the acceptance criteria for all the studies to be performed should be delineated. Sometimes the validation data may fail to meet the acceptance criteria. Appropriate follow-up steps should be identified.

In accord with regulatory guidelines, the author indicated that the validation requirements for analytic methods for new drugs include (1) selectivity, (2) linearity (in the working concentration range), (3) limits of detection (LOD) and quantitation (LOQ), (4) accuracy, (5) precision, and (6) ruggedness.

Selectivity. The selectivity of an analytic procedure is its ability to measure the required analyte in the sample matrix. Selectivity must be demonstrated by testing mixtures of inactive excipients, drug degradation products, and synthesis impurities (if applicable). The author emphasized that at the time of New Drug Application (NDA) the drug degradation products should be well defined through controlled studies. From the selectivity analysis, an appropriate test or procedure can be established for routine analysis.

Linearity. The linearity studies demonstrate the method's ability to obtain test results directly proportional to the analyte concentration in a sample. This is within a given or specified concentration range. The working range (upper and lower levels) needs to be specified. Also, the procedures for testing drug products use, in general, single-point standard calibration. A single concentration of the reference standard is tested to determine the concentration of the test samples. Note that typical calculations assume that the response factor is the same for both the test and the reference samples. The response factor is the ratio of the response to the concentration.

In the working range, three to six measurements should be made for at least six samples of increasing concentration within the range from 25 to 125% of the targeted standard specified in the analytic procedure. Furthermore, replicate testing permits the determination of the precision component of the analysis. Finally, from the mean responses obtained a linear regression best-fit plot of the response curve with the concentration analyzed may be obtained. This helps compare the actual data points with those calculated from the regression values. This typical linearity plot provides an assessment of the proportionality of the response.

Limits of Detection and Quantitation. Linearity should also be carried out to detect impurities and degradation products in the presence of the drug (analyte). LOD and LOQ need to be established. LOD may be defined as that concentration that gives a peak height response three times greater than the baseline noise level. This is for chromatographic analysis. The LOQ may be defined as the lowest amount of an analyte that can be determined quantitatively with precision and accuracy under typical experimental conditions.

Accuracy. As expected, the accuracy of the analytic procedure is critical. The accuracy depicts the closeness of the test results obtained with that of the true value. It is useful to determine the accuracy by comparing the results obtained with another test method. Note that accuracy measurements provide an assessment of the effectiveness of the sample preparation procedure. The recovery studies define the overall range of the method (that is, the concentration range in which the linearity of the response, accuracy, and precision have been demonstrated). The author emphasized that the recovery studies demonstrate and instill confidence in the actual sample preparation procedures.

Precision. Furthermore, the precision of an analytic method is the degree of agreement among individual test results as the procedure is applied over and over again to multiple aliquots of a homogeneous sample. This author emphasized that the precision studies combine the equipment-related aspects of linearity and selectivity with the sample preparation considerations of the accuracy studies.

Ruggedness. Finally, ruggedness studies are essential. Initially, only one analyst may be analyzing the data using the same analytic equipment. Later on, more analysts may be added to the project. Then additional precision data should be obtained. An assessment of method ruggedness is required with regard to sample and standard stability. Later on interanalyst and interlaboratory ruggedness assessments may be required.

Initial validation of the analytic methods yields the required assurance of the reliability of these procedures (Hokanson, 1994). This validation is a dynamic process; and changes are inevitable as the product undergoes the development process, and more information is made available concerning the drug product formulation. Initial information on the validation process can serve as the basis of a later or new validation protocol for subsequent studies.

V. PROCESS VALIDATION OF BULK BIOPHARMACEUTICALS

Lazar (1993) indicated that bulk biopharmaceuticals are different from drug products. Bulk biopharmaceuticals are made by chemical synthesis, by fermentation, and by recovery from natural materials. Drug products are, however, made by formulation of materials of high established quality. This author defined a bulk biopharmaceutical as "an active ingredient that is intended to furnish pharmacological activity." Lazar (1993) indicated that although the principles of validation are universal, the differences between the processes used to produce the bulk biopharmaceutical chemicals and those used to produce drug products may require differences in application. Because, process validation has been presented already for drugs, it will not be repeated here for bulk biopharmaceutical chemicals. Only differences that are relevant or items that may not have been included when talking about drugs may be briefly presented. It is of interest to note that since the 1980s, the FDA has increased its attention to the production of bulk biopharmaceuticals due to episodes in which process failures have eventually led to product recall. Lazar (1993) indicated that the FDA expects to apply the finished dosage forms of the cGMP requirements to all bulk biopharmaceutical areas including development, manufacturing, control, and distribution.

Lazar (1993) mentioned the different types of validation that are used:

Prospective validation is the procedure to establish (or establishing) documented evidence that a system does what it is supposed to do based on a plan.

Concurrent validation is the procedure to establish documented evidence that a system does what it is supposed to do based on information generated during the actual implementation of the system.

Retrospective validation is the procedure to establish documented evidence that a system does what it is supposed to do based on a review and on an analysis of historic information. It has been mentioned that retrospective validation is 20 times more expensive than prospective validation, and that sometimes it is cheaper to replace the old system than to validate it (Rosser, 1994).

Revalidation is the procedure that may be initiated periodically or when changes are made to equipment, systems, or processes. Lazar (1993) emphasized that the revalidation effort will depend on, as expected, the nature and the extent of the changes. This author emphasized that in the presence of changes or process failures, there should be a system for periodic review of validated processes to assess the need for revalidation. Any change to a validated process should be analyzed with respect to potential cost savings, environmental effects, and need to revise a drug master file (DMF) or NDA (Demmer *et al.*, 1994). Also, one needs to determine if revalidation is necessary and how extensive this should be.

VI. VALIDATION OF THE PREPARATION OF CLINICAL MONOCLONAL ANTIBODIES

Monoclonal antibodies (MAbs) from hybridomas are now in clinical use, and must be made safe (Mariani and Tarditi, 1992). These authors emphasized that the new biological products must be made free of viral and nucleic acid contamination. They indicated that virus validation is increasing as the number of biologicals under development increases.

Example 10.7

Briefly describe the validation procedure to purify MAbs from mouse ascites fluid (Mariani and Tarditi, 1992).

Solution

The validation of the preparation of MAbs has been analyzed (Mariani and Tarditi, 1992). Figure 10.4 shows the two-step HPLC method used to purify MAbs. This figure also shows that the viral agent removal was validated for the two chromatographic steps. Protein A chromatography was used along with hydroxylapatite chromatography to provide the required MAb purity, which is equivalent to that obtained with immunoaffinity chromatography (Mariani, 1989).

Mariani and Tarditi (1992) indicated that validation of the process was performed on a research scale version of the purification process (see Figure 10.4). For validating the protein A chromatography step, four 0.5-ml aliquots of murine ascites fluids each containing 4.5 mg of monoclonal immunoglobulin (IgG) were spiked with four different viruses. These authors indicated that these viruses provided a representative panel of contaminating agents with a wide range of physicochemical characteristics. These samples were then applied to each of the four different analytic protein A HPLC cartridges. Each column was then eluted with a buffer.

These authors indicated that the viral reduction factor is the quotient of the virus titer in the spiked samples and that in the emergent IgG samples. Similarly, the reduction factor was evaluated for the hydroxylapatite chromatography step. Except in this case the spiked samples were aliquots of the protein A-purified murine monoclonal IgG. This was the product from the first purification step. These authors indicated that the samples were subjected to

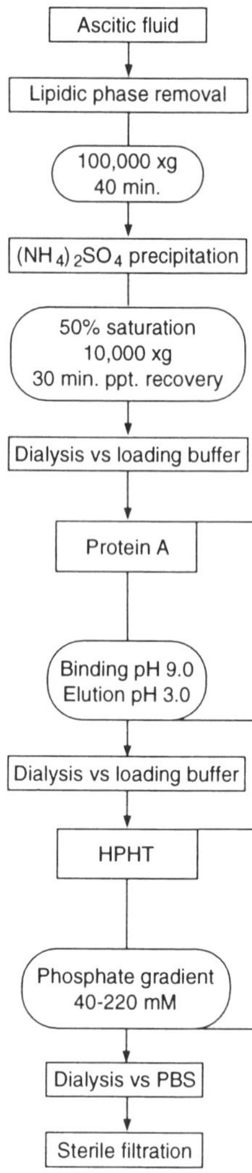

Virus	Polio Sabin 1	Alueszky Herpes	SV40	Mo–MuLV
Nucleic acid	RNA	DNA	DNA	RNA
Envelope	Non enveloped	Enveloped	Non enveloped	Enveloped (retrovirus)
Resistance at room temperature	Very resistant	Sensitive	Very Sensitive	Sensitive
Toxicity of the products on the cells used for the titration	No toxicity	No toxicity	No toxicity	No toxicity

Protein A				
Total titre of infected sample loaded on the column	3.74×10^6 pfu	4.87×10^5 pfu	1.8×10^7 pfu	2.5×10^9 ffu
Total infectious titre of harvested sample	4.55×10^3 pfu	5×10^3 pfu	4.24×10^5 pfu	4.35×10^7 ffu
Protein A Reduction Factor (1) log pfu or log ffu	2.91	1.98	1.63	1.76

HPHT				
Total titre of infected sample loaded on the column	1.80×10^6 pfu	3.12×10^5 pfu	2.18×10^7 pfu	3.2×10^9 ffu
Total infectious titre of harvested sample	150 pfu	<9.5 pfu	1.45×10^3 pfu	<4.6 ffu
HPHT Reduction Factor (2) log pfu or log ffu	6.08	>4.52	4.17	8.84

Overall				
Overall Reduction Factor (1 + 2) log pfu or log ffu	8.99	>8.50	5.80	10.64

pfu: plaque forming units
ffu: focus forming units

FIGURE 10.4 The two-step HPLC method utilized to purify monoclonal antibodies (MAbs) from mouse ascites fluid. Validation was done for viral agent removal by the protein A and hydroxylapatite chromatography steps. [From Mariani, M. and Tarditi, L. (1992). *Biotechnology, 10,* 394, with permission.]

both the hydroxylapatite chromatography step and a 20-min linear gradient from 40 to 200 mM phosphate. The removal of the viral agents by the two-step antibody purification procedure is satisfactory, because there is a log reduction factor of 5.8 to 10.64 depending on the virus strain.

They also demonstrated that their purification procedure was able to remove murine DNA. Table 10.1a shows the removal of murine DNA during the MAb purification. These authors indicated that the validation of the process for murine DNA removal includes three purification steps: ammonium sulfate precipitation, protein A chromatography, and hydroxylapatite chromatogra-

TABLE 10.1a Elimination of Murine DNA during Monoclonal Antibody (MAb) Purification[a]

Purification step	Input ^{32}P-murine DNA (cpm[b])	Recovered IgG fraction (cpm[b])	% Remaining DNA		Reduction factor	
			Single	Cumulative	Single	Cumulative
Ammonium sulfate precipitation	844,916	43,938	5.2	5.2	1.28	1.28
Protein A	4,505,760	4,178	0.09	0.00468	3.05	4.33
Hydroxylapatite	4,424,440	114,054	2.58	0.00127	1.59	5.92

[a] From Mariani, M. and Tarditi, L. (1992). *Biotechnology, 10,* 394, with permission.
[b] cpm = Counts per minute.

phy. For the ammonium sulfate precipitation step the reduction factor was 1.28. This means that there was $10^{1.28}$ times less DNA in the effluent than in the sample input. For the protein A and the hydroxylapatite chromatographic steps, the reduction factors were 3.05 and 1.59, respectively. The cumulative reduction factor was 5.92, that is, the sum of the reduction factors of the three steps. The FDA recommendation for this process was that the final product should not contain more than 10 pg per dose (of 1 mg). The initial concentration of the DNA of 5 to 10 ng/mg in the ascites fluid would be reduced to less than 10 fg per dose of the final product by the purification steps. This is due to the high reduction factor (cumulative) obtained. Table 10.1b shows the results of the spike-off experiments for viral removal.

These authors also validated the removal of protein A from the final product. According to the Federal Code of Regulations (FCR) [U.S. CFR §21, ¶610, 15b], the concentration of protein A contamination must not exceed 1 ppm per dose. They noted that for routine protein A purification, the detection of protein A was always below the detection limit of the enzyme-linked immunosorbent assay (ELISA) kit (Oros System, Cambridge, UK). The detection limit is 40 ng/ml in the presence of mouse IgG at 1 mg/ml. Thus, on an average, 1 mg of IgG (a single dose) purified by protein A chromatography would contain 40-ppm protein A at most. Thus, the hydroxylapatite column needed to reduce protein A by a factor of 50 to meet the federal regulations. In fact, the hydroxylapatite

TABLE 10.1b Murine Spike-Off Experiments[a]

Virus to be removed	Protein A clearance	Hydroxylapatite clearance	Total clearance
Polio Sabin 1	$10^{2.91}$	$10^{6.08}$	$10^{8.99}$ pfu[b]
SV40	$10^{1.63}$	$10^{4.17}$	$10^{5.80}$ pfu
Aujeszky herpesvirus	$10^{1.98}$	$10^{4.52}$	$10^{6.50}$ pfu
Moloney murine	$10^{1.76}$	$10^{8.84}$	$10^{10.64}$ ffu[c]

[a] From Mariani, M. and Tarditi, L. (1992). *Biotechnology, 10,* 394, with permission.
[b] pfu = Plaque-forming units.
[c] ffu = Focus-forming units.

column succeeded in removing the protein A quite effectively, in that the protein A level in the final product was 0.24 ppm. Also, these authors indicated that protein A losses from the first purification step of 40 ng/ml have never been found.

Mariani and Tarditi (1992) indicated that the validation of the purification protocol has been limited to the two chromatographic steps (plus ammonium sulfate precipitation). Other steps such as sterile filtration, dialysis, heat inactivation, and ultracentrifugation of the ascites fluid could also be used to remove these contaminants. These were not evaluated because they were not required.

VII. VALIDATION STUDIES FOR THE REGENERATION OF ION-EXCHANGE CELLULOSE COLUMNS

Ion-exchange chromatography is widely used at different stages in the downstream processing of biological products. Levison *et al.* (1995) indicated that in the manufacture of biopharmaceuticals the regulatory aspects of the chromatographic process is an important aspect. These authors stated that the large-scale ion-exchange processes can be carried out using batch stirred tank or column techniques. As far as validation is concerned, column techniques are better because they are easier to manage and control compared with an open batch stirred tank unit operation. These authors emphasized that for the validation process it is necessary to demonstrate that the product eluted is of the desired quality with regard to its sterility, endotoxin content, and contaminants arising from the chromatographic medium itself (leachables).

Sofer and Nystrom (1991) emphasized that the question of leachables is a key issue in the validation process when chromatographic columns are used. If leachables from the chromatographic medium are coeluted with the product, then this is a serious problem. Levison *et al.* (1995) indicated that there is little published information on leachables, though these leachables have received attention in the field of affinity chromatography. Here this problem is referred to as ligand leakage.

Example 10.8

Show validation studies in the regeneration of ion-exchange cellulose (Levison *et al.*, 1995).

Solution

Levison *et al.* (1995) analyzed the clean-in procedure (CIP) for the process-scale chromatography of hen egg-white proteins on two fast-flowing anion exchange celluloses derivatized with either (diethylamino)ethyl (DEAE) or 2-hydroxypropyltrimethylammonium (QA) functional groups. These authors indicated that the CIP procedure that is effective for column regeneration has been examined for chromatographic performance, sanitization, and media stability in terms of leachables.

They indicated that a CIP using 0.5 M NaOH for 12 to 16 h (overnight) is effective in restoring column performance for DE 52 (Levison *et al.*, 1989),

QA 52 (Levison *et al.*, 1990), DE 92 (Levison *et al.*, 1992), and Express-Ion D (Levison *et al.*, 1994). Figure 10.5 compares the column performance of Express Ion Q for hen egg-white loading: (1) pre-CIP and (2) post-CIP treatment. One notes that the elution profile after CIP treatment is similar, if not slightly improved over that observed for pre-CIP treatment. The analytic loading of hen egg white is shown in Fig. 10.5(a). Thus, the NaOH treatment does not have any detrimental effect on the chromatographic performance of the medium.

The CIP procedure was investigated as an effective bed sanitization step (Levison *et al.*, 1995). Sodium phosphate was used as the mobile phase, because it is more suitable for sustaining the viability of microorganisms than a Tris buffer is. Table 10.2 shows the results obtained following a challenge with a

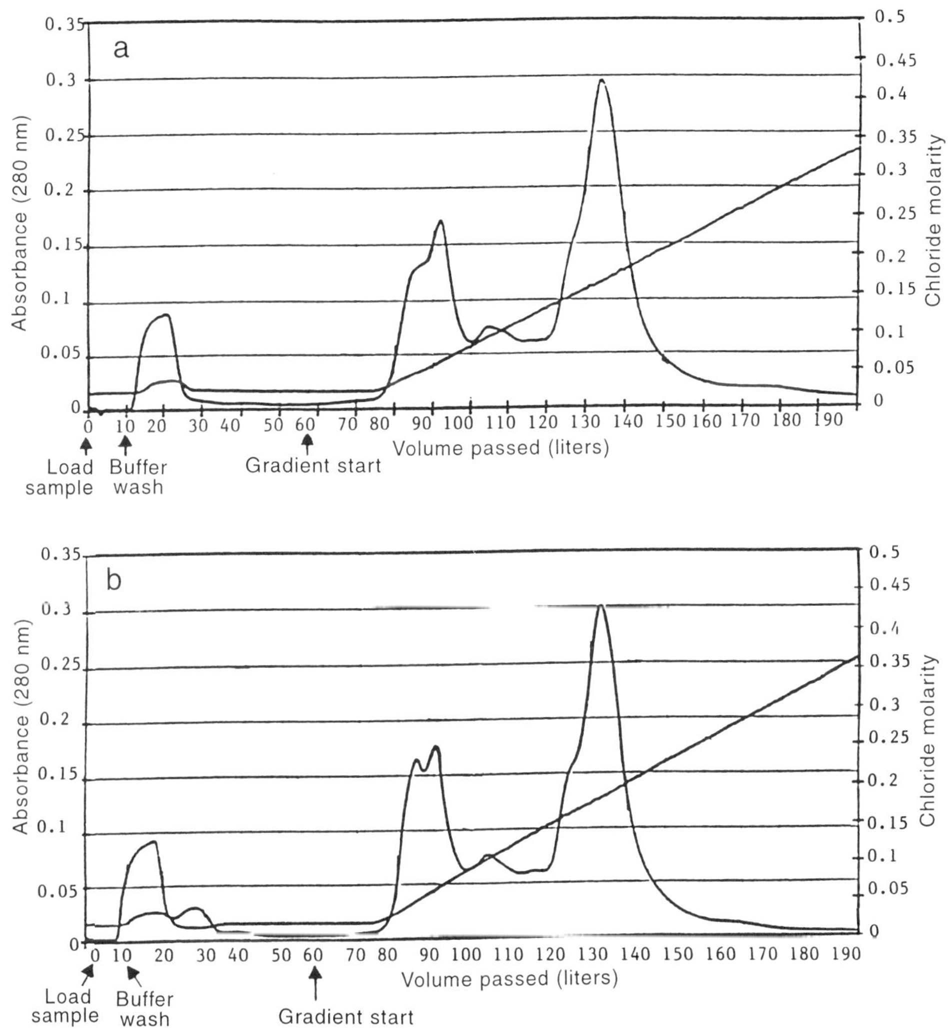

FIGURE 10.5 Column chromatography of hen egg-white proteins on Express-Ion Q on a process scale (16 × 45 cm i.d.) using 0.025 M Tris-HCl buffer, pH 7.5: (a) analytic loading (100 g) before preparative run; (b) analytic loading (100 g) after CIP. [From Levison, P. R. *et al.* (1995). *J. Chromatogr. A, 702,* 59, with permission.]

TABLE 10.2 Sanitization Testing of Columns of Express-Ion D and Express-Ion Q[a]

Stage of investigation	TVC (cfu[b]/ml)	Sterility test	Endotoxin (EU/ml)	Rabbit pyrogen test
Express-Ion D				
Challenge	5.5×10^6	nd[c]	nd	nd
Pre-CIP	7.7×10^4	Fail	>60	nd
Post-CIP	<1	Pass	<0.06	Pass
Express-Ion Q				
Challenge	4.6×10^6	nd	nd	nd
Pre-CIP	7.6×10^3	Fail	>60	nd
Post-CIP	<1	Pass	<0.06	Pass

[a] From Levison, P. R. (1995). *J. Chromatogr. A, 702*, 59, with permission.
[b] cfu = Colony-forming units.
[c] nd = Not determined.

mixed suspension of *E. coli, Staphylococcus aureus, Pseudomonas aeruginosa, Aspergillus niger, Candida albicans,* and *Bacillus subtilis.* These authors indicated that following the CIP and subsequent bed reequilibration using sterile endotoxin buffers, the column gave negative responses in the total viable count test, was sterile, and contained very low levels of endotoxin yielding a negative response in the rabbit pyrogen test. Table 10.2 and this test demonstrate that the 0.5 M NaOH overnight treatment is an effective procedure for the sanitization of packed columns of Express-Ion D and Express-Ion Q following a heavy microbial challenge using a mixed suspension of microorganisms. Sofer and Nystrom (1991) recommended this for validation studies.

Levison *et al.* (1995) indicated that the final aspect of their analysis deals with leachables. More specifically, this refers to the hydrolysis of the functional group from the ion exchangers during the CIP procedure. These authors indicated that Whatman International derivatizes the microgranular cellulose matrix with the DEAE group

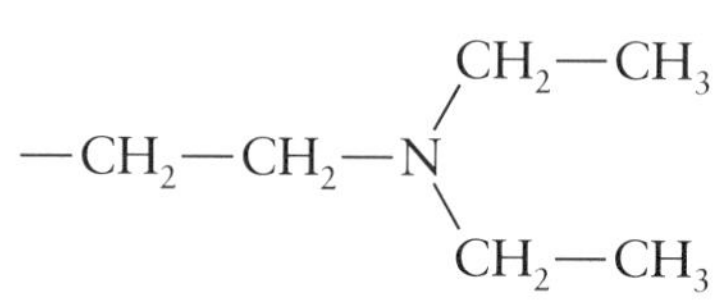

and the QA group

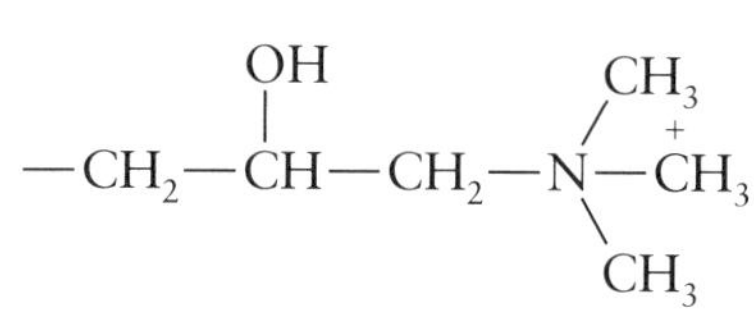

Levison (1993b) indicated that the functional group is covalently attached to the cellulose via ether linkages to the distal carbon atoms. This author indicated that if the hydrolysis of the functional groups were to take place, then it is to be anticipated that the alcohol *N,N*-diethylethanolamine would be liberated from the Express-Ion D column. Similarly, the 2,3-dihydroxypropyltrimethylammonium salt would be liberated from the Express-Ion Q column. Levison *et al.* (1995) noted that no detectable hydrolysis of the functional groups had occurred during either the CIP and the NaOH fraction; and also the reequilibrated media contained no detectable volatile organics under conditions where either the *N,N*-diethylethanolamine (for Express-Ion D), or the 2,3-dihydroxypropyltrimethylammonium salt (for Express-Ion Q) could be identified.

The analysis of Levison *et al.* (1995) is an excellent example of what needs to be done with regard to validation of the regeneration of ion-exchange cellulose columns. This study: (1) should serve as a guideline toward the validation of these media in regulated processes, and (2) may form the basis for extended process-specific validation studies.

VIII. CLEANING VALIDATION AND RESIDUE LIMITS

The quality of a pharmaceutical product is significantly affected by the cleaning process, which involves pipes, equipment, and containers with pharmaceutical products (Zeller, 1993). The FDA is paying attention to the cleaning aspect in validation studies to improve the consistency and uniformity in inspection coverage. The FDA has sought industrial help as far as the cleaning procedures in validation is concerned to help serve as a guide, to assist in the harmonization of the process, and to present and specify acceptable contamination limits.

Example 10.9

Briefly describe some of the important results of cleaning validation and residue limits (Zeller, 1993).

Solution

Some of the aspects of validating CIPs for equipment used to manufacture creams and ointments have been analyzed (Zeller, 1993). One of the basic problems in the validation of cleaning procedures is the sampling process. The author indicated that there are three basic sampling procedures: (1) swabbing the inner surfaces, (2) sampling the final-rinse fluid, and (3) testing the following batch for contamination. This author indicated that the third technique has associated with it risk, cost, FDA attitude, etc. Thus, only the first two options are examined.

Swabbing inner surfaces produces data with problems because this is a random process (Zeller, 1993). This author cautioned that easily accessible swabbing areas could lead to overoptimistic results. Conversely, poorly accessible areas could produce over-pessimistic results. Also, data from rinsing-fluid samples may yield false results if the contaminating substances are either poorly soluble or insoluble, or if the fluid cannot make good contact with all parts of the equipment. The author recommended combining both of these methods to help alleviate the disadvantages in each of these methods. Figure 10.6 shows the flowchart or decision tree to assist in the selection of the sampling proce-

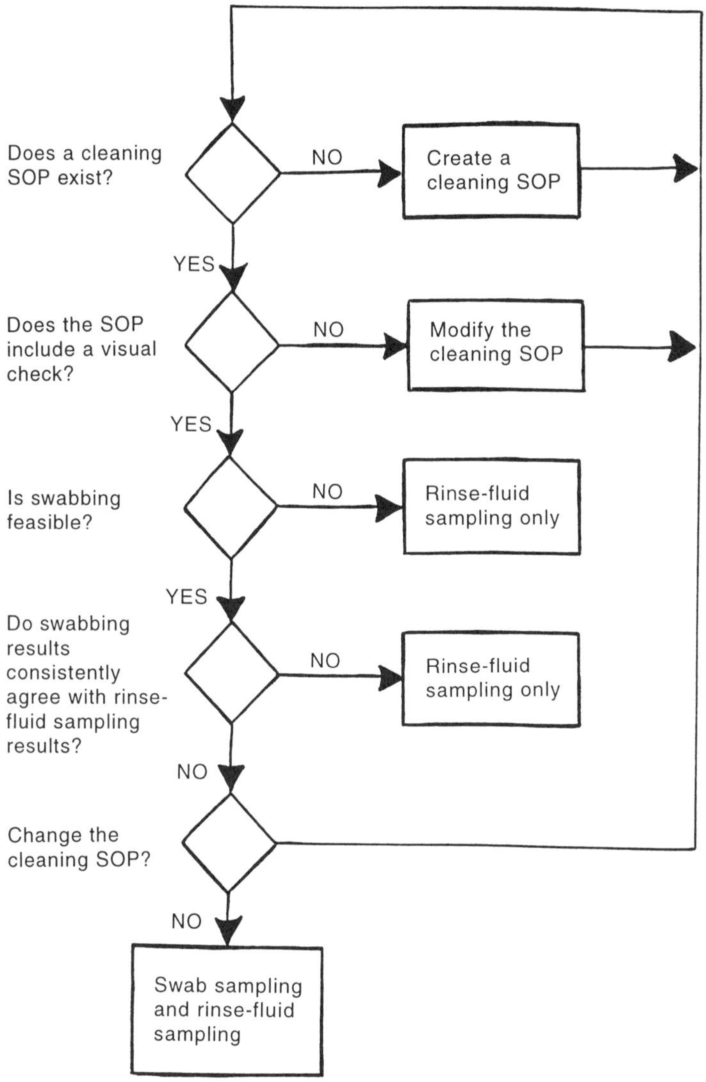

FIGURE 10.6 Flow-chart or decision tree to help select the sampling procedure. [From Zeller, A. O. (1993). *Pharm. Technol.*, *17*(10), 72, with permission.]

dure. The intent of combining the two methods is to make the validation and revalidation process as economical as possible, besides not reducing the significance of the results or the assurance of the product quality. Simplicity requires the elimination of the swab test, wherever it is possible to do so.

The selection of the analytic method should pay attention to both reliability and practicality (Zeller, 1993). This author suggested identifying one of the ingredients in the formulation as the *guiding substance*. In his case the selection of the guiding substance took into consideration the following factors: (1) concentration of the target substance in the product, (2) ultraviolet (UV) absorption characteristics, (3) pharmacological potency, (4) existence of a suitable analytic

TABLE 10.3a Amounts of Guiding Substance Found in the Samples of Rinse Fluids[a]

	3rd Rinse	4th Rinse	5th and Final rinse
Batch 20003:	16.1	3.9	1.4
amount of guiding	16.1	3.5	1.6
substance (mg)	18.2	4.2	1.6
	17.6	4.3	1.2
	mean = 17.0	mean = 4.0	mean = 1.4
Batch 20005:	13.1	4.5	3.4
amount of guiding	14.7	5.0	4.2
substance (mg)	13.7	4.9	4.0
	mean = 13.8	mean = 4.8	mean = 3.9

[a] From Zeller, A. O. (1993). *Pharm. Technol.*, *17*(10), 72, with permission.

method, and (5) solubility. Table 10.3a shows some typical numbers for the amounts of guiding substance found in the samples of rinse fluids. One notes a continuous decrease in the concentration of the guiding substance observed in the rinsing fluids of the last three washes. Table 10.3b shows the amounts of guiding substance found in specific places by swabbing. As expected, the highest concentrations (still an order of magnitude below the limit) were found in narrow corners or angles that are poorly accessible to the water jet from the CIP nozzles.

The detergent selection should also be done carefully, in that: (1) it should be easily detected, and (2) it should not affect the detection of the guiding substance (Zeller, 1993). Furthermore, this author suggested that the level of microbial contamination should also be evaluated. Microbial sampling was performed on the surface of the equipment. Plates were incubated and the number of colony-forming units were counted. For validating their process this author tested for *Staphylococcae*, *Pseudomonas*, *Enterococcae*, yeasts and

TABLE 10.3b Amounts of Guiding Substance Found in Specific Places by Swabbing[a]

Place of swabbing	Area	Amount of guiding substance (mg)
Interior surface of drum	10 × 10	1.50×10^{-4}
Paddle	10 × 10	Below detection limit
PTFE scraper	10 × 5	Below detection limit
Aspiration pipe curvature	12 × 5	Below detection limit
Stirrer	10 × 5	Below detection limit
Screw at the stirrer	2 × 2	1.58×10^{-5}
Pump–pipe exit	3 × 5	2.13×10^{-4}
Curve of drum lid	10 × 5	1.15×10^{-4}
Observation window and wiper	10 × 10	6.10×10^{-5}
Security valve exit	2 × 2	3.79×10^{-4}

[a] From Zeller, A. O. (1993). *Pharm. Technol.*, *17*(10), 72, with permission.

molds, and total organism count. Furthermore, cleaning standard operating procedure (SOP) resulted in no measurable levels of microorganisms in the majority of their equipment. However, a measurable but acceptable level of microorganisms was detected at the recirculation piping and at the solvent feedpipe. An extra cleaning procedure was added at these locations. This modification in the CIP led to the required absence of microorganisms at all locations. This author also indicated that their analytic methods were validated, and also the sampling of the rinsing fluids containing detergent was validated.

Adner and Sofer (1994) indicated that cleaning and the validation of cleaning is a critical issue for the manufacture of recombinant DNA protein products, MAbs, and oligonucleotide therapeutics. These authors further emphasized that the increasing multiuse of facilities leads to an increasing importance to the cleaning aspect and its validation. Chromatography is used quite often during the purification steps. Depending on the feed stream, at least two or three steps may be required to achieve the required purity. These authors emphasized that chromatography is not a sterile process, and cleaning is absolutely essential so that one obtains a reproducible process. They emphasized that inadequate cleaning of columns and systems may lead to further column and product contamination.

Example 10.10

Briefly describe chromatography cleaning validation (Adner and Sofer, 1994).

Solution

Cleaning should be part of every chromatographic cycle (Adner and Sofer, 1994). One estimate is that approximately 20% of each cycle is devoted to cleaning. Berglof (1993) indicated that viruses undetectable after the first cycle can elute in a subsequent cycle with the product if inadequate sanitization steps are performed after each cycle. These authors emphasized that a reproducible and validatable chromatographic process can only be achieved if its cleaning cycle incorporates appropriate detection methods, and is both controlled and reproducible. According to these authors it is critical to work with an actual production stream. Furthermore, challenges performed at small scale are helpful. However, production-scale chromatography equipment must be used for validation. During validation with production batch sizes, they indicated that challenges should include cleaning under the worst-case scenarios.

During validation cleaning, Adner and Sofer (1994) recommended that the detection equipment measure specific contaminants, analyze HPLC profiles, evaluate total organic carbon, determine total protein, and measure pH and conductivity (Baffi, 1993). Though visual inspection is helpful, it is inadequate in itself. Unanticipated feed stream variability may cause fouling of columns and will reduce the column resolvability (Adner and Sofer, 1994). Thus, cleaning methods should be robust enough to take care of this variability. Furthermore, these authors indicated that storage conditions should be adequate (and validated) so as to ensure the absence of microbial contamination, and to ensure the removal of storage solutions before subsequent runs.

Generally, the packed columns used earliest in a process require the most stringent cleaning conditions (Adner and Sofer, 1994). Irreversible binding of specific contaminants to chromatographic media should be evaluated as soon as possible. Often a loss in capacity may occur. Sometimes there may be no loss in capacity of the product if, for example, aromatic compounds are bound. These may just lead to discoloration of the chromatographic media. Then the media may function reproducibly for multiple cycles.

These authors indicated that fermentation feed streams often contain large quantities of nucleic acids. The use of 3 M NaCl and 1 M NaOH is effective in removing nucleic acids from anion exchangers. However, sometimes the 3 M NaCl and 1 M NaOH may not remove all the nucleic acids, because they are so tightly bound. The authors also indicated that some of the radiolabeled nucleic acid maybe retained. Then validation in this case requires demonstrating, at a small scale, that radiolabeled DNA does not elute with the product under operating conditions. If the intended use of the chromatographic medium is for 100 cycles, then these authors recommended that the column be cycled 100 times and the product be tested for the presence of radiolabeled DNA.

It is also suggested that a chromatographic medium be dedicated to one product (Adner and Sofer, 1994). Cleaning after each cycle prevents and minimizes fouling, and extends the lifetime of the medium. Furthermore, cleaning prevents bound denatured proteins and other contaminants from eluting in subsequent runs with the product. Seamon (1993) indicated that the protein buildup in columns is a major concern with columns, because it can affect column performance, contaminate subsequent runs, cause denaturation, and complicate validation.

Table 10.4 shows some typical cleaning agents and their applications. Adner and Sofer (1994) cautioned that each feed stream is unique, and a careful analysis of the suitability and efficacy of the cleaning method should be made for each step in the process.

These authors indicated that one method of evaluating a cleaning agent is by challenge testing. A known contaminant is added to a product. Then it is treated with the cleaning agent. Thereafter, it is tested for contaminating residues. Table 10.5 indicates the effect of 0.5 M NaOH on microorganisms. Note that in all cases, after 30 min the surviving organism concentration was less than three organisms per milliliter. In this case the test was performed in S-Sepharose HP.

The killing efficiency is highly dependent on the type of organism (Adner and Sofer, 1994). Nonspore-forming bacteria and fungi are rapidly killed,

TABLE 10.4 Cleaning Agents[a]

Cleaning agent	Contaminants
Sodium hydroxide	Virus, endotoxins, nucleic acids, proteins
Sodium chloride	Nucleic acids, proteins
Detergents	Lipids, hydrophobic proteins

[a] Adner, N. and Sofer, G. (1994). *Biopharm,* 7(3), 44.

TABLE 10.5 Effect of 0.5 M NaOH on Microorganisms (Performed in S-Sepharose HP)[a]

Test strain	Initial concentration (organisms/ml[b])
Staphylococcus aureus	2.0×10^5
Escherichia coli	6.0×10^5
Pseudomonas aeruginosa	1.4×10^6
Candida albicans	2.6×10^5
Aspergillus niger	4.5×10^6
Gram-negative strain mixture	7.7×10^6

[a] Adner, N. and Sofer, G. (1994). *Biopharm*, 7(3), 44.
[b] Surviving organisms per milliliter after 30 min in all cases was less than three.

whereas bacterial spores require higher concentration and longer contact time. These authors emphasized that FDA expects "cleaning validation protocols will include sampling of equipment surfaces, including those which are more difficult to clean, as well as testing water and/or solvent rinses for residues" (Lord, 1993). Furthermore, the effectiveness of the cleaning steps must be evaluated, and the cleaning procedures should be demonstrated to be consistent.

In conclusion, these authors indicated that cleaning and cleaning validation are an essential part of the production of a process for therapeutics. However, there are no specific requirements. A suitable cleaning program should begin at the start of the development, and must include a validation master plan. Furthermore, cleaning and sanitization helps ensure that the process produces a reproducible product of specified quality. They emphasized that companies need to incorporate the cleaning procedure into the development process. This would assist considerably in validating the process at full scale.

IX. CONCLUSIONS

Validation is an integral part of the development process for therapeutics. Besides, it is mandatory due to FDA cGMP guidelines. Hughes (1994) recommended developing a validation master plan (VMP) that is a valuable internal document. This should contain the company's philosophy and intentions in all operational areas that assist in the manufacture of products. This author emphasized approaching the VMP as an integrated plan for all functional areas.

Resources for validation, of necessity, must compete with the resources required for development. However, these are necessary, and Bevans-Lynch (1994) described how these resources for validation may be effectively allocated. This author recommended the involvement of validation with design review and approval of drawings, defining responsibilities of the company, subcontractors and consultants, equipment shakedown, documentation, calibration, SOPs, test equipment, validation and test technicians, raw data review, personnel training, and final reports and change control. As expected, the role of validation is very intensive. Both Bevans-Lynch (1994) and Burr (1994) emphasized the interactions and the communication involved between validation and different internal as well as external (FDA and other governmental) agencies.

REFERENCES

Adner, N. and Sofer, G. (1994). *Biopharm, 7*(3), 44.

Akers, J., McEntire, J., and Sofer, G. (1994). *Biopharm, 7*(2), 54.

Baffi, R. (1993). Assay Methodology for Multiproduct Facility Validation, *BioPharm Conference Proceedings '93,* Advanstar Communications: Eugene, OR, pp 41, 42.

Berglof, J. (1993). Validating Aspects Relating to the Use of Chromatographic Media: Biotechnology of Blood Products, *Colloq. Inst. Natl. Rech. Med. 227,* Rivat, C. and Stoltz, J. F., Eds., pp 31–36.

Bevans-Lynch, R. (1994). Managing Validation Resources, BioPharm '94, San Francisco, CA, June 13–15.

Bhote, K. R. (1991). *World Class Quality,* American Management Association: New York.

Biotechnology Task Force, PDA Report, (1992). Industry Perspective on the Validation of Column-Based Separation Processes for the Purification of Proteins, *J. Parenteral Sci. Technol., 46*(3), 87.

Brewer, S. J. (1986). *J. Chem. Technol. Biotechnol., 37,* 367.

Briggs, J. and Panfili, P. R. (1991). *Anal. Chem., 63,* 850.

Burr, S. S. (1994). A Validation Success Story—Change and Approval, BioPharm '93, San Francisco, CA, June 13–15.

Demmer, F., Franklin, N. C., Geussenhainer, S., Hausler, H., Kirrstetter, R., Rufer, C., Walter, E., and Zimmermann, F. (1994). *Pharm. Technol., 18*(12), 36.

Eaton, L. C. (1995). *J. Chromatogr. A, 705,* 105.

Garnick, R. L., Solli, N. J., and Papa, P. A. (1988). *Anal. Chem., 60,* 2546.

Garratt, D. C., Hartley, R. E., and Mussett, M. V. (1981). *Pharm. J., 226,* 112.

Guidelines on the Quality, Safety, and Efficacy of Medicinal Products for Human Use, (1990). *The Rules Governing Medicinal Products in the European Community,* Vol. 3, Addendum, July.

Health Protection Branch, Acceptable Methods, (1992). *Drugs Directorate Guidelines,* Minister of National Health and Welfare: Ottawa, Canada.

Hokanson, G. C. (1994). *Pharm. Technol., 18*(9), 118.

Hughes, T. (1993). Integrated Master Plan for Validation, BioPharm '93, San Francisco, June 13–15, 1993.

Kieffer, R. and Nally, J. (1991). *Pharm. Technol., 15*(9), 131.

Lazar, M. S. (1993). *Pharm. Technol., 17*(12), 32.

Levison, P. R. (1993a). In *Preparation and Production Scale Chromatography,* Ganetsos, G. and Barker, P. E., Eds., Marcel Dekker: New York, p 617.

Levison, P. R. (1993b). In *Celluloses: Materials for Selective Separations and Other Technologies,* Ellis Horwood: Chichester, UK, p 25.

Levison, P. R., Badger, S. E., Toome, D. W., Carcary, D., and Butts, E. T. (1989). In *Downstream Processing in Botechnology II,* de Bruyne, R., and Hughebaert, A., Eds., Royal Flemish Society of Engineers: Antwerp, The Netherlands, p 2.11.

Levison, P. R., Koscieiny, M. L., and Butts, E. T. (1990). *Bioseparation, 1,* 59.

Levison, P. R., Badger, S. E., Toome, D. W., Koscieiny, M. L., Lane, L., and Butts, E. T. (1992). *J. Chromatogr., 590,* 49.

Levison, P. R., Badger, S. E., Toome, D. W., Streater, M., and Cox, J. A. (1994). *J. Chromatogr., 658,* 419.

Levison, P. R., Badger, S. E., Jones, R. M. H., Toone, D. W., Streater, M., Pathirana, N. D., and Wheeler, S. (1995) *J. Chromatogr. A, 702,* 59.

Lord, A. (1993). International Society for Pharmaceutical Manufacturing (ISPE) Meeting, March.

Mahar, J. T. (1993). *Pharm. Technol., 17*(10), 46.

Mariani, M., (1989). *Biochromatography, 4,* 149.

Mariani, M. and Tarditi, L. (1992). *Biotechnology, 10,* 394.

Nally, J. and Kieffer, R. (1993). *Pharm. Technol., 17*(10), 106.

Paul, W. L. (1991). *Pharm. Technol., 15*(3), 130.
Rosser, M. (1994). *Biopharm, 7*(3), 28.
Seamon, K. B. (1993). Technical Issues/FDA Concerns, Presented at BioPharm Conference '93, Cambridge, MA, June 21.
Sofer, G. K. and Nystrom, L. E. (1991). *Process Chromatography A Guide to Validation*, Academic: London, Chapter 2, p 5.
Zeller, A. O. (1993). *Pharm. Technol., 17*(10), 72.

INDEX